COASTAL DUNE MANAGEMENT
Shared Experience of European Conservation Practice

COASTAL DUNE MANAGEMENT

Shared Experience of European Conservation Practice

Proceedings of the European Symposium
Coastal Dunes of the Atlantic Biogeographical Region
Southport, northwest England, September 1998

Edited by
J.A. HOUSTON, S.E. EDMONDSON and P.J. ROONEY

LIVERPOOL UNIVERSITY PRESS

LIVERPOOL HOPE UNIVERSITY COLLEGE

Sefton Council

ENGLISH
NATURE

THE NATIONAL TRUST

EUCC
European Union for
Coastal Conservation

First published 2001 by
Liverpool University Press
4 Cambridge Street
Liverpool
L69 7ZU

Typeset in 10/13pt Meridian by
XL Publishing Services, Lurley, Tiverton, Devon
Printed in the European Community by
The Alden Press, Oxford

CONTENTS

KEYNOTE PAPER

Section 1
WORKING WITH GEOMORPHOLOGICAL PROCESSES

Section 4
THE SEFTON COAST LIFE PROJECT

APPENDIX

PREFACE

In 1987 managers of sand dune coasts from throughout Europe came together in Leiden for the first time to discuss future directions for conservation management (van der Meulen et al., 1989). In the previous decades ecologists such as Westhoff and Ranwell had significantly increased our understanding of the dunes while management attention had primarily been focused on the traditional problems of developing techniques and strategies to combat erosion and instability of sand dunes. Historically the problems of shifting sands inundating settlements and agricultural land were well understood, and more recent problems caused by recreation pressure on fragile mobile dune environments had been addressed. In the UK the publication of the British Trust for Conservation Volunteers handbook on sand dunes (Agate, 1986) and the Institute of Terrestrial Ecology *Coast Dune Management Guide* (Ranwell and Boar, 1986) represented a compilation of this accumulated experience. The role of sand dunes for coastal defence, particularly in countries such as The Netherlands, had prioritized the need to maintain fixed and stable sand dunes.

The increased understanding of dune ecology, however, led to a growing awareness of the significance of natural dynamics of the dunes. Dunes are geomorphologically active environments and the plants and animals typical of coastal dune environments have evolved strategies to cope with the stress and disturbance that characterizes their habitat. The disturbance mosaics created by cyclical patterns of erosion and recolonization of dune surfaces provide the conditions required for organisms from the whole range of seral communties of the dune and slack environment. This mobility within the dune system is even more important on systems where accretion is failing to provide early successional stages at the coastal fringe.

Other factors have also been significant in driving the trend towards stability and late successional stages in coastal dune systems. The arrival of myxomatosis in the 1950s severely reduced rabbit populations, causing rapid succession to high biomass plant communities. Development on the dunes requires stability; the shifting sands of a dynamic environment are not compatible with alternative land uses such as housing, transport, industry and even golf courses. Water extraction caused changes in water table levels, thus adversely affecting slack environments. Influx of nutrients from a variety of sources released the nutrient stress typical of dune habitats and further accelerated the rate of succession.

By the 1987 Leiden meeting, therefore, dune specialists were beginning to realize that rather than stabilizing dunes, their responsibilities might be to do just the reverse – to promote bare and shifting sand by adopting a 'dynamic approach'.

Just over ten years later in Southport dune managers came together once more with renewed confidence in their adoption of a dynamic approach and with experience of implementing policies to promote dynamics, bare sand and early seral communities of dunes and slacks. This experience is reported in this volume, which represents the proceedings of the Southport symposium. It includes discussion of techniques such as stock grazing, mowing, scrub clearance, removal of pine plantations, turf-stripping and destabilization management. The need to adopt a 'whole system' approach has also been reinforced; the techniques must be employed within a strategy which takes the broadest possible spatial and temporal view.

At the same time as nature managers of dune coasts have been developing their understanding of the importance of dynamics and of reversing the negative trends on our dunes, environmental awareness and understanding have been growing at all levels of society. The environmental movement, which blossomed in the 1960s in response to growing environmental problems, has now matured. Sustaining a viable and healthy environment is now recognized as an international imperative that must be implemented at all spatial scales. Public awareness of the importance of issues such as maintaining habitats and ecosystems and protecting rare species is high. People are an important part of the landscape ecology of dune systems. They live, work and spend their leisure time on the dunes and are thus an integral part of the 'whole system'. Consequently they have views on the appearance and management of the dunes which are, at least in part, enlightened by enhanced awareness of environmental problems and the need for sustainability. Without public information and consultation, therefore, the appearance of bulldozers, diggers, tractors, teams of contractors with chainsaws clearing scrub and trees and huge bonfires is not likely to be compatible with people's understanding of environmental conservation. Nature managers need to raise public awareness of the threats to the special plants and animals of the dune systems. In addition they must always question their aims, be able to justify their decisions and, even more importantly, take the views of the dune public into consideration. Thus, although conservation strategy may be driven by national and international trends and initiatives, it can only operate effectively in practice by working together with local people. This conclusion was recognized by many delegates at the Southport symposium and the message was an important outcome of the meeting.

In the 1990s nature managers throughout Europe have been forced to question their aims and values in the implementation of EU Habitats Directive 92/43/EEC (the 'Habitats Directive'). Firstly, implementation of the Directive requires an evaluation of wildlife resources in the context of the annexed habitats and species so as to propose sites for inclusion in the Natura 2000 network. Secondly, managers have to collectively achieve 'favourable conservation status' for the habitats and species so recognized, requiring local definitions of this 'favourable' condition. This requires a careful justification of conservation objectives which must also recognize the role of people as users of the land and observers of nature. The Southport meeting specifically aimed to bring together people involved in implementing the Habitats Directive for sand dunes throughout the Atlantic Biogeographical Region of the European Union to discuss their experience. This volume therefore is in part a case study of the implementation of the Directive for one landscape type.

The Sefton Coast Life Project has grappled with the implementation of the Directive on a large, hindshore dune system on the northwest coast of England. The system has multiple landownership, public and private, and is fragmented by the development of large residential areas and transport routes. Negative trends of enhanced succession, stability and loss of rare dune species have been tackled by implementing a range of techniques to promote dune dynamics. Large numbers of people were consulted in the development of a conservation strategy for the coast. Partnership with major landowners, including the Ministry of Defence, the Territorial Army and the golf clubs (there are seven links golf courses on the coast) has been a major feature of the project. Despite past criticisms of the impacts of golf courses on the wildlife value of the dunes, there is now an acknowledgment of their role in conserving important dune areas which might otherwise have been lost to more destructive development. A unique outcome of the Sefton Coast Life Project has been to demonstrate effective collaboration between nature managers and the golf clubs to maintain and enhance the wildlife value of the golf courses on the dunes. The Sefton Coast Life Project has also networked with colleagues throughout Europe to develop good practice, culminating in the symposium which brought this network together in Southport in 1998.

The Southport symposium has demonstrated that sand dune managers continue to consult, collaborate, question, criticize, evaluate and innovate in developing their management strategies. Importantly, they also monitor the results of implementing their management decisions. Geomorphologists and ecologists continue to improve our understanding of sand dune systems. Their shared experience is presented in this volume.

S.E. Edmondson, May 1999

References

Agate, E. (1986), *Sand Dunes*, British Trust for Conservation Volunteers, Wallingford.

Meulen, F. van der, Jungerius, P.D. and Visser, J.H. (1989), *Perspectives in Coastal Dune Management: Towards a Dynamic Approach*, SPB Academic Publishing, The Hague.

Ranwell, D.S. and Boar, R. (1986), *Coast Dune Management Guide*, Natural Environment Research Council, Institute of Terrestrial Ecology, Huntingdon.

ACKNOWLEDGMENTS

Support, help, hard work and encouragement have come from many sources in the preparation of this book. Without the support of DGXI of the European Commission there would have been no Sefton Coast Life Project. Without the commitment of Lorna Lander, Dan Wrench, David Simpson and Rachel Flannery at the Life Project, the conference and thus the book would never have happened. Throughout we have had support from all the partners in the Life Project and the Sefton Coast Management Scheme. Colleagues at Liverpool Hope University College have made a significant contribution, particularly Ian Evans, Liz Evans and Dr Stephen McKinnell. Finally, our sincere thanks go to Amanda, Mike and Jane.

COASTAL DUNES: RESULTANT DYNAMIC POSITION AS A CONSERVATIONAL MANAGERIAL OBJECTIVE

PROFESSOR WILLIAM RITCHIE
Lancaster University, England

ABSTRACT

Coastal sand dunes are resources for a variety of interacting and often conflicting purposes. In Britain most systems are topographically mature and subjected to coastal erosion. Most erosion is a symptom of a substantial reduction in primary sand supply and the processes of change are therefore natural responses to attain a new resultant dynamic position. In reality, change can only occur in non-vegetated areas, normally the coastal edge and in other specific locations inland.

Conservation management must include the beach and nearshore zone within its terms of reference, for coastal erosion is the dominant process affecting most contemporary dune systems.

Dynamic equilibrium demands movement and most consequent topographical changes often appear either to be incompatible with, or to threaten, existing and proposed land use practices. Hence there can be a tension between management which encourages change and those land uses which require fixed, stable environments.

Managerial responses, for conservation purposes, should be based on the necessity of encouraging natural sand movements. Managerial decisions to allow mobility can be applied most easily in areas where the use of the dunes is inherently 'natural', but become increasingly inappropriate where the land use requirement is for fixed topography and vegetation. In the latter case, however, these restrictive or preventative actions to ensure permanent stability are running counter to natural need and could, in the end, be counterproductive.

The 'Problems'

Coastal sand dunes are prime resources for a variety of interacting and occasionally conflicting purposes. Especially for those systems which remain relatively unmodified by utilitarian use, the need for effective management has increased as a consequence of a greater public concern for environmental matters. In addition, the concept of appropriate sustainable exploitation has become an explicit planning requirement.

The abundance or availability at local, regional and national scales of provision can also pose problems. In Britain, for example, coastal dunes are widespread but their spatial distribution is more or less inverse to the distribution of population. Tensions can therefore occur between agencies with national (and now European) responsibilities and local interests, notably in relation to conservation status and managerial priorities. This can be exemplified best by recognizing the number of coastal sites with conservation status in peripheral areas. This distribution is appropriate at national and European scales but to local inhabitants, the occurrence of so many sites under conservation designation seems both excessive and unnecessary. Elsewhere, formal political and administration boundaries may create difficulties if they run across the natural extent of the dune system.

Coastal dunes are not fixed or finite resources. By most standards their physical and ecological evolution is rapid, especially in the dune ridges which lie close to their contributory sources of sand, and their value alters according to their environmental maturity. Thus from a conservation perspective, the use and management of coastal dunes must take into account a requirement to accommodate change, having first assessed the stage of contemporary development.

Stage of Development

In general, in Britain almost all coastal sand dune systems are mature. They are almost fully vegetated with large areas having developed far beyond the phase of initial sand accumulation and progradation. Many have deep multiple soil horizons; many have complex patterns of mature vegetation; many have a history of semi-continuous land uses which extend for a century or more. Broadly, the absence of primary accretion and growth has long been ascribed to the ending of the long phase of excessive quantities of sand-sized sediment in shallow coastal waters at the conclusion of the post-glacial period. At some indeterminate date, possibly measured in thousands of years, the phase of sediment abundance ended and a long transition period, which also corresponded to a period of relative changes in sea level, ensued. Closer to the present, but again unspecified, and with considerable regional variations, the balance altered from sediment availability to deficiency.

During these historical times of abundant sand on the shallow sea floor, onshore-driving marine processes and numerous low-angle, low-altitude reception surfaces ensured that coastal dunes became ubiquitous in all but the steep cliff-bound sections of the coasts of Britain and Europe. Now these systems are mature and in an erosional phase. There are exceptions, for example at river mouths where sand banks and bars occur, or where fluvio-glacial sand terraces (or other deposits) continue to be eroded and carried downstream to the coast, or where, as in parts of northern Europe, sea levels continue to fall and offshore sand banks become available. History of land use in river catchments, local variations in sea level and other factors may give rise to exceptional examples of continuing 'young' accretion but the overwhelming evidence is for geomorphological senility. However, this mature topographic stage does not have its parallel in concomitant ecological climax vegetation. Grazing, habitation and other interferences appear to have held most coastal dune systems to a sub-climax or plagio-climax stage of useful grassland with varying amounts of shrub and tree cover. Along the European littoral the cultural progression from Stone to Bronze to Iron to Modern Ages can be traced in many coastal dune areas with the implication of persistent anthropic intervention in natural ecological evolution. In earlier cultural stages it is doubtful if many activities had a direct effect on geomorphological development but there would be considerable indirect effects on vegetation and soils.

Landform maturity does not, however, mean total geomorphological inactivity. One area, the coastal edge, cannot be other than active, but similar sites of geomorphological activity normally occur opportunistically further inland, almost entirely for unnatural reasons.

Stability and Dynamic Equilibrium

In its simplest form, the coastal sand dune system is a wave- and wind-driven system which removes excess sand from the marine area (which at any point in time is defined as being landwards from the water depth of the wave-base) to lodge in a beach, part of which dries sufficiently to enable aeolian processes (assuming sufficient wind velocity in relation to sand size) to carry the sand beyond the reach of claw-back by the sea. Sand size, cohesion, vagaries of tide and wave, cycles of wetness and dryness and changes in wind and wave vectors will all vary locally and regionally but, in general, given sufficient time, sand moves from sea to beach to dune where the well-known vegetation sequence (psammosere) of pioneer through to late seral stages is followed, probably discontinuously, to a theoretical climax association.

The evolution of coastal dunes is also a particular example of the simple geomorphological principle of applying energy (waves and wind) to materials (sand) to produce landforms (beaches and dunes). But what happens when the source materials cease to be provided? Unless there is some continuation of supply, and assuming a relatively stable sea level, energy levels are likely to remain more or less available to continue to move materials and to reshape the landforms, but the balance will alter from deposition and accretion with the land prograding seawards to erosion and redeposition with net coastline retreat.

On various timescales, resultant dynamic position is a key principle of coastal geomorphology whereby unprotected materials seek to move to a resting location where they will be disturbed least by all the potential transporting forces (including gravity) which operate on them. This concept is most easily appreciated in the morphodynamics of sand beaches where offshore-onshore and longshore movements continuously adjust the beach shape and profile to incident wave energy as it rises and falls with every tide, but is harder to discern in the day-to-day evolution of coastal dunes, especially where a powerful new factor is introduced – vegetation, which stabilizes the landform and prevents most dynamic change other than the sand-trapping function of long dune grasses.

At the scale of a single dune system, the reduction of sand supply should produce general coastal retreat at the beach–dune interface because there will be a net transfer of sand from the dune to the beach zone. This allows the system as a whole to migrate landwards as the beach attempts to maintain an energy-absorbing constant sand volume. In theory, at this late stage, with little or no supply the total volume of sand in the system, including the beach and nearshore zone, is more or less constant; hence although some areas of sand accumulation are visible, these are due to redeposition not primary deposition. Historically, with excess sand supply equilibrium was achieved by an expansion in the area of dunes, as manifested by the addition of successive shore-parallel dune ridges; now the lack of sand means that equilibrium is achieved by retreat and reworking of the coastal edge by waves. Within any dune system, however, local areas can appear to accrete, but on a longer timescale the system as a whole will be seeking a new equilibrium and the locations where these local adjustments are taking place are normally very small compared with the total area of the dune system. These local areas will usually be defined as zones of instability. They are zones of change and all have two factors in common, which are the prerequisites at all scales and for all time sequences: the sand must be unvegetated and accessible to the two prime energy sources – wind and waves. Bare sand surfaces are created by destroying, removing or burying the sheltering layer of vegetation and, on occasions, upper soil horizon. Thus the processes of contemporary dune evolution are concentrated in zones of actual or potential devegetation.

Thus coastal sand dunes at a mature stage of development and in a condition of inadequate primary sand supply evolve mainly as a consequence of marine rather than aeolian processes. There will be zones of aeolian erosion and redeposition but these are relatively unimportant compared with net coastline retreat at the beach–dune interface. Most dune management is concerned with the control of aeolian processes and this is understandable, but it could also be a symptom of a lack of awareness of the totality of dunes as coastal systems which must include the beach and nearshore zone. Although there is little that can be done to compensate for a negative sand budget, other than by expensive massive beach nourishment projects, management should have as its prime concern the beach and nearshore zone if only to ensure that activities there are not exacerbating the natural regressive evolution of the coastline.

Erosion and Deposition

The movement of sand by waves or wind flow (both of which are extremely complex to analyse) requires free access to individual sediment particles. For depositional processes the upper beach and the embryo dunes are the critical zones. The sand must be dry; it must be uncovered; it must be free to move. Movement, which is mainly by saltation, will be directional according to wind vectors during the specified time interval. Although other factors need to be considered, the prime distinction in any sand dune system is between vegetated and unvegetated areas. For unvegetated areas relative wetness and dryness are also of great importance, especially if wetness is a function of seasonality. Dynamic response is normally a direct function of the degree of 'bareness' and 'dryness' of the sand surface. Thus bare or partially vegetated sand surfaces are properly unstable. 'Unstable' is often perceived in a managerial sense as adverse because it is linked directly to erosion which often carries a negative connotation. This perception is false. In all natural evolutionary stages, whether prograding or retreating, the presence of most types of bare sand areas is not only normal but essential, since volumes of sand must be allowed to move to achieve geomorphological equilibrium.

Normally, unstable areas of activity in most British and European dune systems occur in four areas: the upper beach, the seaward side of the first coastal dune ridge, blowouts and other deflation areas and areas of non-natural surface damage. If the management of coastal dunes espouses concepts of working with nature then these zones require more detailed consideration of the mechanisms of their geomorphological development.

The Importance of the Upper Beach and Coastal Edge

As discussed previously, management must extend to the beach at least as far as the wave base, a depth of water which is defined by the height of the incoming waves. As a rule of thumb the breaker zone would be a reasonable seawards limit. A full analysis of beach processes cannot be given here, but for any understanding of the development of coastal dunes, in addition to an assessment of wave climate, an understanding of the location and continuity of sand supply must be attempted, including a concern for offshore sand dredging or other similar activities. Unfortunately, due to the legal definition of land in British law as extending to the high tide level below which is Crown foreshore, effective management can be hampered by formal rights. Nevertheless, bearing in mind seasonal changes, the crucial zone for wind transport and for direct wave attack under high tide and storm conditions lies above the high-water mark.

The key variables in this upper beach zone relate to the width, thickness, dryness, amount of vegetation and evidence of lateral shifts of material. Some dune systems have a ridge or a veneer or shingle or cobbles at the high tidal or storm wave level which adds to the complexity of the marine and aeolian processes affecting this crucial zone. Often artificial constructions, usually described as beach defences, exist and normally alter natural processes substantially. These constructions normally have a negative effect because they inhibit natural retreat and consequent system resilience. In extreme cases they can encourage enhanced backwash and sand removal downbeach. Groynes deliberately inhibit longshore sediment movement and distort any further natural patterns of development. The construction of a solid concrete or

Plate 1. *The process of undercut and slump is illustrated by this winter photograph of the eroding dune coastline at St Fergus on the northeast coast of Scotland. (W. Ritchie)*

stone sea wall along the face of the frontal dune brings a complete change to the nature of coastal geomorphological processes. Suffice to say that this is little more than introducing a solid, reflective cliff into a hitherto totally mobile and dynamic system and is therefore incompatible with any 'natural' managerial approach.

The upper beach and the coastal edge merge together as a zone of interchange and, as such, this is normally the most active part of most dune systems. The empirical evidence which is used for an assessment of activity is the profile from the crest of the main frontal dune ridge to high tide level. At one extreme, a constructive profile is wide, hummocky and low angle with increasing vegetation cover landwards. At the other extreme, a severely eroding coastline will resemble a cliff with a narrow flat upper beach, especially in winter or after high energy wave events. In a condition of erosion a process of undercut and slump occurs linearly and shore-parallel along the toe of the dune ridge. Waves undercut the dune face which slumps to the upper beach and sand will be removed by waves. Slopes in excess of the angle of rest of sand are common due to the cohesion of buried soil or humic layers, roots or ground water. Binding by moisture in the sand is temporary and selective wind erosion often occurs, especially near the top of the profile. Block slumping is common and, temporarily, vegetation may regenerate on the upper beach from the talus blocks. Free sand at the

Plate 2. *Young, hummocky backshore dunes are unusual but occur in favourable locations such as the accreting spit at the east end of the Culbin dune system on the south side of the Moray Firth in Scotland. (W. Ritchie)*

toe of the dune may be lifted by the wind to be redeposited on the crest or leading edge of the dune where weight is added to the column of sand in the sand cliff and this can exacerbate the slumping process. Upper beach sand can also be funnelled into the inner dunes through coastal blowouts. Moderate edge accumulation can stimulate the growth of most dune grasses and the edge can have a rejuvenated appearance, but this is a product of redeposition rather than primary accumulation. Erosion is shore parallel and linear and often extensive, and dune cliffs are indicative of rapid shoreline retreat. In large dune systems, especially where the beach is divided into cells by rip-currents, these erosional sectors may shift laterally according to changes in the pattern of waves. Point erosion on the seawards edge of the coastal dune is the other mechanism of natural adjustment to a diminished sand supply on the upper beach; point erosion is more commonly described as a blowout, and is transverse or oblique to the alignment of the beach.

The transverse alignment of blowout features makes it possible to contrast the morphology of young accreting systems (where there is a succession of shore-parallel dune ridges which are progressively younger seawards) with mature dune topography. In mature systems this early shore-parallel topography is compounded by transverse secondary and tertiary hollows and ridges which correspond to the axes of blowouts.

Plate 3. *The juxtaposition of a wide sand beach with remnants of a severely eroded coastal dune ridge is not unusual. At Sandwood Bay in Sutherland (Scotland), storm waves reach to and through the foredune ridge. (W. Ritchie)*

These transverse features are often aligned with dominant wind directions. In larger mature dune systems, deflation plains and parabolic dune arcs also reflect dominant or resultant wind forces. Mature dune systems are therefore diverse systems with more recent geomorphic features superimposed on the older prograding, shore-parallel ridge and hollow topography. Similarly, redeposition from blowouts and deflation hollows rejuvenates vegetation, especially long dune grasses, and these areas can lie quite far inland from the coastline.

Blowouts and Deflation Areas

The origin and development of blowouts in coastal dunes have been well documented in geomorphological literature. Essentially a weakness, which is initially small, in a vegetated dune ridge or surface is created by natural (including biotic) or anthropic reasons. The removal of vegetation allows the wind to transport material from the bare surface. Once initiated, recolonization by vegetation can occur if there is a long time interval between periods of suitably strong transporting winds, or if there are weather changes, or if the bare sand becomes waterlogged due to a rise in the water table. Normally, consolidation by revegetation also requires the cessation of the original cause of surface breakage. Alternatively the blowout will grow. The wind will enlarge the erosional area, especially if a funnelling effect increases the frequency and effective velocity of the wind. After a critical stage, most blowouts become unstoppable and pass through corridor, widening and breakthrough stages with an ultimate end-point of removing large volumes of sand from and through the original ridge or surface. This reduces the topography to a near-plain or undulating surface (a deflation surface) which will be either the water table or a buried cohesive layer such as a humic soil or beach gravel or shingle horizon. Once revegetated, however, it remains possible to detect the axis of the blowout and the parallel side-ridges. The redepositional ridge at the exit to the blowout often has a comet-tail shape with the same alignment.

Blowouts are normally classified according to shape and stage (e.g. corridor, V-shaped, cauldron), and by location (e.g. frontal dune, inner ridge, hillside). All blowouts have complementary redepositional features normally downwind, usually but not necessarily as discrete secondary features. In coastal blowouts with variable winds, some blowouts can feed sand from the dunes to the beach , but more frequently

Plate 4. *Irvine Bay in Ayrshire, Scotland, illustrates a typical series of dune rehabilitation and protection measures, including attempts to trap sand on the backshore and to corridor a footpath to the beach. (A.S. Mather)*

they are conduits for sand from the upper beach to other parts of the dune system. Thus blowouts can be not only consumptive and redepositional and therefore mediums of internal transfer but also, especially at the coastal edge, transport corridors for sand from elsewhere. At a later stage of deflation, blowouts at the coastal edge can have floor altitudes below high storm-water level and can contain beach deposits.

In relation to dune management blowouts are important locations for natural dynamic change but, more often, their origins and development are triggered by anthropic damage such as footpaths, sand extraction, burning and grazing pressures (including rabbits). In fact, examination of hundreds of dune systems in Britain suggests that only a few blowouts can be described as natural since most have an iden- tifiable cause of initiation and aggravation. The only exceptions to this assertion are blowouts on the coastal edge where the natural process of wave attack can expose a zone of relative weakness for further concentration of wave and/or wind energy, a process which becomes self-accelerating. Thus unlike shore-parallel or other forms of erosion at the coastal edge, most blowouts and other areas of sand deflation are likely to be non-natural either for biotic or anthropic reasons, and, therefore, in theory controllable. Any consideration of triggering or aggravation of erosion by grazing

Plate 5. *Footpaths often trigger small-scale erosional development. Regeneration of dune hummocks on the lower bare sand slope is typical. Bridge of Don, Aberdeen. (W. Ritchie)*

pressures poses the question about the importance of natural grazing as part of the ecology of the system, especially in relation to rabbit populations which, throughout the historic period, have found dunes to be particularly attractive habitats. The importance of rabbit grazing has varied over time according to such factors as natural predation, their value as a food source, the effects of myxomatosis cycles and competition from other species such as sheep. Further, field evidence indicates that the general effect of over-grazing is not the only issue: burrowing, which creates the initial sand exposure, is of greater significance.

From a managerial perspective, it is likely that small populations of grazing and burrowing species can be tolerated, possibly welcomed as being part of the natural ecosystem, but at some density threshold a step-change occurs where erosion escalates rapidly to unacceptable levels. It is when this step-change occurs that the hypothesis that erosion due to over-grazing is non-natural and therefore controllable becomes valid. If the erosion is caused by other non-natural factors such as sand quarrying or tracking then the hypothesis is clearly valid. During the surveys of all the dune systems of Scotland an attempt was made to obtain the order of magnitude for acceptable ('natural') bare sand areas as a percentage of the total area (dunes, dune pasture and transitional zones) and a value of 2.5 per cent was obtained. The value of 5 per cent was also found to be the point at which the system was perceived as 'having an erosion problem'. These ratios might therefore be suggested as approximate indices of 'normal' erosion for similar mature systems in temperate latitude dune systems.

Whether natural or not, blowouts occur in most dune systems. They have a predictable evolution which will reach an equilibrium of a reduced surface altitude, undulating relief, full revegetation and geomorphological inactivity. Before a blowout cycle is complete some stages produce diverse, highly dissected and vigorous zones of rapid topographic and vegetational change. If managerial action to control the blowout is required, it should be designed to recognize the end point of the cycle and accelerate that condition. Fixing a blowout as a steep-sided deflation axis by attempting revegetation or sand trapping within the feature has a high probability of failure and can only be justified if erosion and redeposition are destroying a fixed asset or threatening a viable alternative use of the feature or adjacent areas. Some users of sand dunes regard blowout zones as topographically attractive as a consequence of their landscape diversity and their value as sites which demonstrate on a small scale the essential principles of dune development. Nevertheless they are important mechanisms for transferring sand further landwards beyond the reach of wave redistribution and are therefore prime methods of the net landward shift of the system as a whole.

Conclusion: Managerial Options

Managerial options for coastal sand dunes are often severely constrained by ownership, planning regulations, financial pressure and historical precedents, none of which need

Table 1. Managerial summary

Key geomorphological considerations

- Conservational management of coastal dunes must include the beach and nearshore zone
- The natural development of coastal dunes is dependent on movement and change
- Any engineering works in dune areas especially at the coastline should use 'soft' solutions
- Knowing the sand budget is as important as the understanding of biotic and abiotic processes
- Most coastal dune systems in Europe are mature and in erosional and recycling phases
- Control of the water table is often overlooked in understanding and, potentially, in managing coastal dunes
- Spatial analysis of dune ridge alignments gives insight into historical and ongoing sand movements
- Linear, shore-parallel erosion at the beach–dune interface is normal but most blowouts are initiated for abnormal reasons

Managerial framework

- Relatively few dune systems remain free from utilitarian pressures – past, present and future
- Fragmentation of ownership and managerial objectives can render conservational management problematical

Managerial constraints – summary

- Ownership
- Planning regulations
- Financial imperatives
- Historical legacy
- Preservational attitudes
- Fear of consequences of 'natural' changes
- Inability to include beach and nearshore zone

Plate 6. *Coniferous afforestation is common along the Moray Firth coastline of Scotland. At Findhorn, the westwards migration of the river is producing severe erosion of the massive dune system. Eroded sand feeds local sand bars and beaches which are producing redepositional backshore dunes several kilometres downdrift to the west. (W. Ritchie)*

to take any significant account of natural conditions or processes. In reality, although there are regional variations, especially in parts of Wales and Scotland, management of most dune systems is often designed to preserve existing land use and economic return. Nature Reserves and Sites of Special Scientific Interest also tend to have management systems which take into account and often permit existing land uses and therefore economic benefit to continue – sometimes for pragmatic reasons and sometimes because the existing land use helps to preserve the scientific interests of the system. Any inventory of all the dunes of the British Isles would demonstrate how few systems remain free from some form of commercial activity in all or part of the area. Thus, in reality, operational management in all but a few areas must continue to give priority to the likely economic impact of any proposed action. Therefore, for the most part, management which is based on scientific, conservational or even sustainable principles is at best rare, often irrelevant and in a few instances in conflict with the existing forms of exploitation. Accordingly and typically, in many dune areas, management takes the form of remedial action; for example, to stop erosion because sand drift is damaging agricultural land or blocking a drain or closing an access road.

Such management is both piecemeal and reactive. Holistic proactive management is almost impossible where there is fragmented ownership, patchworks of different types of land use and different regulatory frameworks for different parts of the system. It is almost impossible to be effective if the beach zone does not fall within the managerial compass. Most sand dune systems in Britain fall into the category of having little opportunity for holistic management, therefore the optimum managerial situation that might be achievable for these areas is an overarching forum or board which might agree to manage the system within conservational principles for the greater good and, in the end, sustainable exploitation of the resource.

A few sand dune systems exist under a single managerial structure. These areas are usually some type of conservational area and are either owned by the region or the state or by an agency which has appropriate managerial control over a single owner, or can control multiple landowners' individual actions. Ideally these might become demonstration areas for good practice. Good practice for a mobile, changing, responsive environment must have at its core dynamic conservation, not static preservation. The distinction between conservation and preservation is not merely semantic. It should encapsulate the essential difference between a managerial approach which tries to proactively develop the resource by encouraging natural evolution and a holding approach that resists habitat and ecological change and seeks to prevent those free processes which, in reality, create long-term resilience and protective adaptability.

On a wide range of timescales, change is inherent in all natural systems. Coastal sand dunes, on the whole and in part, change relatively quickly. Thus management processes should be designed to work *with* change. The concept of resultant dynamic position is an effective working principle which, along with an understanding of the consequential succession of soils, vegetation and other biota, provides the basis for effective conservation and therefore sustainable exploitation for those dune systems that remain sufficiently unmodified to have primarily ecological value. Elsewhere, with respect to general coastal management, decades of effort by coastal geomorphologists and other environmental scientists to advocate 'soft' solutions are beginning to show some sign of success with those who are charged to take decisions on coastal protection and maintenance. Although the past cannot be undone, and many unsuitable developments occupy many coastal areas, 'soft' engineering solutions are now much more common and coastal sand dunes by their nature are prime examples of the benefits of 'soft' solutions. 'Hard' solutions may be justifiable in extreme cases or as a recurrent requirement to preserve the status quo, but for a dune system with a primarily conservation function such methods are intellectually very difficult to justify and, over time, are increasingly likely to produce more problems than solutions.

Section 1

WORKING WITH
GEOMORPHOLOGICAL PROCESSES

LONG-TERM GEOMORPHOLOGICAL CHANGES AND HOW THEY MAY AFFECT THE DUNE COASTS OF EUROPE

K. PYE

Department of Geology, Royal Holloway, University of London, England

ABSTRACT

Coastal dunes are naturally dynamic systems which respond to fluctuations in wind/wave energy regime, sea level and sediment supply on timescales ranging from a few years to millennia. The likely effects of possible future climate and sea level change on the dune systems of northwest Europe are difficult to predict given the current limitations of available coupled physico-ecological models and the diversity of environments which exist. However, important indications can be obtained from studies of past changes during earlier periods of the Holocene. Based on the experience of past changes, it would seem likely that changes in wind/wave climate are likely to have a more important effect than sea level change during the course of the next century. However, under even the most extreme Intergovernmental Panel on Climate Change (IPCC) climate change predictions, the effects on dune systems are likely to be modest on this timescale. The most significant geomorphological consequences will be increased frontal dune erosion, blowout development, and transgressive dune development. These changes could be partly controlled by management measures, but in the interest of both geomorphological and ecological diversity it would be better to allow these systems to evolve naturally wherever possible. Losses of habitat in some areas will be compensated by gains in other areas downdrift, and by the enhanced value of a dynamic landscape.

Introduction

Many of the major coastal dune systems in northwest Europe were initiated during the mid to later part of the Holocene period (i.e. after 6000 years BP). Remnants of older Holocene and Pleistocene coastal dune systems are found in some localities, but are of limited extent compared with older Quaternary dune complexes in other parts of the world, such as eastern and southern Australia, the west coast of the United

States, and southeastern Brazil. Since their initiation, the later Holocene dune systems of northwest Europe have experienced significant fluctuations in climate, sea level and sediment supply which have had major impacts on the dune morphology and ecology. Additionally, almost all European dune systems have been significantly affected by human activities, including grazing, afforestation, sand mining and re-profiling for recreational purposes such as development of golf courses and tourist facilities. Future changes in environmental forcing factors, both natural and anthro-pogenic, are certain to occur, and accommodation of the resulting geomorphological and ecological responses presents a major challenge for dune managers.

Unfortunately, the magnitude and timing of future changes in environmental forcing factors are surrounded by considerable uncertainty. The most recent predic-tions by the IPCC suggest that global mean temperature may increase by between 0.8 and 4 degrees centigrade, with a best estimate of about 2 degrees centigrade, by the year 2100. In addition, the weather in many areas is likely to become more extreme and more unpredictable, and global mean sea level is likely to rise by between 20cm and 96cm, with a best estimate of 49–55cm, by 2100. However, at the present time, global climate models are unable to predict the nature and magnitude of changes of climate at the sub-regional scale with any degree of confidence. Furthermore, predic-tive models which link changes in environmental forcing factors to changes in dune morphology and ecology are at present in a very early stage of development, and of doubtful reliability. For this reason, there is considerable value in undertaking studies of past changes which may provide important clues as to the nature of possible future changes.

Potential Effects of Climate Change on Dune Systems

Climate changes can affect coastal dune systems in a variety of ways. Changes in wind strength and direction are likely to increase or decrease potential rates of aeolian sand transport, both from the beach to the dunes and within the dunes themselves, leading to changes in rates of dune growth and migration. Other things remaining equal, an increase in wind strength/duration will increase the likelihood of damage to vegeta-tion, development of blowouts, and enhance the migration rate of transgressive dunes and sand sheets. Conversely, a reduction in wind speed or duration will tend to encourage sand stabilization and soil development. An increase in the frequency and/or magnitude of storm events, in particular, is likely to have a major impact on frontal dune systems, mainly in the form of wave scarping, dune cliff formation and recession, and blowout development. While frontal dune erosion inevitably leads to loss of habitat area, it may also bring benefits in the form of geomorphological and ecological dynamism, and may be very positive from an ecological point of view. Increased sand flux from the beach to the hind-dune areas may stimulate floristic growth and diversity.

At a global scale, the size of individual dune forms shows a direct relationship with wind energy and sand availability; hence an increase in mean wind speed, or in the frequency of extreme wind events, may well lead to the formation of larger, more active dunes, in areas which currently experience low to moderate wind energy regimes.

Changes in wind directional variability are likely to have an effect on the morphology of dunes and blowouts (Jungerius et al., 1991). Unidirectional wind regimes tend to produce elongated deflation features and dunes, whereas bimodal and multi-modal wind regimes produce more complex and equi-dimensional forms (Pye and Tsoar, 1990). Analysis of historical wind records from locations in northwest Europe has clearly demonstrated that there have been marked variations in wind speed and direction over the last 100–150 years (e.g. Anthonsen et al., 1996; Clemmensen et al., in press). These changes, combined with fluctuations in precipitation, have led to changes in dune morphology and degree of aeolian mobility. For example, Anthonsen et al. (1996) documented the change in morphology of a large barchanoid dune in northern Denmark to one with more parabolic form during the period 1887–1986.

Changes in actual or (perhaps more importantly) effective precipitation total and seasonal distribution can have a major effect on aeolian sand transport rates, since moisture content of near surface sand both the threshold wind velocity required to initiate aeolian transport and influences the vigour of dune vegetation. On timescales of weeks to decades, changes in groundwater table levels have a major effect on dune stability/instability, since in active dune systems the base level of deflation is controlled essentially by the position of the water table. Other things remaining constant, a decrease in effective precipitation will lead to a fall in water table levels, greater stress on the vegetation, and increased likelihood of blowout development, while an increase in effective precipitation will encourage vegetation growth and dune or sand sheet stabilization. However, high precipitation alone is not sufficient to prevent a high degree of dune mobility if wind energy levels are high enough. Some of the world's largest dune systems are found in regions which receive more than 2000mm of rainfall per annum, including North Queensland (Pye, 1993) and Oregon (Cooper, 1958).

Changes in temperature and humidity have an effect on aeolian sand transport rates by controlling the rate at which surface sand dries out after rainfall, and on rates of plant growth (Sherman and Hotta, 1990). Under some circumstances, an increase in mean temperature may be expected to enhance potential rates of aeolian transport on bare surfaces but may retard it on vegetated surfaces owing to more vigorous plant growth. Evidence from the Holocene stratigraphic record indicates that periods of warmer, slightly wetter climate in the geologically recent past resulted in widespread dune stabilization and soil development in many parts of northwest Europe, while relatively cold periods, with lower total precipitation, were associated with increased sand blowing and dune or sand sheet mobility (Pye and Neal, 1993; Clemmensen et al., in press). The Little Ice Age, in particular, is well documented as a time of widespread aeolian sand instability in many parts of northwest Europe. Warmer, drier intervals, with marked summer droughts, can encourage sand instability, notably in semi-arid areas. Droughts directly affect vegetation growth through changes in soil moisture regime, and may also lead to increased risk of fire damage.

Coastal dunes along the Atlantic coast of northwest Europe occur along a climatic gradient with marked differences in temperature, precipitation, humidity, cloud cover, wind regime and fire incidence, and within each dune system there is a wide variety of microclimates (Wartena et al., 1991). Each of these different microclimate associations is likely to respond slightly differently to changes in climatic parameters, and the resulting geomorphological and ecological consequences are likely to be correspondingly varied.

Potential Effects of Changes in Sea Level

Changes in sea level can also have significant effects on coastal dune systems. Mean sea level and tidal range directly influence beach width and morphology, and thereby the area exposed to wind action at low tide. An increase in mean sea level is likely to lead to beach narrowing and foredune erosion, while a fall in relative sea level will encourage beach widening and foredune progradation. There has been much debate about whether changes in sea level encourage or hinder the development of transgressive dunes and sand sheets. Several alternative models were discussed by Pye (1984) and Christiansen and Bowman (1986). As pointed out by Pye and Bowman (1984), on coasts with high wind energy, a large proportion of the sand on eroding beaches is transferred landwards to form transgressive dunes and sand sheets, rather than being moved offshore as predicted by the Bruun model of shore erosion (Bruun, 1962). On low wind energy coasts, rising sea level results in submergence or seaward transfer of much of the eroded sand. Falling sea levels tend to favour development of beach ridges and low foredune ridges on low to moderate energy beaches, but on high energy shores, especially where rainfall is low and vegetation cover sparse, transgressive dunes and sand sheets are likely to develop. However, changes to the shore and frontal dunes due to sea level rise need to be considered in three dimensions rather than two, since beach and foredune erosion in one area may provide a source of sand which allows net accretion in other areas downdrift (Carter, 1991). Examples of this situation are widely found at the present day; for example, on the Sefton Coast, in northwest England, erosion of frontal dunes at Formby Point provides sand for beach and foredune accretion both to the south at Ravenmeols and to the north at Ainsdale and Birkdale (Pye and Neal, 1993).

Fluctuations in sea level and resulting shoreline position will also have an effect on groundwater levels in the adjacent dune system and will therefore affect the base level for deflation. In general, rising sea level results in rising water table levels and an increase in thickness of accumulated aeolian sediment. Falling sea level causes falling water table levels and a lower base level of deflation. Where changes in shoreline position are marked, however, the situation may be spatially complex. Much depends on the width and height of the dune sand body, and on the hydraulic conductivity of the sand (van der Meulen, 1990; Noest, 1991). Shoreline changes also lead to a shift in the extent of saline spray deposition, with implications for soil conditions and vegetation stability. In practice, changes in sea level and climate will have a combined effect on groundwater levels which can only be modelled on a site-specific basis.

Lessons from Historical Studies

In the past ten to fifteen years a considerable number of historical studies have been undertaken on the dune systems of northwest Europe. These studies have provided chrono-stratigraphical and palaeoecological information which can be correlated with climate-proxy evidence and which allows testing of hypotheses about the effects of changes in environmental forcing factors on the geomorphology and ecology of dune systems in different environmental settings (e.g. Bakker et al., 1990; Pye and Neal, 1993; Clemmensen et al., in press). The results of these studies have shown that there have been major variations in the scale and distribution of aeolian activity in the historically recent past. At the present time, many dune areas in northwest Europe are characterized by an unusually high degree of stability which reflects the effectiveness of artificial stabilization measures in recent decades.

Periods of warmer, wetter conditions during the Holocene, such as the Atlantic period, were associated generally with dune stabilization and soil development, while colder periods, such as the Little Ice Age, were stormier and characterized by widespread sand blowing. The evidence from the Holocene record suggests that fluctuations in wind/wave climate, shoreline erosion/accretion, and frontal dune stability appear to have been far more important than changes in sea level as a cause of episodic dune activity. Beach-dune systems can quite easily cope with rates of sea level rise of several millimetres per year provided that a supply of sediment is maintained. If sediment supply is not maintained, frontal dune erosion will result even if sea level is stable or falling. Changes in beach/frontal dune sediment budget are highly sensitive to fluctuations in wind/wave regime, and consequently frontal dune systems are those which are likely to experience the earliest and most pronounced consequences of future climate change. If current IPCC predictions prove to be accurate, the most obvious effect will be an increase in beach and foredune erosion rates, and enhanced blowout development and transgressive dune/sand sheet instability related to shoreline recession, but enhanced progradation is likely in some downdrift areas. Such changes should not be viewed as a threat, however, but more as an opportunity to enhance ecological and geomorphological diversity. Consequently, management intervention should be kept to a minimum and systems should be allowed to evolve naturally wherever possible.

In many areas, sand dunes will be able to act as effective dynamic flood defences if permitted to evolve naturally in the face of sea level rise and shoreline erosion. However, in some problem areas, where dunes provide protection for high-value land assets, and where there is low onshore wind energy and high rates of littoral marine transport, it may be necessary to sustain the integrity of the dune systems by traditional dune management techniques and other beach management measures, including beach recharge, groynes and offshore breakwaters. Rates of frontal dune erosion in northwest Europe typically lie in the range 0.5–3m per year. These rates are not likely to increase dramatically during the next century even under the worst case IPCC scenarios. Although there will inevitably be habitat losses in some areas, these will be largely offset by gains in other areas provided that natural sediment transport cells are allowed to function in an uninterrupted manner. Shore and frontal dune erosion, together with instability in back dune areas, should be seen as natural processes which enhance, rather than detract from, the landscape quality of dune areas, and temptations to stabilize or preserve them in their present position should be resisted.

References

Anthonsen, K.L., Clemmensen, L.B. and Jensen, J.H. (1996), 'Evolution of a dune from crescentic to parabolic form in response to short-term climatic changes: Rabjerg Mile, Skagen Odde, Denmark', *Geomorphology*, **17**, 63–77.

Bakker, T.W., Jungerius, P.D. and Klijn, J.A. (1990), 'Dunes of the European coasts. Geomorphology, hydrology, soils', *Catena Supplement 18*, Catena Verlag, Heidelberg.

Bruun, P. (1962), 'Sea level rise as a cause of shore erosion', *American Society of Civil Engineers, Journal of Waterways and Harbours Division*, **88**, 263–342.

Carter, R.W.G. (1991), 'Near-future sea level impacts on coastal dune landscapes', *Landscape Ecology*, **6**, 29–39.

Christiansen, C. and Bowman, D. (1986), 'Sea level changes, coastal dune building and sand drift, northwestern Jutland', *Geografisk Tidskrift*, **86**, 28–31.

Clemmensen, L.B., Pye, K. and Murray, A. (in press), 'Stratigraphy, sedimentology and landscape evolution of a Holocene dune system, Lodbjerg, N.W. Jutland, Denmark', *Sedimentology*.

Cooper, W.S. (1958), 'Coastal dunes of Oregon and Washington', *Geological Society of America Memoir*, **72**.

Jungerius, P.D., Witter, J.V. and van Boxel, J.H. (1991), 'The effects of changing wind regimes on the development of blowouts in the coastal dunes of The Netherlands', *Landscape Ecology*, **6**, 41–48.

Meulen, F. van der (1990) 'European dunes: consequences of climate change and sea level rise', in T.W. Bakker, P.D. Jungerius and J. A. Klijn (eds), *Dunes of the European Coasts. Catena Supplement 18*, Catena Verlag, Cremlingen, 183–95.

Noest, V. (1991), 'Simulated impact of sea level rise on phreatic level and vegetation of dune slacks in the Voorne area (The Netherlands)', *Landscape Ecology*, **6**, 89–97.

Pye, K. (1984), 'Models of transgressive dune building episodes and their relationship to Quaternary sea level changes: a discussion with reference to evidence from eastern Australia', in M. Clark (ed.), *Coastal Research: UK Perspectives*, Geo Books, Norwich, 81–104.

Pye, K. (1993), 'Late Quaternary development of coastal parabolic megadune complexes in northeastern Australia', in K. Pye and N. Lancaster (eds.), *Aeolian Sediments Ancient and Modern*,

International Association of Sedimentologists Special Publication 16, Blackwell Scientific Publications, Oxford, 23–44.

Pye, K. and Bowman, G.M. (1984), 'The Holocene marine transgression as a forcing function in episodic dune activity on the eastern Australian coast', in B.G. Thom (ed.), *Coastal Geomorphology in Australia*, Academic Press, Sydney, 179–96.

Pye, K. and Tsoar, H. (1990), *Aeolian Sand and Sand Dunes*, Unwin-Hyman, London.

Pye, K. and Neal, A. (1993), 'Late Holocene dune formation on the Sefton Coast, northwest England', in K. Pye (ed.) *The Dynamics and Environmental Context of Aeolian Sedimentary Systems*, *Geological Society Special Publication No. 72*, Geological Society Publishing House, Bath, 201–17.

Sherman, D.J. and Hotta, S. (1990), 'Aeolian sediment transport: theory and measurement', in K.F. Nordstrom, N.P. Psuty and R.W.G. Carter (eds.), *Coastal Dunes Form and Process*, John Wiley and Sons Ltd, Chichester, 11–37.

Wartena, L., van Boxel, J.H. and Veenhuysen, D. (1991), 'Macroclimate, microclimate and dune formation along the West European coast', *Landscape Ecology*, **6**, 15–27.

A LIFE-ICZM DEMONSTRATION PROJECT FOR IRISH BEACHES AND SAND DUNES

J. POWER, J.A.G. COOPER, J. McGOURTY, J. McKENNA, M.J. MacLEOD
Coastal Research Group, University of Ulster, Northern Ireland
and G. CONVIE
Donegal County Council, Ireland

ABSTRACT

This three-year project, part of the EU Integrated Coastal Zone Management (ICZM) network, will design, implement and demonstrate sustainable management strategies for Irish soft coasts. Seven demonstration sites in Donegal (Republic of Ireland) were selected by the partner organizations. Three inter-related phases are being carried out at each site: scientific description, plan production and plan implementation. A fundamental aspect is the participation of coastal communities at every stage of the project. In addition, the experience of the National Trust in long-term management of beach and dune systems in Northern Ireland is being used to inform the decision-making process.

This paper will briefly review the project structure, progress and preliminary findings, using a single site as a case study. The Lisfannon dune system, located on the eastern shore of Lough Swilly, is of mainly recent origin (post-1950s). The site contains significant botanical and ornithological interest – habitats include mud flats, (fixed) grey dune, saltmarsh and a brackish lagoon. However, long-shore sediment supply has now largely ceased, and the beach and dunes are under wave attack. This directly threatens the access road, car park and toilet block at a Blue Flag beach used by large numbers of day visitors from nearby urban centres. A *do-nothing* strategy is likely to lead to a loss of these facilities, and also large areas of terrestrial habitat. At the other extreme, engineered coastal protection measures are unlikely to be economically feasible or environmentally desirable.

Armed with an understanding of how each system functions, the project staff are working closely with competent authorities, interest groups, beach users and the local community to arrive at management strategies which are both acceptable and sustainable.

Introduction

Approximately 20 per cent (1000km) of the Irish coast is protected by sandy beaches and dunes (Carter, 1990). Traditionally, these areas were considered of limited economic value, providing sand for construction, marram grass for animal bedding, rabbits and rough grazing. However, with the increasing popularity of leisure activities, such as walking, bathing, surfing and sailing, tourism and recreation have become the principal uses of beaches and dunes. Indeed, in the mountainous, high rainfall areas of the Irish west coast, many dune systems have been developed as leisure and tourism facilities. These include golf courses, sports pitches, caravan sites, and at Carrickfin in County Donegal, the regional airport. Beaches and sand dunes are now recognized as important contributors to biodiversity in the Irish landscape. This is reflected in the large number of Irish dune systems which are candidate Natura 2000 sites, designated primarily in response to the Habitats Directive (European Community, 1992). The Directive identifies Irish machair as a priority Annex I habitat, reflecting its particular international significance.

In spite of their importance for tourism, conservation and coastal defence, the environmental quality of many Irish dune systems has deteriorated in recent years. There are numerous examples of beach lowering, shoreline erosion and dune loss, often resulting directly from ill-conceived engineering works (e.g. Wexford Harbour, Portrush and Portballintrae, County Antrim). Dune habitats, in a semi-natural state because of centuries of grazing management, have been modified further. They have been degraded by the intensification of livestock farming, which has encouraged farmers with coastal land to over-winter large numbers of animals on dune pasture. Recreational uses (e.g. walking, parking, motorcycle scrambling) and the development of leisure facilities have also damaged or destroyed dune habitats, and have led to a general decline in the aesthetic quality of the coastal landscape. Taking a wider perspective, the increasing pressure placed on the coastal zone by both agriculture and tourism has led to reductions in the quality of estuarine and inshore waters, and locally, to the supply of mains drinking water. The expansion of tourism in Ireland has not been of positive benefit for all coastal communities. For example, the short, weather-dependent holiday season does not tend to encourage investment in permanent infrastructure, full-time employment or wider economic development.

It is widely recognized that many of the current management problems associated with dune systems are the result of decisions made in ignorance. There is a general lack of understanding of coastal processes, and of the wider consequences of individual actions. This transcends all levels of society (administrators, developers, land owners, beach users). In Ireland, there is also currently little co-ordination of policy or management between different sectoral interests (e.g. planning, conservation, fisheries), and between organizations with terrestrial and maritime responsibilities (Brady Shipman Martin, 1997).

It is in this context that a LIFE-funded project was launched to investigate alternative management strategies for dune systems in the north of Ireland. This study forms part of the EU 1997–99 Demonstration Programme in Integrated Coastal Zone Management, designed to assist the development of ICZM policy in the Community (European Commission, 1997).

Methodology

Donegal County Council and the University of Ulster (Coleraine) are currently developing management plans for beaches and dune systems at Culdaff, Lisfannon, Portsalon, Downings, Magheraroarty, Narin and Rossnowlagh. These are located on the north and west coasts of County Donegal (Republic of Ireland). They occupy a diversity of environmental and cultural settings, ranging from a remote Atlantic beach in the Gaeltacht (Irish-speaking area) to an intensively used estuarine beach close to population centres. The project structure is summarized in Figure 1.

Site assessment
The management plans are to be based on a sound understanding of natural processes and current uses. Unfortunately, there is little environmental information available for Irish coastal areas at a scale appropriate for management. To address this deficiency, an ongoing programme of interdisciplinary scientific assessment is being carried out at each site. This can be split into three related areas: geomorphology, ecology and

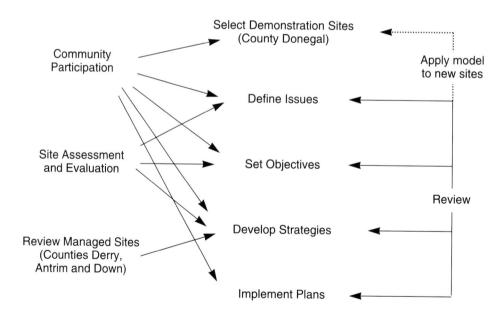

Figure 1. *Operational structure of the LIFE-ICZM project.*

human utilisation. Coastal landforms and physical processes are being studied by historical analysis, mathematical modelling and topographic survey. All published maps and aerial photographs are being used to trace the development of each system from the 1830s to the present. This is performed by overlaying images using a computerized rectification system (ARC-INFO). Simulation of wave dynamics and associated morphological response using the HISWA modelling package, and based on bathymetric data from published charts, will simulate the effects of storms on the beach and dune systems. Topographic and seasonal variations in beach and dune profiles are being studied using a high-resolution Global Positioning System (GPS). The ecological component includes surveys of plant, beetle and bird communities along the dune profiles. The biodiversity and management of the dune systems is also being assessed at the macro-scale by site habitat mapping (using Natura 2000 categories). Beach utilization is being studied by questionnaire and observational surveys in order to understand the factors that lead people to visit the beaches, and to describe and interpret the ways people behave once they are there. Assessing the distribution of visitors, vehicles and activities is of great importance, both in describing current pressures and in zoning uses in future management plans. This process will be aided by the installation of remote video cameras linked to the Internet.

Community participation

The management plans will attempt to strike a balance between utilization and conservation of the beach and dune resource. The plans must also represent the views of landowners, tenants, beach users, the local community, and organizations with an interest in the coastal zone. Furthermore, they must be politically acceptable, both locally and at county level. Open dialogue and public participation are therefore fundamental aspects of every stage of the project. This is to ensure the widest possible support, and to avoid potential conflicts. Regular public meetings and postal consultation exercises are to be held at each site to obtain information, and to canvass opinion on proposed strategies prior to the implementation of the plans.

Review of managed sites in Northern Ireland

Assessment of the experiences (both positive and negative) of long-term management of Atlantic dune systems by the National Trust in Northern Ireland will be used to inform the decision-making process. The National Trust have been monitoring the effectiveness of management initiatives at dune sites which combine high visitor numbers with nature conservation objectives, e.g. Portstewart (County Derry) and Murlough (County Down).

Plan implementation

Implementation began with awareness-raising at a number of levels (e.g. interpretation boards at demonstration sites, articles in the news media, scientific presentations). This continues with the production of a brochure and a web-site. The management plans will be part-implemented in 1999 and 2000 using LIFE project resources, and it is intended to implement them fully by seeking additional funds at regional, national and international levels. It is envisaged that beach/dune plans will be reviewed within the framework and timescale of the Donegal County Development Plan, thus integrating them with the existing planning framework. Such an approach is intended to ensure that the plans are implemented within a formal administrative structure. The project also intends to produce a good practice guide for community-based sustainable coastal management, which will be launched at an ICZM workshop to be held in Donegal in 2000.

Case Study

The project has largely completed the assessment phase at all sites. The process of designing and implementing the management plans is moving from the identification of issues to the definition of objectives and strategies. In order to illustrate the key

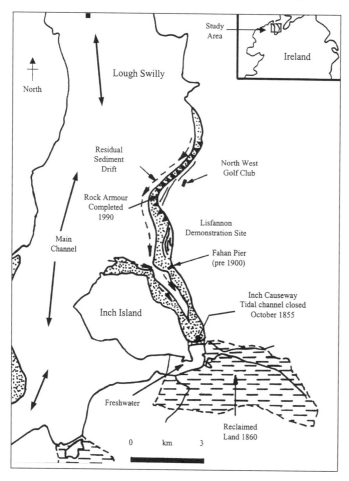

Figure 2.
Geomorphological processes and historical changes in Lough Swilly affecting dune development at Lisfannon (after Carter, 1989). Stippled areas are inter-tidal sediments.

features of the demonstration project, and to highlight experiences to date, a single case study will be presented.

Assessment

The Lisfannon dune complex is one of the most dynamic sites on the Irish coast (Figs. 2 and 3, Plate 1). Examination of 1834 and 1905 Ordnance Survey sheets (scale: ≤6 inches to the mile) shows a wide inter-tidal area, and a mean high-water mark closely following a rocky shore. This is now marked by an overgrown walled embankment protecting the disused Buncrana to Derry railway from wave attack. These sea walls are up to 300m from the present shoreline. The intervening area is now occupied by sand dune, saltmarsh and wetland habitats.

Sediment accumulation at the site has taken place under conditions of strong southerly drift, driven by locally generated estuary waves. A limited sediment supply in the area with little contemporary input has caused successive development of drift-aligned forelands. As sediment supply begins to deplete, drift alignment breaks down

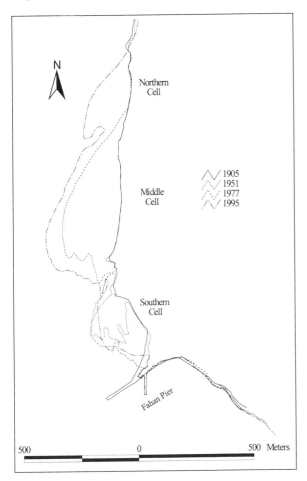

Figure 3. Shoreline changes at Lisfannon 1905–1995 determined by ARC-INFO comparison of maps and aerial photographs.

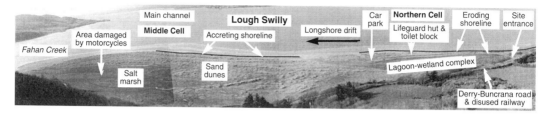

Plate 1. *Middle and northern cells at Lisfannon (1998). The dunes seaward of the Derry–Buncrana road have developed since the early 1950s.*

and the foreland erodes, with sediment deposited further south (downdrift). An initial foreland, on which the North West golf course is now located, developed during the early 1800s. It continued to accumulate into the mid-1900s, but when sediment supply was exhausted, it began to break down and experienced continuous erosion.

As the golf course eroded, much of the sediment removed accumulated on a foreland immediately to the south on which modern-day recreation takes place. Deposition of some of this eroded material in Fahan Creek has been aided by reclamation in its upper reaches, which reduced the tidal prism and created conditions favourable to siltation of the tidal channel (Fig. 2).

At the Lisfannon demonstration site, the process of dune accretion began under conditions of positive sediment supply by expansion of small areas of older dunes at Fahan Pier. A developing series of shore-parallel dune ridges gradually isolated the back-dune area from the sea. This led, over a period of 40–50 years, to the development of a varied dune topography and wide diversity of habitats – grey fixed dune, saltmarsh, brackish lagoon, freshwater wetland and alder scrub. Their present species composition is a function of complex interactions of factors, e.g. time, sediment supply, hydrology, marine inundation, rabbit grazing and human disturbance. The site now consists of three distinct sediment cells. The southern cell is the oldest part of the system. Here foredune (*Elytrigia juncea* and *Ammophila arenaria* communities) and yellow dune (*Ammophila*-dominated communities) are absent. Species-rich grey dune, often with over 20 plant taxa per square metre, extends to the high water mark. This is dominated by *Ammophila arenaria, Festuca rubra*, herbs and mosses. Examples include lady's bedstraw *Galium verum*, thyme *Thymus praecox*, cat's ear *Hypochaeris radicata*, plantain *Plantago lanceolata*, and the mosses *Rhytidiadelphus squarrosus, Calliergon cuspidatum* and *Hylocomium splendens*. The seaward edge of the middle cell and the boundary with the northern cell contain extensive foredunes. These areas are backed by grey dune, saltmarsh and dune/saltmarsh mosaics (with species such as sea milkwort *Glaux maritima* and thrift *Armeria maritima*). The younger northern cell has few foredunes, a narrow strip of consolidated dunes, and is dominated by a shallow, seasonally fluctuating brackish lagoon-wetland complex.

Faced with progressive shoreline erosion, the golf club to the north of the site installed a fringing wall of rock armour (completed in 1990). This dramatically reduced the amount of sediment moving south to maintain the northern part of the Lisfannon system. It also coincided with the formal development of this area as a Blue Flag tourist beach. The northern cell is favoured by visitors because of its wide inter-tidal zone and lack of stones in the beach sand. A shore-parallel access road and car park were installed on the newly consolidated dunes. Here, the immature dune grassland between the beach and the access road has recently been severely damaged by trampling, nutrient enrichment and marine erosion. This area was subject to wave attack, breaching and marine overwash during a minor storm in February 1998. Other recent developments at Lisfannon have been trimming of the seaward edge of the southern cell by wave attack, continued accretion of the middle cell, and a large marina currently under construction on the foreshore of the southern cell. Lisfannon was included in the Lough Swilly Natural Heritage Area in 1994. However, it was not designated as either a candidate Special Area of Conservation for its dune habitats, or as a Special Protection Area for its bird populations (which include Brent geese *Branta bernicla* and oystercatcher *Haematopus ostralegus*). It is often used for environmental education projects, as the process of sand dune accretion and succession can be graphically illustrated.

A survey of visitors to Lisfannon indicated that informal family recreation, walking, swimming and pet exercising are the most common activities on this beach. It tends to be used mainly for short duration visits by local people, and day-trippers from nearby Derry City. Lisfannon differs from many other recreational beaches in Ireland in that cars are not permitted to park on the strand.

Issues

The following list of management issues was compiled by the project team and the local community at a public meeting held at Lisfannon in June 1998:

1. Breach of the dune line of the northern cell by storms;
2. Recreational damage to dune front;
3. Use of motorcycles and dune buggies on the dunes;
4. Marina development;
5. Dumping of spoil around the periphery of the site;
6. Car access and parking facilities;
7. Litter;
8. Toilets;
9. Dog fouling;
10. Provision of lifeguards;
11. Public access to foreshore and dunes;
12. Bathing water quality and appearance;
13. Public information and awareness;
14. Retention of Blue Flag status.

Objectives and strategies

Measures to address the range of management issues have not yet been defined for any of the LIFE demonstration sites in Donegal. It is anticipated that a hierarchical set of interrelated objectives and associated strategies will be most appropriate. At Lisfannon, the first-order decision for management is whether to retain the dune system or to permit it to erode. If the first option is selected, an objective might be to address the recent destabilization of the northern cell, as this threatens the existence of the Blue Flag beach.

The range of potential strategies in response to the erosion of the northern cell may include:

1. Protect the site by hard engineering to maintain the current shoreline;
2. Use environmentally sensitive soft engineering methods (e.g. renourishment, revegetation) to sustain beach and dune development. Control access by vehicles and pedestrians through the dunes. Accept the possible loss of the site;
3. Control access by vehicles and pedestrians through the dunes. Accept the probable loss of the site;
4. Do not implement erosion control measures and do not maintain beach facilities. Accept the inevitable loss of the site.

As at all the demonstration sites, ongoing management continued during the period of assessment, and will undoubtedly influence emerging strategies. For example, as a result of damage to the frontal dune ridge at Lisfannon, a temporary wire fence was erected parallel to the shoreline to prevent access by vehicles, and to control visitor movements. In addition, a recent inter-departmental project meeting within the County Council revealed the availability of sediment to be dredged from Buncrana Harbour for a new ferry port. Grain size analysis suggests that this material would be suitable to renourish the beach at Lisfannon.

Full scientific findings will be presented at a series of meetings in early 1999. At this stage, the project team, authorities with a coastal management function, landowners and beach/dune users will jointly develop objectives and strategies for Lisfannon and the other demonstration sites. At this stage, all potential strategies will be considered. However, given the rationale behind the project, and the funds available for implementation, large-scale engineering work to stabilize dune shorelines is unlikely to be acceptable on scientific, environmental or economic grounds.

Concluding Remarks

The issue of shoreline instability dominates all discussions on the management of the Lisfannon site. This must be addressed before other strategies can be implemented. Public perception that all property and infrastructure must be protected from coastal erosion may be an impediment to rational decision-making at Lisfannon. Another important factor is the apparent lack of statutory protection afforded to a non-SPA/cSAC coastal site in the face of proposed development (such as the marina on the southern cell at Lisfannon). Confusion surrounding land ownership, jurisdiction and legal responsibility on such a dynamic coastline is a further constraint to sustainable coastal management.

It is clear that a scientifically informed decision-making forum is necessary for effective coastal management. However, scientific input alone is not sufficient, and decisions will ultimately be based on cultural, political, economic and social considerations, in addition to models and databases. We believe, however, that it is only by incorporating scientific approaches into the process that the consequences of management options can be fully appreciated in advance.

References

Brady Shipman Martin Ltd (1997), *Coastal Zone Management: A Draft Policy for Ireland,* Government of Ireland, Dublin.

Carter, R.W.G. (1989), 'Resources and management of Irish coastal waters and adjacent coasts', in R.W.G. Carter and A.J. Parker (eds.), *Ireland: A Contemporary Geographical Perspective,* Routledge, London, 393–420.

Carter, R.W.G. (1990), 'The geomorphology of coastal dunes in Ireland', in T.W. Bakker, P.D. Jungerius and J.A. Klijn, *Dunes of the European Coast,* Catena-Verlag, Cremlingen, 31–40.

European Commission (1997), *Better Management of Coastal Resources,* Office for Official Publications of the European Communities, Luxembourg.

European Community (1992), *EC Council Directive 92/43/EEC Conservation of Natural Habitats and of Wild Flora and Fauna,* European Community, Brussels.

DUNE MANAGEMENT ON THE ATLANTIC COAST OF FRANCE: A CASE STUDY

ROLAND PASKOFF
Lumière University, France

Introduction

The southwestern Atlantic coast of France, the Aquitaine coast, is an almost rectilinear shoreline, 230km long, the only interruption being in its middle and corresponding to the inlet of the Arcachon lagoon (see Fig. 1). In its natural state, still prevailing in the eighteenth century, this coast was characterized by mobile long ridges of barchan type dunes, up to 7km wide and often exceeding 50m in height (Buffault, 1942). At that time foredunes did not exist because of strong onshore winds removing much more sand than spontaneous vegetation was able to trap (Duparc, 1983). Mobile dunes were extensively stabilized in the first half of the nineteenth century through planting of maritime pines *Pinus maritima*. However, this new forest, which rapidly became the main economic resource of the region, was threatened, at least on its seaward margin, by continuing sand arrival. To protect the forest from the impact of blowing grains and the corrosive effect of salt spray, it was decided to create an artificial dune, acting as a rampart, along all the length of the coast. The quartzitic sand coming from the beach was held in place by palissades which were raised as they were buried, and by planting of marram *Ammophila arenaria*.

Fragility of the Artificial Dune

The profile of the artificial dune was designed by an engineer, J.S. Goury. Known as the *ideal profile* (see Fig. 2), it shows an outer slope on the seaward side with a gentle gradient of about 20 per cent, a subhorizontal plateau, both being planted, and a lee slope or slipface with a steep gradient of 60 per cent which corresponds to the angle of repose of dry sand. The artificial dune, about 150–200m wide and 12–15m high, represents a fragile construction. It needs permanent maintenance which is insured by a State agency, the *Office national des forêts*.

Three main factors account for the fragility of the artificial dune: rapid marine erosion, intense aeolian dynamics and touristic pressure.

On the Aquitaine coast, characterized by a high energy wave regime, marine erosion

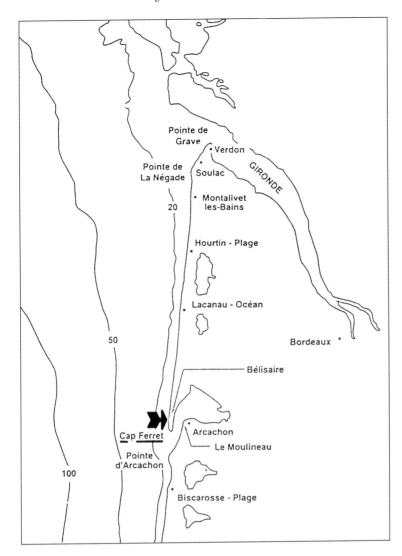

Figure 1. *The Aquitaine coast: location map.*

is severe because the sediment budget is negative and it has been so probably since the beginning of the Christian era. The retreat of the shoreline has nowhere been less than 1m/yr on average and often exceeds 2m/yr (Lorin and Viguier, 1987). Marine erosion does not occur uniformly along the coast. In sectors affected by rapid erosion, the outer slope of the dune is cut by a sandy cliff whose top corresponds to the 'plateau' (see Plate 1). Between these receding sectors, areas are found where the shoreline is temporarily stabilized or even prograding. As a matter of fact, the same sector of coast is alternately eroding and accreting, a situation due to sand waves migrating southward. However, in the long run, erosion is more important than accretion. German blockhouses of the defensive Atlantic wall dating back to World War II, originally built up on the edge of the plateau of the dune and now tumbled onto the beach

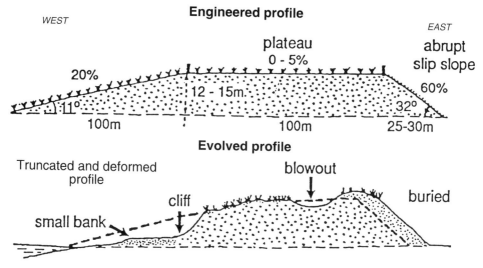

Figure 2. *Profiles across the artificial dune of the Aquitaine coast.*

or even onto the foreshore, give evidence of a rapid retreat of the coastline.

High and frequent northwesterly winds are blowing on the Aquitaine coast (Froidefond and Prud'homme, 1991). Sandy cliffs cut into the dune by marine erosion increase wind velocity and turbulence leading to the excavation of breaches and corridors which in course of time may enlarge into wide blowouts, allowing sand tongues to migrate landward. As a result, the artificially created dune is not only getting out of shape, but also moving inland (see Fig. 2). The average quantity of sand arriving to the rear of the dune may exceed 25m³/m/yr. In some instances, the landward migration of the dune is spectacular since it reaches a velocity greater than 10m/yr (Barrère, 1992).

Near urbanized areas, uncontrolled human access, which has been steadily increasing during the last decade in relation to the touristic development of the Aquitaine coast, leads to the degradation of the vegetation cover of the dune. Consequently, an enhanced wind action has accelerated the destabilization of the dune and its landward migration.

The Site of Cap-Ferret

The site of Cap-Ferret may be chosen as a significant example of the difficulties which arise in the management of the artificial dune in a highly visited part of the Aquitaine coast. It also gives a good illustration of an intervention by the *Conservatoire du littoral,* a State agency whose main objective is to buy coastal sites to remove them permanently from the threat of urbanization, in close partnership with the *Office national des forêts.* In France both agencies are the main actors as far as dune management is concerned.

Plate 1. *Truncated profile of the artificial dune at Cap-Ferret; the outer slope has been eroded and a cliff is cut into the plateau.*

The site of Cap-Ferret lies at the southernmost part of the sand spit which isolates the Arcachon lagoon from the ocean (see Fig. 1). Over more than two centuries the spit had been lengthening at a rate of about 3m/yr. However, over the last 30 years a reverse trend is prevailing and erosion is taking place at a rate of 2.5m/yr. As elsewhere on the Aquitaine coast, an artificial dune was erected here, but the transfer of land to private ownership at the beginning of the century resulted in a lack of maintenance of the fixing vegetation which virtually disappeared due to uncontrolled access. The present phase of severe marine erosion combined with the touristic boom of the last decades made the situation worse. This resulted in the sand becoming mobile again from deflation hollows. Blowouts eventually increased in size and coalescing tongues of sand advanced toward the pine forest. Between 1989 and 1992 work was carried out to control aeolian activity, but without stopping it completely as the rejuvenation it gives to the landscape is worth maintaining. Vehicular access also had to be controlled and was reduced by 50 per cent by limiting the number of car parks.

Here we will consider the restoration works in the northern part of the site, which is 2km long. In this area advancing sand was not only invading the forest but also engulfing dwellings which had been imprudently built up just behind the dune (see Plate 2). The aim was not only to control the aeolian processes by the reinstatement of a plant cover, but also to reconcile the protection of the restored environment with public access through fencing, footpaths and information panels, since the latter is one of the objectives of the policy conducted by the *Conservatoire du littoral*. Prior to the work, ecodynamic mapping at a scale of 1:5,000 was carried out by geographers from the University of Bordeaux under the direction of Professor P. Barrère. Such a document, combining geomorphological features and plant characteristics, was used as a guideline for the technicians of the *Office national des forêts* which was in charge of the restoration works and is still responsible for site management.

Works involving different types of operations started in 1994. First of all, some 100,000m³ of sand had to be extracted from the slipface invading the houses and then transported to the seaward side of the dune whose sandy cliff was reprofiled. Nylon nets acting as windbreaks were used to fill up blowouts. Deflation areas were covered with branches to stop large volumes of sand being blown away (see Plate 3). The density of branches, which were spread mechanically, was about 900kg/ha on average. Their slow decomposition, providing a supply of organic matter, encourages spontaneous plant colonization. However, this technique has the disadvantage of introducing foreign species into the dune environment. Extensive surfaces were planted with *Ammophila arenaria* and *Elytrigia juncea* (see Plate 4).

It appears that the situation is now stabilized. As a matter of fact, the success of the operation is in part due to a temporary respite in marine erosion in relation to the arrival of a sand wave in this sector. However, because the coastline is receding in the long run, there still exists the risk of the urbanized area being covered with sand again.

Plate 2. *Inner slope of the artificial dune advancing and engulfing houses at Cap-Ferret.*

Plate 3. *Branches covering the plateau of the artificial dune at Cap-Ferret.*

Plate 4. Ammophila arenaria *planting on the gentle seaward slope of the artificial dune at Cap-Ferret.*

Conclusion: A New Approach in Dune Management

In France as in other developed countries, during the 1980s and 1990s, thinking on dune management underwent a notable change due to a better dialogue between technicians and scientists. The *Office national des forêts*, whose historical mission in dune areas was to stop completely all moving sands through afforestation, nowadays allows the movement of dunes to a certain extent. Along the Aquitaine coast, since the retreat of the shoreline is unavoidable, the landward migration of the artificial dune is accommodated rather than opposed, intervention being aimed at reducing to a minimum natural degradation and deterioration from trampling pressure (Favennec, 1998). A smoother, more undulating topography is taking the place of the geometrical *ideal profile*. Traditional engineering techniques, such as the reshaping of the dune with heavy machinery, are no longer systematically used. Instead of brutal and traumatic mechanical operations, soft ecodynamic methods, based on Nature's ability to self-regenerate and on the assistance of geomorphic processes through fencing and planting, are now in practice. In this respect, the *Conservatoire du littoral*, which owns many dune sites, has acquired a great expertise and attaches great importance to the dissemination of environmental techniques in dune restoration and management (Paskoff, 1996). Since the policy of the *Conservatoire* is to avoid an artificialization of the environment, sand movements are accepted on its sites when they result from natural phenomena, except where they may threaten to bury human settlements which cannot be displaced. Maintenance of biodiversity is also a concern: efforts are made to eradicate introduced invasive plants for the benefit of native species. The necessary respect for natural processes is progressively gaining ground among those who are involved in dune management in France.

References

Barrère, P. (1992), 'Dynamics and management of the coastal dunes of the Lande, Gascony, France', in R.W.G. Carter, T.G.F. Curtis and M.J. Sheehy-Skeffington (eds.), *Coastal Dunes*, Rotterdam, Balkema, 25–32.

Buffault, P. (1942), *Histoire des dunes maritimes de Gascogne*, Bordeaux, Delmas.

Duparc, J.L. (1983), 'La restauration des dunes littorales de Gascogne', unpublished PhD thesis, Bordeaux III University.

Favennec, J. (1998), 'The dunes of the Atlantic coast of France, typology and management', *Coastline*, **1**, 14–16.

Froidefond, J.-M. and Prud'homme, R. (1991), 'Coastal erosion and aeolian sand transport on the Aquitaine coast', *Acta Mechanica*, **Suppl. 2**, 147–59.

Lorin, J. and Viguier, J. (1987), 'Régime hydro-sédimentaire et évolution actuelle du littoral aquitain', *Bulletin de l'Institut de Géologie du Bassin d'Aquitaine*, **41**, 95–108.

Paskoff, R. (1996), 'Aim for the dune! Natural recovery?', *Annales du Conservatoire du Littoral*, **96**, 156–62.

COMPARATIVE CHARACTERISTICS OF OLD AND NEW GENERATIONS OF CURONIAN SPIT DUNES

R. MORKUNAITE

Institute of Geography, Vilnius, Lithuania

and A. CESNULEVICIUS

Vilnius University, Lithuania

Introduction

The Curonian spit separates the Curonian lagoon from the Baltic Sea. It is composed of both marine and aeolian sand. Of a total length of 98km, 50km belongs to Lithuania. Preliminary investigations of the Lithuanian coastal dunes were launched as early as the middle of the nineteenth century. More complex investigations of the coastal zone began in the 1940s–1950s. They included the questions of dune lithology, litho-dynamics, morphology, dune dynamics, buried soils of the old dunes, genesis and evolution of the new and old dunes. Using the principle of complex methodology (morphological, lithological and morphometrical analysis) in this work we make an attempt to reveal the differences between the old and new generations of dunes. This is intended to contribute to future work in determining the duration of historical sand-blow in the Lithuanian coastal (and continental) dunes.

Methods

Using 1:10,000 topographic maps we made cross profiles of relief which reveal the morphological differences between the old parabolic and new dunes (see Figs. 1 and 2). A morphometric evaluation of the relief of the Curonian spit was carried out using cartographic measurements on the maps and morphometric classification. Using the material published by other authors (Gudelis and Michaliukaite, 1959, 1976; Michaliukaite, 1962; Gudelis et al., 1993) and data from our own investigations, we analysed the granulometric composition of the sands. Using the data obtained we made cumulative curves of the granulometric composition of sands and from these, using a graphical method, we calculated quartile values (Q_1, Q_2 (Md), Q_3), and a measure of sorting (S_o) using the Trask formula (Shvanov, 1968). The data obtained were interpreted using the classification of granulometric classes of sediments proposed by Malinauskas (1980) and the classification of sorting suggested by Fuechtfaner (Racinowski and Szczypek, 1958).

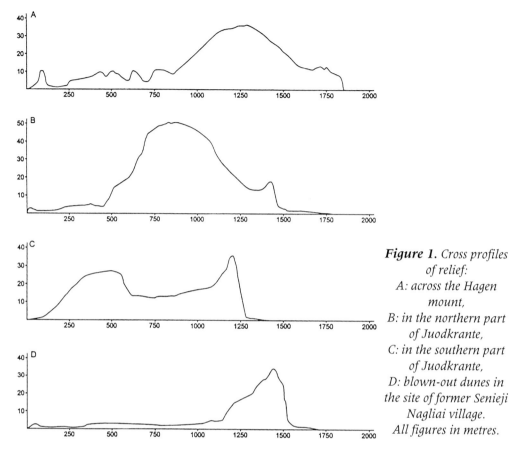

Figure 1. Cross profiles
of relief:
A: across the Hagen
mount,
B: in the northern part
of Juodkrante,
C: in the southern part
of Juodkrante,
D: blown-out dunes in
the site of former Senieji
Nagliai village.
All figures in metres.

Dunes Morphology

Old parabolic dunes are particularly well expressed in Juodkrante, where their length exceeds 4km and width ranges between 200 and 300m. They have a rather complicated form: many branches of dunes are linked together and they dissect the relief. Deep and complicated hollows and depressions are situated on the western slope of parabolic dunes. A small area of parabolic dunes in Nida–Skruzdyne is not so well morphologically developed. To the north of Pervalka, 15–20 years ago there still existed 6–8m high blowout residuals of parabolic dunes. At present no blowout dunes can be found. There are only exposures of old buried dune soils (Gudelis, 1989–1990).

New generation dunes stretch between Juodkrante and Pervalka, Nida–Pilkopa–Rasyte and to the north from Rasyte. The greater part of the ridge has been planted. The 9km long sector of blown-out dunes between Juodkrante and Pervalka differs considerably from other blown-out dunes of the Curonian spit. Its windward slope is dissected by buried soils and sand hillocks as well as by blowout forms, blowout valleys and depressions (Minkevicius, 1982). The dunes of this sector have flat tops formed

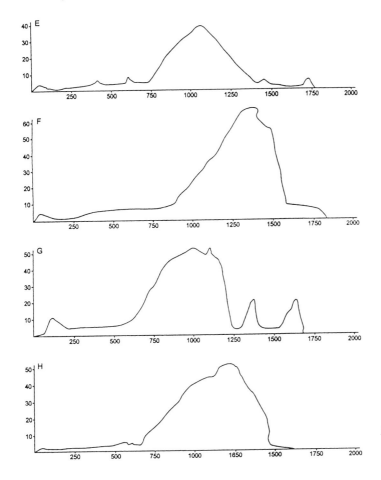

Figure 2. Cross profiles
of relief: E: Pervalka
dune ridge,
F: Vecekrugo mount.,
G: Nida (Skruzdyne),
H: Nida (Parnidis dune).
All figures in metres.

by deflation and vegetation. By contrast, the windward slope of the dune chain to the north of Nida is not dissected by exposures of buried soils and blowout remnants. The crest of the main dune ridge abounds in blown-out forms which are less stable.

In many places the ridge of old parabolic dunes coincides with new (active) dunes. This is reflected by the internal structure of the dune ridge; the layers of blown-out sands are separated by old forest soils. Soil analysis by paleobotanical and radiocarbon methods revealed that the lower soil horizon formed at the end of Atlantic/beginning of Subboreal period, i.e. approximately 4390±110 years ago. By comparison, the longshore ridge of high dunes on the Curonian spit is comparatively young, not older than 150 years (Gudelis and Michaliukaite, 1959).

Morphometric Types

Wind is the main relief-forming factor. Therefore, blowout formations prevail on the Curonian spit. The blowout relief is characterized by a variety of forms, ranging from

Figure 3. *Morphometric types of Curonian spit:*
1 = small low hills; 3 = small high hills; 5 = average size and height hills;
6 = average size high hills; 10 = wavy plain; 11 = flat plain.

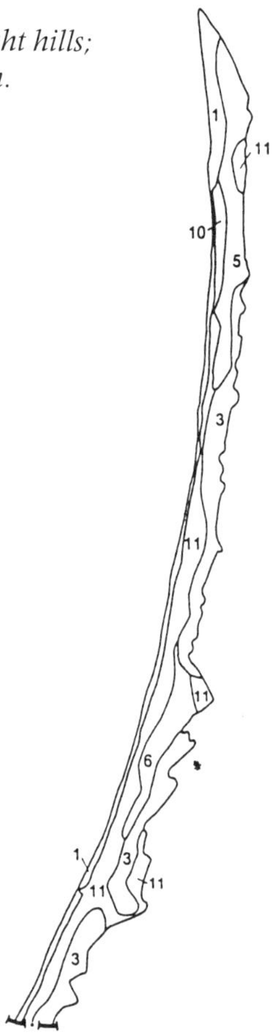

small, low hillocks to large, high dunes. A morphometric evaluation has identified six types of relief (see Fig. 3). The western side of the Curonian spit is characterized by an artificial foredune ridge which can be attributed to the type of relief of *small low hills*. It is immediately followed by a wide blown sand plain composed of *flat plain* and *wavy plain* types of relief. The approaches to the lagoon are occupied by the main dune ridge. The sector between Smiltyne and Juodkrante belongs to two types of relief, *average size and height hills* and *average size high hills*. Between Juodkrante and Pervalka the ridge is composed of *small high hills*. To the south of Pervalka, the ridge loses its uniform character: its western part is composed of *average size high hills*, the eastern part of *small high hills*. The capes of the spit are dominated by a *flat plain* type of relief (see Table 1).

Figures 1 and 2, cross profiles of the Curonian spit, show the main differences along the dune ridge. They are particularly obvious when comparing the old parabolic dunes and blown-out ridge. In the environs of Juodkrante and Nida, an area of parabolic dunes, in front of the main dune there stretches a sector of blowouts and blowout depressions, whereas the remaining part of the spit has asymmetric ridge slopes (not everywhere). The old parabolic dunes are attributed to the type of relief of *average size high hills*, whereas the new ones belong to the type of relief *average size and height hills*.

Table 1. Morphometric types of Curonian spit relief

Type of relief	Area		Number of areas
	km²	(%)	
Flat plain	44.10	42	3
Wavy plain	2.97	3	1
Small low hills	12.17	12	1
Small high hills	20.40	21	3
Average size and height hills	12.90	13	1
Average size high hills	8.31	9	1

Grain-Size Characteristics

The grain-size characteristics reflect the conditions of sand accumulation. Thus, by an interpretation of the data obtained by grain-size analysis it is possible to reconstruct the conditions of sand accumulation. The present investigation is based on published data of grain-size composition of the blown-out sands (Gudelis and Michaliukaite, 1959). However, its graphical interpretation is different. Additionally, a special grain-size classification of sands was used (Malinauskas, 1980). The authors also made use of material obtained during geomorphological–paleogeographical investigations of old buried soils in the environs of Pervalka and Pilkopa–Nida carried out in 1988–1989 and individual data collected in Juodkrante in 1998 (see Table 2).

The cumulative curves (see Fig. 4) reveal what fractions dominate and what their amounts are in the windward slope of the Agila dune (mixed sands prevail), the western slope of a nameless dune situated to the north of Pervalka (mixed sands), the foot of the southwestern slope of Vingiakope dune (fine-grained sands), Pilkopa environs (Grobstas gate) (mixed sands) and the top of Ieva mount (fine sands).

The grain-size composition of new generation dunes (most samples taken during the expeditions of 1984, some from 1997–1998) is somewhat different: the surface layer of the windward slope of Agila mount is composed of fine-grained sand, the middle of the Parnidis dune slope is dominated by medium-grained sand, the windward slope

Table 2. Granulometric indices of aeolian sands in the Curonian spit

No.	Sampling sites	Sampling depth (cm)	M_d (mm)	S_o
Old dunes				
1	Western slope of Agila mount	150–200	0.093	1.55
2	Western slope of nameless dune	210–220	0.092	1.53
3	Foot of northwest slope of nameless dune	130–180	0.148	1.59
4	Foot of southwest slope of Vingiakope dune	180–200	0.088	1.49
5	Pilkopa. Grobstas cape	>120	0.145	1.54
6	Ieva dune	20	0.194	1.12
New dunes				
7	Windward slope of Agila mount	3	0.085	1.30
8	Middle of Parnidis dune slope	3	0.183	1.93
9	Windward slope of Liepa mount	3	0.130	2.39
10	Top of Skirpstas dune	20	0.280	1.29
11	Western slope of Vingiakope	20	1.413	1.14

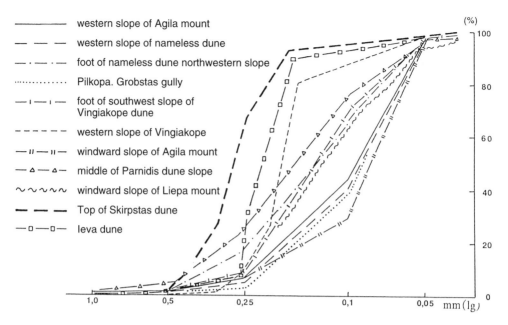

western slope of Agila mount

western slope of nameless dune

foot of nameless dune northwestern slope

Pilkopa. Grobstas gully

foot of southwest slope of
Vingiakope dune

western slope of Vingiakope

windward slope of Agila mount

middle of Parnidis dune slope

windward slope of Liepa mount

Top of Skirpstas dune

Ieva dune

Figure 4. *Cumulative curves of granulometric composition of Curonian spit aeolian sands.*

of Liepa mount, medium-grained sand and Vingiakope dune (western slope), medium-grained sand. The cumulative curves of grain-size composition of blown-out sand reveal that sands of most new dunes tend to accumulate in coarser fractions than in the old dunes. Graphic calculations of grain-size quartile measures Q_1, Q_2 (Md), and Q_3 and degree of sorting determined by the method of Trask (Shvanov, 1968) show that S_o in the old dunes is (on average) 1.47, and in the new dunes 1.61. Thus according to our data, the sands of the old generation dunes are only slightly better sorted.

To the south of Nida there is a clear tendency that the sands of the new dunes are poorly and very poorly sorted, whereas the sands of old dunes are intermediate (according to the Fuechtfaner classification). According to the data of other authors (Jarmalavicius and Zilinskas, 1996) the most fine-grained sand has accumulated in the foredune ridge (its average S_o value is 1.21), whereas the coarsest and least sorted sand can be found near the dynamic shoreline (S_o – 1.44). It should be noted that local deflation is still active in the artificial foredune ridge and the sector of blown-out dunes. It results in the formation of blowout depressions, open areas (particularly in the artificial foredune ridge), open blown-out southern dune slopes in depressions and blown-out trampled paths.

The differences between the old parabolic and younger dunes show that the old dunes have an older lithology. These forms are already preserved by the forest. By comparison, the new dunes have a coarser lithology and their surface is less transformed. These dunes are subject to localized erosion.

Conclusions

1. The old and the new dunes differ in their form and morphometric indices. The foot of the windward slope of the old parabolic dunes contains large deflation depressions and blowouts. The slopes are almost symmetrical (their inclinations differ only slightly). The windward slope of younger dunes begins immediately after the blowout remnants area. The slopes are clearly asymmetrical and inclinations differ.

2. The sands of the old generation dunes are mixed and fine-grained, whereas the sands of new generation dunes are mixed and medium-grained. A comparison of cumulative curves revealed that sands of the new dunes tend to be dominated by coarser fractions than the old dunes.

3. While determining the degree of sorting S_o of the blown-out sands we should differentiate with regard to location from Nida. To the south of Nida the new dune sands are badly to very badly sorted, while the old dune sands are moderately sorted.

4. The shape of the old parabolic dunes is preserved under the forest. Younger dunes are blown out and present conditions for the development of geomorphological processes.

References

Gudelis, V. (1989–1990), 'Lithology of old parabolic dunes of Curonian spit and dynamics of Litorina sea coastal processes', *The Geography Yearbook*, **25–26**, 13–17 (in Lithuanian).

Gudelis, V., Klimaviciene, V. and Savukyniene, N. (1993), 'The old soils of Curonian spit and their palynological characteristics', *Problems of Coastal Dynamics and Paleogeography of the Baltic Sea*, **2**, 93–98 (in Lithuanian).

Gudelis, V. and Michaliukaite, E. (1959), 'On the question of lithology and eolodynamic differentiation of new eolian sands of Curonian lagoon', *The Geography Yearbook*, **2**, 535–59 (in Lithuanian).

Gudelis, V. and Michaliukaite, E. (1976), 'Old parabolic dunes of Curonian spit', *Geographia Lithuanica*, Vilnius, 59–63 (in Russian).

Jarmalavicius, D. and Zilinskas, G. (1996), 'Distribution peculiarities of granulometric composition of surface sediment on the Lithuanian coast of the Baltic Sea', *Geography*, **32**, 77–84 (in Lithuanian).

Malinauskas, Z. (1980), 'Structure and composition of intramorainic complexes of Pleistocene', dissertation for cand. geol. sc. degree (in Russian).

Michaliukaite, E. (1962), 'Old dunes of Curonian lagoon and their soils', *The Geography Yearbook*, **5**, 377–89 (in Lithuanian).

Minkevicius, V. (1982), 'Eolian relief of the Baltic coastal area', *The Geography Yearbook*, **20**, 156–61 (in Lithuanian).

Racinowski, R. and Szczypek, T. (1958), *Prezentacja i interpretacja wynikow badan uziarnenia osadow czwartorzedowysch*, Warshawa (in Polish).

Shvanov, V.N. (1968), *Sand Rocks and Methods of their Investigation*, Moscow (in Russian).

THE DEVELOPMENT OF A GREEN BEACH ON THE SEFTON COAST, MERSEYSIDE, UK

S.E. EDMONDSON, H. TRAYNOR and S. McKINNELL
Environmental and Biological Studies, Liverpool Hope University College, England

ABSTRACT

The coastline of the Sefton Coast sand dune system is eroding rapidly along the more open, central section, and accreting at the northern and southern ends of the system where it is transitional to estuarine systems. The development of a green beach feature at the accreting northern end of the system is described by using time-series aerial photography (1989–1997). In this period the feature grew from approximately 2 to 12ha. In addition, field survey shows the profile of the feature and the development from saltmarsh to dune-building vegetation to build a low sand dune ridge on the beach. The significant influence of human impact, particularly vehicles driving along the beach, on the development of the feature is also demonstrated. The process of dune formation from saltmarsh vegetation, together with the incipient primary dune slack which is formed behind it, is an important mechanism for accretion on this northern end of the sand dune coast.

Introduction

This paper aims to describe and evaluate the green beach that has formed at the northern end of the Sefton Coast sand dune system. The feature is a shore-parallel ridge formed on the beach plain and is initiated by colonization of the saltmarsh grass *Puccinellia maritima*. This promotes enhanced sand accretion rates which eventually allow the establishment of sand dune grasses.

The open coast of the central part of the dune system grades into estuarine conditions of the Mersey estuary in the south, and the Ribble in the north. Between 1845 and 1906, the coastline was prograding along its entire length, but since 1906 the central section, centred around Formby Point, has been rapidly eroding, a trend associated with the establishment of a negative beach sediment budget and reduction in backshore width. The northern and southern ends of the system, however, have continued to prograde (Plater et al., 1993) (see Fig. 1).

Payne (1983) reports the first appearance of *Puccinellia maritima* on the beach at

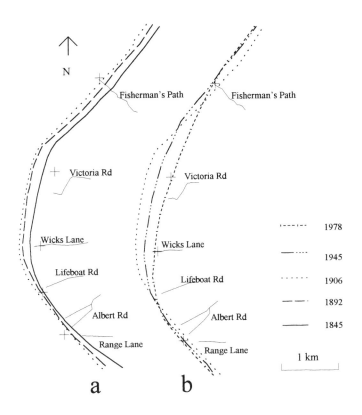

N

Fisherman's Path

Fisherman's Path

Victoria Rd

Victoria Rd

Wicks Lane

Wicks Lane

Lifeboat Rd

Lifeboat Rd

Albert Rd

Albert Rd

Range Lane

Range Lane

a   b

˙----˙---	1978
——˙····	1945
······	1906
——˙—	1892
——	1845

1 km

Figure 1. *Erosion and accretion on the Sefton Coast. The study area is at the northern accreting end of the system; (a) growth of Formby Point 1845–1906, (b) erosion of Formby Point 1906–1978 (after Turner, 1984).*

Ainsdale in 1982. At Birkdale, a little further north, the sward of *P. maritima* which initiated the current green beach first began to appear in 1986 (Sturgess, 1988). A similar feature had occurred in the area between Ainsdale and Freshfield (now part of the erosional front) in the 1930s. Allen (1932) reported a line of *P. maritima* on the sandy beach '10 yards seaward of the base of the outermost dunes' which first appeared in 1930. By the early summer of 1931, there was a 350 metre long ridge beyond the reach of ordinary tides. Other saltmarsh species had colonized the area together with some freshwater marsh and ruderals. The species lists include *Aster tripolium, Ranunculus sceleratus, Plantago maritima, Bolboschoenus maritima, Triglochin* spp., *Cochlearia* sp., *Atriplex* sp., *Tripleurospermum inodorum, Phragmites australis, Puccinellia distans, Alopecurus geniculatus* and *Lolium perenne.* By July of the same year two more ridges had formed seaward of the first, the primary colonizer in each case being *Puccinellia maritima.* A large pool of 'practically fresh water' colonized by a range of algal species separated the first and second ridges, indicating the formation of an incipient primary dune slack. This feature reported by Allen (who reports the third ridge as being only vaguely discernible after the winter of 1931) is likely to have been relatively transient, being on a section of the coast reported by Plater et al. (1993) to be eroding in the 1930s. The current green beach, however, has now persisted for more than ten years and is on an accreting section of the coast.

Methodology

The green beach was mapped by tracing the area of beach vegetation detectable by eye from stereoscopic pairs of vertical aerial photographs from five dates between 1989 and 1992 (see Table 1). The maps were imported into a GIS (Geographical Information System) and geo-registered, to show changes in the shape of the feature and to allow measurement of the total area.

Table 1. Vertical aerial photograph sources used to map development of the green beach. All photographs were loaned by Sefton Metropolitan Borough Council.

Date	Type of photograph	Scale
1989	Black and white	1:10000
1992	True colour	1:10000
1993	True colour	1:10000
1995	False colour infra-red	1:10000
1997	True colour	1:10000

In October 1997 three profiles (a northern, middle and southern section) across the feature, from the embryo dunes to the seaward limit of the vegetation, were surveyed at 10m intervals, mapped by three point bearings onto prominent landmarks, and levelled to Ordnance Datum. A 10m-wide belt transect along the line of the profiles was used to record the vegetation in $100m^2$ quadrats. Twinspan (using the Vespan III package, Malloch, 1995) was used to classify the vegetation recorded. Species lists for the whole site were recorded by Smith (1997; 1998).

Results

Mapping

Maps drawn from the aerial photographs show an increase in size and some changes in morphology of the beach over the study period (see Fig. 2). The extent of the green beach on these maps increased from 2ha to more than 12ha in the study period (see Table 2 for exact figures). Ground-truthing of these maps, however, shows that these figures significantly under-record the maximum extent of the green beach. The position of the southern profile across the green beach (see 1997 map, Fig. 2) is nearly 1km south of the southern tip of the mapped area. Although there will have been some extension of the green beach in the four-month period from the time of the 1997 air photograph to the time of the field survey, the maps are assumed to represent the extent of a more established phase of green beach development, after the initial colonization by *Puccinellia maritima*.

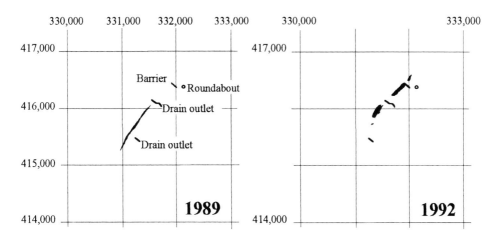

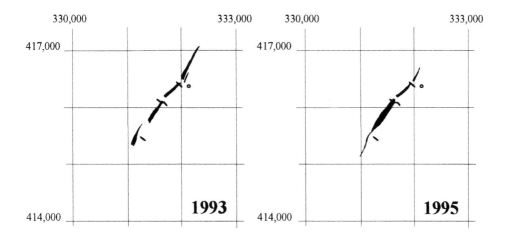

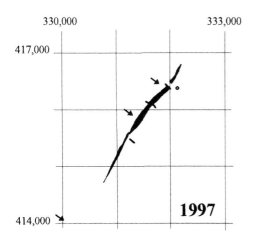

Figure 2. *Maps drawn from time-series aerial photographs showing the development of the green beach. The positions of the beach profiles are indicated by arrows on the 1997 map.*

Table 2. Area of the green beach mapped from air photographs

Date	Area of the green beach (m²)
1989	20030
1992	36940
1993	81191
1995	81161
1997	124840

In 1989, the mapped area was approximately 1km in length and covered an area of just over 2ha. This is very similar to Sturgess' 1988 description of the green beach which was just over 1km in length and varying in width from 10 to 25m. Between 1989 and 1992 the beach continued to increase in total area but significant changes in the shape and position of the green beach occurred. The beach was centred along a more northerly position and was wider but with two major gaps along its length. This change in position and shape is presumed to be the result of the storm surge of 26 February 1990 and subsequent backshore recovery (see Pye, 1991). The frontage of the entire dune system suffered severe erosion during this storm, amounting to 6m between Ainsdale and Southport, the area of the green beach (Plater et al., 1993). The more northerly gap in the line of the beach is due to a car park and vehicle and pedestrian access from the roundabout shown in Figure 2. The more southerly gap is associated with a drain outlet.

Between 1992 and 1993, the measured area of the green beach more than doubled and the line had extended both northwards and southwards, the southerly extension being separated from the main line of the beach by another large gap again associated with a drain. In 1995, apart from the site of the car park and access point, the gaps had become narrow but the overall size of the beach had declined slightly due largely to a contraction of the northern end. By 1997 the green beach had reached its maximum measured area, but both the narrow gaps associated with the drains and the larger gap around the car park had persisted.

Vehicles, walkers and also beach cleaning play a major role in ultimately defining the outline of the green beach (see Fig. 3). A heavily used, sparsely vegetated footpath runs between the green beach and the foredune communities. A vehicle track is associated with the most southerly persistent gap in the green beach. The car park (defined by a barrier) and beach access is clearly responsible for maintaining the larger, northern gap. A line of posts has been installed along a section of the western edge of the green beach immediately seaward of which is bare sand affected by vehicles, horse riding and mechanical beach cleaning (for amenity purposes).

Vegetation and profiles

The southern profile is at an early stage of development and shows a slightly raised

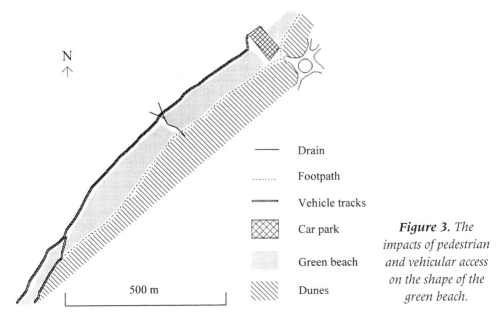

N
↑

Drain

Footpath

Vehicle tracks

Car park

Green beach

Dunes

500 m

Figure 3. The impacts of pedestrian and vehicular access on the shape of the green beach.

profile relative to the adjacent beach level. The older profiles show a clear ridge raised above the level of the adjacent beach to a maximum height of 1.54m on the central profile (see Fig. 4). Both profiles also show a depression between the green beach and the dunes which is mostly wet and muddy at the surface. The shelter of the green beach has produced a low energy environment where finer sediment has been deposited from water brought in on abnormal high tides. For much of the length, however this depression is not washed by normal tides and the feature is thus a large, incipient, primary dune slack.

Twinspan analysis classified the vegetation data from 33 quadrats into 4 types (see

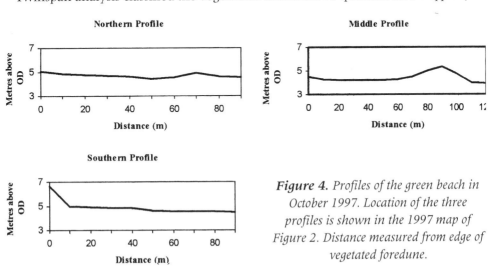

Northern Profile

Middle Profile

Southern Profile

Figure 4. Profiles of the green beach in October 1997. Location of the three profiles is shown in the 1997 map of Figure 2. Distance measured from edge of vegetated foredune.

Table 3). The first two are saltmarsh communities with *Puccinellia maritima* constant. Type 1 was found on sites with shallow, standing water with filamentous algae also constant. Type 2 is the most common vegetation type on the green beach and includes the early stages of development where only *P. maritima* occurs as well as sites where *Aster tripolium* and *Bolboschoenus maritima* are common, other species being rare. Types 3 and 4 are sand dune communities found on the higher parts of the green beach. These communities are found mostly at the seaward edge of the feature where the dune-building grasses trap sand blowing up from the beach.

Discussion

Accretion at the northern and southerly ends of the dune system is vital in balancing erosional losses in the centre around Formby Point. The green beach development is an important mechanism for coastal accretion at the northerly end, which has continued now for 13 years to produce a significant coastal landform. The formation of new dune and slack communities is also significant given the lack of dynamics and early seral communities elsewhere on the dune system (Edmondson et al., 1993).

The process of the development is clearly transitional from saltmarsh to sand dune. Sturgess (1988) suggested that the high nutrient and sediment content of water from the Mersey estuary was significant in initiating the growth of large microbial populations,

Plate 1. *Sand accumulation around clones of* Puccinellia maritima *forming hummocks on the beach.*

Table 3. Floristic tables for the four vegetation types identified by Twinspan analysis from 33 10 × 10m quadrats. Roman numerals are the frequency with which the species is found within the quadrats (maximum V), numbers in brackets are the maximum percentage cover values in the quadrats.

Type 1, n = 3	
Puccinellia sp.	V (60)
Filamentous algae	V (10)
Salicornia europaea	II (10)
Triglochin maritimum	II (10)

Type 2, n = 22	
Puccinellia maritima	V (90)
Aster tripolium	III (90)
Bolboschoenus maritimus	II (50)
Glaux maritima	I (80)
Spartina anglica	I (70)
Beta vulgaris maritima	I (10)
Atriplex sp.	I (15)
Elytrigia juncea	I (30)

Type 3, n = 6	
Ammophila arenaria	IV (30)
Aster tripolium	IV (50)
Leymus arenarius	III (50)
Elytrigia repens	I (40)

Type 4, n = 2	
Elytrigia juncea	V (10)
Ammophila arenaria	III (50)
Atriplex sp.	III (1)
Leymus arenarius	III (30)
Rumex crispus	III (5)

particularly of blue-green algae, diatoms and euglenids, on the wide, flat shore at the northern end of the Sefton Coast. The micro-organisms would bind surface sediment particles together, providing a more stable surface suitable for colonization by *Puccinellia maritima. P. maritima* has an even greater stabilizing effect, its roots inhibiting movement of sediment in the substrate and its leafy shoots trapping wind-blown sand in the sub-aerial environment. Upward growth of the shoots through the accumu-lating sand creates small hummocks on the beach surface which gradually coalesce to form a low ridge on the beach (see Plate 1). Other saltmarsh species are able to colonize the ridge and also the line of former beach plain inland of the new ridge. Transition to sand dune vegetation occurs when the height of the ridge is above the level of normal tides. The taller sand dune grasses, firstly *Elytrigia juncea* then *Leymus arenarius* and *Ammophila aenaria*, accelerate sand accretion, reaching maximum growth rates at the seaward edge of the ridge where the flux of sand from the beach is trapped. The species recorded in the majority of the quadrats in October 1997 are very similar to those recorded by Allen (1932) on the ridges which occurred in 1931, indicating that the feature was formed by a similar process. He concluded that the vegetation was a mix of saltmarsh and sand dune vegetation.

Transitions between saltmarsh and sand dune environments are often of great botanical interest (Radley, 1994). Smith (1997; 1998) recorded 93 higher plant species on the green beach (including the incipient dune slack behind the ridge) in the summers of 1997 and 1998 (see Table 4) and described the area as having 'outstanding botanical interest'. The species are a mix of saltmarsh, sand dune, freshwater marsh and ruderal plants, some of which are regionally rare or do not occur elsewhere on the sand dune system. The depression behind the green beach is the only incipient primary dune slack on the coast.

The presence of saltmarsh species and muddy substrates on the foreshore of a seaside resort area is considered by some local people, however, as aesthetically unattractive. The centre of the Victorian seaside resort of Southport is less than 1km from the northern end of the green beach. Despite considerable public concern about the erosion around Formby Point, there appears to be some disquiet about accretion of the green beach near Southport. Given the climate of ongoing controversy about management to promote dune dynamics elsewhere on the coast (see Edmondson and Velmans, this volume), interpretation of the green beach and public consultation on its future management present a significant challenge to coastal managers.

Conclusions

The green beach is a relatively stable, accreting landform which is of high conserva-tion significance both geomorphologically and ecologically. The transition from saltmarsh to sand dune is uncommon, and is associated with regionally rare species, an unusual species composition and an increasing species diversity. Within the whole coast context it is important in balancing the budget of erosional loss at Formby Point

Table 4. Species list for the green beach (Source: Smith, 1997; 1998)

Agrostis stolonifera	Glaux maritima	Puccinellia distans
Ammophila arenaria	Hippophae rhamnoides	P. maritima
Apium nodiflorum	Honkenya peploides	Pulicaria dysenterica
Arctium minus	Hordeum sp.	Ranunculus sceleratus
Aster tripolium	Hypochoeris radicata	Rumex crispus
Atriplex glabriuscula	Impatiens glandulifera	R. obtusifolius
A. laciniata	Juncus ambiguus	Raphanus raphanistrum
A. littoralis	J. articulatus	maritimum
A. portulacoides	J. bufonis	Sagina nodosa
A. prostrata	J. gerardii	Salicornia europaea agg.
Avena sativa	J. inflexus	Salsola kali kali
Beta vulgaris maritima	J. maritimus	Samolus valerandii
Bolboschoenus maritima	Leontodon saxatile	Schoenoplectus tabernae-
Cakile maritima	Leymus arenarius	montani
Carex extensa	Lolium perenne	Senecio jacobaea
C. flacca	Lycopersicon esculentum	S. vulgaris
C. otrubae	Lycopus europaeus	Silene uniflora
Centaureum littorale	Mentha aquatica	Sonchus apser
C. pulchellum	Myosotis laxa	S. arvensis
Chenopodium album	Oenanthe crocata	Spartina anglica
C. rubrum	O. fistulosa	Spergularia marina
Cirsium arvense	Parapholis incurva	S. media
Cochlearia anglica	Pesicaria maculosa	Stellaria media
Elytrigia juncea	Phragmites australis	Suaeda maritima
E. repens	Plantago coronopus	Taraxacum sect. Ruderalia
Epilobium hirsutum	P. lanceolata	Trifolium fragiferum
Erygium maritimum	P. major	Triglochin maritima
Euphorbia paralias	P. maritima	T. palustris
Festuca rubra	Polygonum aviculare	Tripleurospermum
F. arundinacea	P. lapathifolium	maritimum
Glaucium flavum	P. oxyspermum raii	Typha angustifolia

and in providing substrates for the development of early seral communities. Human influences, especially vehicles and beach cleaning, define the exact shape of the beach and limit accretion rates. There are indications that some local people view the green beach as unattractive and a threat to the amenity and recreational value of the area. Interpretation and public consultation must therefore be an important aspect of the future management of the green beach.

Acknowledgments

The authors would like to thank the Sefton Coast Life Project for help and for loan of the air photographs. In addition, thanks go to Dr Phil Smith for his meticulous recording of the plant species on the green beach.

References

Allen, M.J. (1932), 'Recent changes in the sea beach flora at Ainsdale, Lancashire', *North West Naturalist*, **7(2)**, 114–17.

Edmondson, S.E., Gateley, P.S., Rooney, P. and Sturgess, P.W. (1993), 'Plant communities and succession', in D. Atkinson and J.A. Houston (eds.), *The Sand Dunes of the Sefton Coast*, National Museums and Galleries on Merseyside, 65–84.

Malloch, A.J.C. (1995), *Vespan III: A Computer Package to Handle and Analyse Multivariate and Species Distribution Data*, Institute of Environmental and Biological Sciences, University of Lancaster.

Payne, K. (1983), 'The Vegetation of the Ainsdale Dunes', unpublished report, English Nature, Aindsale Sand Dunes NNR.

Plater, A., Huddart, D., Innes, J.B., Pye, K., Smith, A.J., and Tooley, M.J. (1993), 'Coastal and sea level changes', in D. Atkinson and J.A. Houston (eds.), *The Sand Dunes of the Sefton Coast*, National Museums and Galleries on Merseyside, 23–34.

Pye, K. (1991), 'Beach deflation and backshore dune formation following erosion under storm surge condition: an example from Northwest England', in O.E. Barndorff-Nielsen and B.B. Willetts (eds.), *Sand, Dust and Soil in their Relation to Aeolian and Littoral Processes*, Acta Mechanica, Supplementum 2.

Radley, G.P. (1994), *Sand Dune Vegetation Survey of Great Britain, Part I: England*, Joint Nature Conservation Committee.

Smith, P.H. (1997), 'A Vascular Plant List for the Green Beach, 4 and 17 July 1997', unpublished report to Sefton Coast Life Project.

Smith, P.H. (1998), 'Additions to the Vascular Flora of the Birkdale-Ainsdale Green Beach', unpublished report to Sefton Coast Life Project.

Sturgess, P. (1988), 'Tagg's Island – A Survey', unpublished report, Sefton Metropolitan Borough Council.

Turner, D.A. (1984), *A Guide to the Sefton Coast Data Base*, Sefton Metropolitan Borough Council, Engineers Department.

EXPECTED POSITIVE EFFECTS OF SHOREFACE NOURISHMENT ON THE VEGETATION OF CALCIUM-POOR DUNES AT TERSCHELLING (THE NETHERLANDS)

R. KETNER-OOSTRA

Freelance vegetation ecologist, Bennekom, The Netherlands

ABSTRACT

Sand suppletion on the shoreface may result in sand-drift of base-rich sand over dry dune vegetation. This will counteract the negative effects of nitrogen deposition and favour the continuation of lichen-rich vegetation types in these dunes.

Introduction

Since the 1980s a significant increase of grasses and mosses has been observed in the dry calcium-poor dunes at Terschelling (The Netherlands). These grey dunes were formerly characterized by a lichen-rich grey hairgrass *Corynephorus canescens* community *Violo-Corynephoretum dunense*. Now sand sedge *Carex arenaria* and marram *Ammophila arenaria* are dominant and the number of species has decreased considerably (see Table 1 and Ketner-Oostra, 1992; 1993). Partly this is a result of natural succession, but soil acidification has accelerated due to acid rain and nitrogen deposition (national average 40kg.ha^{-1}.yr^{-1}; the Wadden Sea islands 20kg.ha^{-1}.yr^{-1}). Fluctuations in rabbit populations also play a role.

Dynamic Coastal Management

At Terschelling the coastline between Hoorn and Oosterend has retreated considerably. Between 1965 and 1990 Rijkswaterstaat (National Institute for Coastal and Marine Management) straightened the coastline as a coastal protection measure. At beachpole 15,6 (one of the markers on the coastline) the foot of a large, protruding dune was levelled. This created a more landward position for the base of the dune and a wider beach. At the landward side of beachpole 15,6 stabilized dunes have been exposed to reactivated sand from the foredunes. In 1993, in order to stop further loss of sand at the coastline, a pilot project of sand nourishment under water on the shoreface was implemented. This Innovative Nourishment Techniques Evaluation

Table 1. Number of plant species in the grey hairgrass community at Terschelling in 1966 and 1990, in relation to soil pH. Location: R.D.-dune at Oosterend; average numbers of 6 plots of 4 × 4m

	Western slope		Southern slope	
	1966	1990	1966	1990
Distance to foot of dune	380m	380m	450m	450m
Number of grass-like plant species[a]	5	4	4	7
Mean cover of grass-like plants[a]	10%	90%	10%	90%
Number of herb species	9	5	8	5
Number of moss species	6	3	4	3
Number of lichen species	14	4	12	5
Mean cover of lichens	50%	3%	15%	1%
pH.H_2O	5.2[b]	4.8/5.3[c]	5.6[b]	3.9/4.0[d]

[a] including Carex arenaria
[b] in 1966: in soil layer 0–8 cm
[c] in 1991: in Ah1 = 0–4 cm / Ah2 = 4–10 cm;
[d] in 1991: in Ah1 = 0–5 cm / Ah2 = 5–16 cm

(NOURTEC) project involved three coastal sites. In Germany and Denmark similar projects were implemented (Spanhoff, 1998). At Terschelling two million cubic metres of sand was added to the shoreface between beachpoles 14 and 18. The beach volume increased at a rate unprecedented in the last 35 years (Biegel and Spanhoff, 1996).

Methodology

Vegetation relevés and chemical soil analyses were part of a monitoring programme in the lichen-rich grey hairgrass community started in 1995 (Ketner-Oostra, 1997).

Positive Effects of Sand-Drift on the Vegetation *(see Table 2)*

As a result of soil dynamics and mineral content, marram-dominated vegetation is characteristic of the seaward and landward side of the frontal dune at beachpole 15,6. The western slope of the adjoining large, landward dune has a vegetation rich in herbs, including a population of *Pilosella peleteriana* (Gadella, 1981).

The moss and lichen cover is rich in species typical of a calcium-rich soil. Soil analysis showed the sand to be almost neutral with a high base content (see Tables 3 and 4). On the southern slope of the large, landward dune herb cover is less and the average lichen cover is 40 per cent, but locally is 90 per cent. Species characteristic of neutral and acid soil grow together. Results of soil analysis indicate drift of base-rich sand to

Table 2. *Number of plant species at the large dune, landward of the frontal dune ridge, near beachpole 15,6 north of Hoorn; average numbers of 7 relevés of 2 × 2m, in the period 1993–95*

	Western slope	Southern slope
Distance to foot of dune	100m	200m
Number of grass species	9	8
Mean cover of grass species	10%	17%
Number of herb species	12	13
Mean cover of herb species	30%	10%
Number of moss species	5	7
Number of lichens	9	12
Mean cover of lichens	30%	40%

raise pH values; the surface soil layer, 0–2cm, was found to be pH.H_2O 6.1 (pH.KCl 5.4), whilst beneath this, 2–10cm, lower values of pH.H_2O 5.8 (pH.KCl 4.4) were recorded. Without sand drift the surface layer would be of lichens typical of acid, decalcified conditions (see Table 1).

Conclusion

The process of soil acidification seems to be slowed down by the incoming sand which has a high base content. As a result, lichen-rich vegetation types in the investigated dune area are maintained. To monitor the further development of vegetation and soil, several permanent plots were laid out as part of a monitoring programme in 1995 (Ketner-Oostra, 1997).

The National Institute for Coastal and Marine Management intends to maintain

Table 3. *Results of soil analyses in the 0–10cm soil layer of the seaward and landward sides of the frontal dune and the two slopes of the large adjoining landward dune at beachpole 15,6 north of Hoorn, in the period 1995–96*

	pH.H_2O	pH.KCl	% $CaCO_3$	% N	% C	% P
Seaward side frontal dune	–	9.2	0.6	–	–	–
Landward side frontal dune	7.2	7.4	0.8	0.004	0.12	0.014
Large dune, western slope	6.5	6.0	0.1	0.050	0.64	0.027
Large dune, southern slope	6.0	4.9	0.1	0.039	0.51	0.023

Table 4. Cation exchange capacity (CEC) and base content of the 0–10cm soil layer of the landward side of the frontal dune and two slopes of the large adjoining landward dune at beachpole 15,6 and of a lichen-rich inner dune ridge at Midsland; in 1995

	CEC (meq.kg^{-1})	Base content (%)
Dunes at beachpole 15,6		
Landward side, frontal dune	8.0	92
Large dune, western slope	31.0	79
Large dune, southern slope	17.2	43
Inner dune ridge at Midsland		
Northern slope of inner dune ridge	12.7	13

active sand drift management in the future. The continuity of incoming calcium-rich sand may counteract the negative effects of nitrogen deposition, which will have positive effects on the survival of lichen-rich dune grasslands in the Wadden Sea area.

References

Biegel, E.J. and Spanhoff, R. (1996), *Effectiveness of a Shoreface Nourishment, Terschelling, the Netherlands*, Rijkswaterstaat, National Institute of Coastal and Marine Management Report.

Gadella, T.W.J. (1981), 'Het Schellings havikskruid *(Hieracium peleterianum* Mérat)', *Gorteria* **10**, 121–29.

Ketner-Oostra, R. (1992), 'Vegetational changes between 1966–1990 in lichen-rich coastal dunes on the island of Terschelling (The Netherlands)', *Int. J. Mycol. Lichenol.*, **5**, 63–66.

Ketner-Oostra, R. (1993), 'Buntgrasduin op Terschelling na 25 jaar weer onderzocht', *De Levende Natuur*, **94**, 10–16.

Ketner-Oostra, R. (1997), *De korstmosrijke Duin-Buntgras-gemeenschap op Terschelling. De periode 1990–1995 vergeleken met de periode 1966–1972. Onderbouwing lange termijn monitoring-programma.* Staatsbosbeheer Fryslân, Report.

Spanhoff, R. (1998), 'Success of the shoreface nourishment at Terschelling', *Wadden Sea Newsletter 1998*, 9–15.

Section 2

NATURE MANAGEMENT

BRAUNTON BURROWS IN CONTEXT:
A COMPARATIVE MANAGEMENT STUDY

J.R. PACKHAM

School of Applied Sciences, University of Wolverhampton, England
and A.J. WILLIS
Department of Animal and Plant Sciences, University of Sheffield, England

ABSTRACT

The vegetation of Braunton Burrows in southwest England has coarsened considerably since World War II and since the myxomatosis epidemic which greatly reduced rabbit grazing from 1954 onwards. In the last few decades the area covered by scrub has greatly increased, with a consequent local loss of biodiversity, particularly amongst smaller plants of low relative growth rate (RGR). The water regime, with its strongly domed water table, appears to operate much as it did in the 1950s but the deepening, in 1983, of a drain of an adjoining marshland caused a lowering of the adjacent water table. Scrub eradication by mechanical means has proved both difficult and expensive. Grazing produces better quality vegetation than mowing but it is important to select the most appropriate grazing animal. Cattle control some coarse vegetation well, as do ponies; sheep and ponies are more effective in the creation of a short sward. Rabbits again exert an important influence in some areas.

After an outline of new management procedures, planned to improve the condition of Braunton Burrows, there is a more general discussion of management and conservation problems in other British dune systems. These include woodland, scrub and invasive species (often *Hippophae rhamnoides* with consequent eutrophication problems), visitor pressure, trampling and management control of successional processes. A distinction is made between recurring and restoration management, and the importance of monitoring with the continued use of control plots is emphasized.

Introduction

Coastal sand dunes are of great importance as biological and ecological systems (Willis, 1989), in which plant–water relations are often critical (Willis and Jefferies, 1963), grazing is important (Boorman and Fuller, 1982), and mycorrhizas play a major role in nutrient cycling (Read, 1989). Braunton Burrows, on the north coast of Devon in southwest England, is one of the best known, most studied and diverse sand dune systems in the UK. Although no longer a statutory National Nature Reserve (a consequence of management problems), the site is a candidate Special Area of Conservation under the Habitats Directive containing priority habitats.

This paper starts with a brief description of Braunton Burrows and the causes of its high biodiversity. Present problems are related to recent history, including the military use of the system during World War II and especially the changes consequent upon the diminution of rabbit grazing after the myxomatosis epidemic of the 1950s.

A discussion of the immediate management issues at Braunton Burrows, and the strategies being employed to address them, is followed by a brief review of dune management problems in general, with particular reference to a number of other British dune systems.

The conclusion stresses the value of well-controlled experiments in unravelling the complex biotic and abiotic systems which drive sand dune dynamics. It emphasizes the importance of retaining the whole range of successional stages within coastal systems, but also notes the stochastic nature of many events influencing sand dune conservation. No matter how carefully considered the original management plan, it will serve its purpose only if the plant and animal communities are adequately monitored and management adjusted accordingly.

Braunton Burrows:
Description and Causes of High Biodiversity

The coastal dunes at Braunton Burrows are described in some detail by Willis et al. (1959a; 1959b). They cover over 800ha, are over 2000 years old, show a great diversity in water regime, are constantly changing under the influence of the wind, and are composed almost everywhere of highly calcareous sands and, in the foredune area, are relatively high in magnesium. They extend for 5km northward from the joint estuary of the Rivers Taw and Torridge and form one of the highest dune systems in the British Isles, rising to over 30m OD in places. On the seaward side the dunes are fully exposed to the prevailing westerly winds and at low tide the large, gently graded expanse of Saunton Sands provides a plentiful supply of wind-blown sand. The western part of the Burrows is a complex of dunes whose irregular ridges run approximately north-south parallel to the coastline; the highest crests are mostly to landward. The inland part of the dune system is irregular but generally fairly low-lying with prominent sand hills only locally; the low areas of this region are flooded in wet seasons, as may be numerous slacks between the high dunes. The gentle seaward slope of the inter-dunal slacks is stabilized by the water table. To the east is the even lower-lying Braunton marsh, most of which has been drained and is used for agriculture.

Transect data for some dune slacks in the southern part of the system show a clear distinction between the pioneer species established in the first ten years before *Salix repens* has achieved dominance, and the secondary species which have subsequently colonized (Hope-Simpson and Yemm, 1979). Hope-Simpson (1997) outlined the work done to stabilize bare areas of the dunes during the period 1952–1961, also describing seral trends noted at Braunton Burrows for the period up to 1987. *Elytrigia juncea* had by then become much more abundant than it was in 1959.

A lowering of the water table in recent years on the most inland part of the dune system, probably connected with drainage operations, has been indicated by Anthony (1998). She compared the present state of the Burrows with that described by Willis et al. (1959a; 1959b). In respect of the southern transect T4 (Willis et al., 1959a; Fig. 6), the profile had changed only little, but the dune crest was flatter and the slopes less pronounced. Increase of *Salix repens* var. *argentea, Rubus fruticosus* agg., *Ononis repens, Carex arenaria* and *Festuca rubra* were noted, with pH values not as strongly alkaline as previously. With diminution of the rabbits, increased plant growth and prolonged leaching, the system appears to have progressed further towards a climax.

Since 1947 more than 350 taxa of angiosperms, some ten species of pteridophytes, over 70 species of bryophytes and a substantial lichen flora have been recorded from the dune system. The large number of vascular plants in the southern and middle sectors is about 80 more than expected from the Arrhenius regression $\log y = \log a + n \log x$, where y = number of species, x = area, a is a constant and n an index of regression (Dony, 1977; see also Connor and McCoy, 1979) for continuous exponential increase of number of species with area (Willis, 1985a). Naturalized and alien species form an interesting but minor element of the flora, while there also occur a considerable number of flowering plants rare or uncommon in Britain, e.g. *Teucrium scordium* and *Scirpoides holoschoenus*, some in substantial quantity.

There are numerous reasons for the species richness of Braunton Burrows (Willis, 1985a). As in other dune systems, its dynamic nature ensures the constant provision of habitats ranging from the bare sand favourable to the growth of annuals to more stable terrain bearing perennial herbs or shrubs. The influence of the water regime, including the domed water table, on niche differentiation is described by Packham and Willis (1997) while features such as aspect, slope, exposure, and protection further widen the variety of environments present. On the Burrows the high soil pH and calcium levels have precluded the development of the extensive dune heath seen, for example, at the South Haven Peninsula in Dorset on the south coast of England. Moreover, the low levels of soil nitrogen, phosphorus and potassium considerably diminish the competitive ability of large species of high RGR and tend to allow the continued survival of many 'stress-tolerant' forms (Willis, 1963; 1965; Willis and Yemm, 1961). High levels of rabbit grazing exert a similar influence, but rabbit populations are now more localized. The distributions of *Juncus acutus* and *Scirpoides holoschoenus* appear to be controlled by climate (Willis, 1985a) in this favourably located area. Some flowering populations, such as those of *Dactylorhiza praetermissa* and *D. incarnata*, vary substantially from year to year, perhaps owing to 'runs' of weather.

Recent History and Management Problems

The main management problems of Braunton Burrows in recent years have been concerned with a general coarsening of the vegetation and a much greater need for scrub control. As already mentioned, the high diversity of its vegetation arose largely from a combination of relatively low nutrient status and the grazing control exercised, until the outbreak of myxomatosis, by rabbits. Army activities during World War II, when the Burrows were used to practise amphibious landings by tracked vehicles, left what Breeds and Rogers (1998) describe as a 'semi-desert of shifting sand'. Subsequent sand stabilization by fencing and marram planting and the much reduced rabbit grazing greatly restricted the extent of bare sand areas available for plant colonization, an essential feature in this dynamic environment. There has also been considerable concern about the level of the water table, particularly beneath the slacks (Breeds and Rogers, 1998). Especially towards the landward part, there has since 1980 been a notable increase in woody plants, including *Salix cinerea* ssp. *oleifolia, S. repens* var. *argentea, Ligustrum vulgare, Rubus fruticosus* agg., *Betula* spp., *Crataegus monogyna* and *Prunus spinosa. Hippophae rhamnoides*, which formed large thickets up to 2ha in extent, was tackled in the 1970s, when this shrub was effectively controlled by the use of scrub-clearing saws, the cuttings being burnt. In some areas the coarse bush grass, *Calamagrostis epigejos*, and *Brachypodium sylvaticum* have spread. These changes are adversely affecting biodiversity, leading to a reduction of small-growing species which tend to be shaded out by tall vegetation.

The dunes have been owned and managed by the Christie family since 1856 and there was concern that alterations in the existing management, in particular the extensive grazing by cattle proposed by English Nature, might result in unfavourable changes in the vegetation and the water table. Between 1964 and 1996 the Burrows had been managed as a National Nature Reserve by English Nature and its predecessor bodies. The site was de-declared in 1996 because the lack of grazing was causing its nature conservation interest to decline and English Nature was unable to introduce measures to counteract this problem. Subsequently, however, the Christie family, English Nature and the Ministry of Defence (MOD) agreed jointly to sponsor a 3-year grazing experiment in a 16ha enclosed paddock with controlled grazing by cattle and sheep. This commenced in the spring of 1998. The area remains a Site of Special Scientific Interest, and, as mentioned previously, is a candidate Special Area of Conservation under the Habitats Directive.

Previous attempts to control woody plants by mechanical means were made from about 1985 to 1996. Not surprisingly in the uneven terrain, and some areas with wartime relics (hidden trenches, ironwork and concrete), the use of machinery, including chainsaws, flails and even a mini forage harvester, proved difficult and uneconomic, maintenance costs being very high (Breeds and Rogers, 1998). There were also problems in disposing of cuttings, especially as eutrophication is to be avoided in the interest of biodiversity.

Effects of Sheep Grazing and Mowing on the Vegetation of Braunton Burrows

From 1987 to 1996, Soay sheep (Plate 1) were intermittently grazed on landward areas of the dune system; another area was mown. Plots of at least 25 × 25m were monitored (Mr P.C. Robinson, personal communication). Some were fenced to exclude sheep grazing while others were grazed by Soay sheep, or mown. All plots were in an area named Soay Plain on the Braunton Burrows Habitat Map issued by English Nature in 1990. The vegetation of seven plots was recorded in 1991 and 1996, presence/absence data being obtained for 30 randomly placed (1m²) quadrats for each plot on both occasions (Fig. 1). Two of the monitored plots (not mown ones) were in dry dune pasture, three in damp dune pasture and two in wet dune pasture, both of which were sheep-grazed.

In *dry dune pasture,* species diversity (as shown by quadrat records) declined significantly (P<0.001) to a similar extent between 1991 and 1996 in a sheep-grazed plot and one with only rabbit grazing. It is of note, however, that the number of species of flowering plants found within the thirty quadrats sampled in each plot varied rather little from 1991 to 1996, suggesting that niches suitable for most species still remained though the areas of such niches were diminishing in many cases. In both plots there was little change in the abundance of *Galium verum, Lotus corniculatus* and *Thymus poly-*

Plate 1. *Soay sheep grazing at Braunton Burrows. (Photograph by J.R. Packham.)*

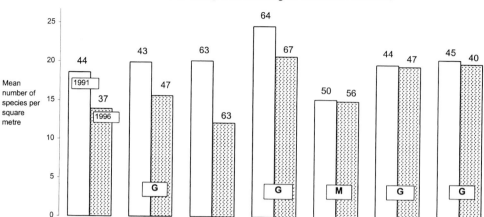

Figure 1. *Change between 1991 and 1996 in the mean number of species of flowering plants per m² in plots which were subject to minor rabbit grazing only, sheep-grazed (G), or mown (M). The first of each pair of columns represents the figure for 1991, and the second that for 1996. The figures at the top of each bar represent the total number of flowering plant species found within the thirty quadrats sampled on each occasion. Note that sheep grazing and mowing had begun before 1991.*

trichus, all common species. In the sheep-grazed area *Agrimonia eupatoria* was frequent in 1996 but not recorded in 1991; in the rabbit-grazed area *Vulpia fasciculata* was frequent in 1996 but unrecorded in 1991 (however, rabbits were scarce during the experimental period).

In *damp dune pasture*, where the only full contrast of experimental regimes is available, there were significant declines in average numbers of species in both sheep-grazed and only rabbit-grazed plots, but a greater decrease in the latter. In the sheep-grazed area there were gains of *Agrimonia eupatoria, Anacamptis pyramidalis* and *Pulicaria dysenterica*. In the rabbit-grazed plot, associated with much greater scrub growth, there was no spread of *Calamagrostis epigejos,* but a reduction of *Festuca rubra* and loss of *Dactylorhiza praetermissa*.

In the mown plot, grazed only by rabbits, there was very little change in the average number of species per quadrat from 1991 to 1996. However, as might be expected, there were decreases of *Cirsium arvense, Eupatorium cannabinum* and *Pulicaria dysenterica,* but increases of low-growing species such as *Hydrocotyle vulgaris, Potentilla reptans* and *Ranunculus repens* as well as of *Anacamptis pyramidalis,* some grasses and the semi-parasitic *Parentucellia viscosa* and *Rhinanthus minor.*

Grazing of the damp dune plot ceased after 1996, but the experimental treatment of the mown plot continued into the summer of 1998. There was a striking difference between the general appearance of the three damp dune pasture plots in July 1998, whereas their vegetation cover was judged visually to be very similar in height and species composition when monitoring of these plots began in 1987. Some 65 per cent of the exclosure was now covered by quite tall scrub, where the number of species per square metre tended to be much lower than in the species-rich rabbit-grazed gaps. In the formerly grazed plot, where *Festuca rubra* was very abundant, there was a single large bush of *Ligustrum vulgare* and much low *Salix repens; Crataegus* was prominent amongst the shrub species which had often been recorded as juveniles but which were now developing into small saplings. The mown plot now (1998) has little *Salix repens,* and the less than 1 per cent scrub cover consisted of rapidly grown shoots of *S. cinerea* which had regrown after the mowing of 1998. (In contrast, a regularly mown plot seen at Ainsdale NNR on the Sefton Coast in September 1998 was dominated by *Salix repens.*)

Species diversity was well maintained in *wet dune pasture* in sheep-grazed plots from 1991 to 1996, *Epipactis palustris* and Dactylorchids surviving well, but *Galium palustre* was much reduced.

It is concluded from these and other data that, as previously reported at Newborough Warren (Hewett, 1985), both sheep grazing and mowing help to prevent further loss of plant diversity in vegetation which has become rank. Sheep grazing also tends to create regeneration niches, and to restore the species-rich areas of short turf that are of particular conservation interest on Braunton Burrows.

Hydrology

The water regime of dune systems has a very strong influence on the vegetation. Plants not only of the slacks, but also of many areas of low dune pasture, have their root systems within reach of the water table. In contrast, the vegetation of the higher dunes is well above the level of the water table and its capillary rise (the latter is usually some 30–40cm). Indeed, many plants have characteristic distributions relative to the water table (Willis et al., 1959b; Willis, 1985b). Some species can establish from seedlings only at about the level of the water table, but can later trap sand and build up hummocks or dunelets, e.g. *Salix repens, Scirpoides holoschoenus.*

Large dune systems, especially where the hinterland is low and the water regime autonomous, have a distinctly domed water table. At Braunton Burrows the water table at the crest of the dome is some 6m higher than at the shoreline or inland bordering Braunton Marsh (Willis et al., 1959a, Fig. 8).

Information about the movement and shape of the water table may be gained from suitably placed wells or water holes at known altitudes. Recent measurements of the water table made by Mr J.M. Breeds (personal communication), using steel pipes set flush with the ground, show that its fluctuations continue to be strongly correlated with the rainfall, as found by Willis et al. (1959b). The amplitude of the long-term changes (over many years) may be some 1.5m but annual changes are usually much smaller. The height of the water table tracks the previous rainfall, most notably that of a preceding three-month period, as shown in Figure 2. It is noteworthy that sampling point 3N, with a ground level of 10.20m OD and situated in a dune slack central to the Burrows, was flooded for some four months in 1994 and again at the beginning of 1995. The water regime at the crest of the dune consequently appears to be very similar now to that in the 1950s. Indeed much of the hydrological system of the Burrows appears to operate as it did in the past.

Following several consecutive years of below average precipitation (1983–92), however, concerns expressed by English Nature led to a hydrological investigation directed by Dr A.G. Williams, University of Plymouth. Ground water recharge is by precipitation in this hydrologically near-autonomous system, and the general overall decline in the elevation of the water table from 1983 to mid-1992 was undoubtedly owing to below average precipitation during this period (Burden, 1997). This was exacerbated by the deepening, in 1983, of the West Boundary Drain between Braunton Marsh and Braunton Burrows, which caused a lowering of the adjoining water table. Drainage of an adjoining golf course and a large field close to the Burrows may have had an additional influence. There is also the possibility that loss of water through evapo-transpiration from the coarser vegetation now extant, including the scrub whose area has greatly increased since 1954, is greater than from the lower vegetation previously present.

Whereas much of the lower dune system flooded regularly during the winter, this seldom occurs now and the black-headed gulls, curlews and lapwings which bred on Braunton Burrows in the 1950s are now only occasional winter visitors. This change, the loss of the shore dock *(Rumex rupestris),* and a 45 per cent decline of water germander *(Teucrium scordium)* since 1982, have all been ascribed by English Nature to a fall in the water table (Dr R. Wolton, personal communication).

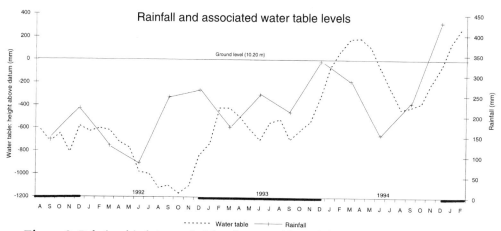

Figure 2. *Relationship between height of water table and the rainfall which it tracks at site 3N, in a slack central to the dune system at Braunton Burrows for the period 1991–95. Each symbol on the rainfall record is shown at the end of the three-monthly period whose total rainfall it represents. (See also Fig. 16, Willis et al., 1959b.)*

Plate 2. Ononis repens *and* Hydrocotyle vulgaris *revealed by cattle grazing of* Calamagrostis epigejos. *Note the heavy* C. epigejos *litter remaining from the previous year. (Photograph by J.R. Packham.)*

Future Management

It is now intended to monitor with some precision the effect of cattle, together with sheep, on the dune system, to try to establish whether they will give adequate control of the rank grasses and woody species. An extensive new trial has now begun in a fenced-off area in the central part of the Burrows, with replicated plots grazed and ungrazed, and the effects will be monitored over a three-year period. Cattle browse scrub effectively and also graze *Calamagrostis epigejos* whose litter tends to eliminate smaller species (Plate 2). Soay sheep help to produce a low herb-rich sward and also graze *Juncus acutus* which is becoming prolific in places. A new ten-year MOD Braunton Burrows Management Plan, jointly agreed with English Nature and the Christie Devon Estates Trust, commenced in 1998. It is hoped that this plan will advance one of the long-term objectives of the MOD, the restoration of NNR status to Braunton Burrows.

Management Problems in Some Other British Dune Systems

Ranwell and Boar (1986) produced a management guide for coastal dunes, and recent conservation management practice on British dune systems is discussed by Houston

(1997). The influence of grazing on British sand dunes has been investigated by Boorman (1989a; 1989b), who has also studied the impact of paths on sand-dune vegetation (Boorman and Fuller, 1977).

Many British dune systems suffer from management problems, including those concerned with hydrology, similar to those at Braunton. The dune system on the South Haven Peninsula, Dorset, is compared with that of Braunton Burrows by Packham and Willis (1997). It has an area about a quarter that of Braunton Burrows, has soils distinctly low in calcium (Wilson, 1960) and also in some micro-nutrients (Willis, 1985b); it supports an acidophilic vegetation on its older dunes. Its origin is as recent as the early seventeenth century and it is still prograding, while the shoreline at Braunton has changed very little for hundreds of years. Despite their differences, both systems have highly diverse floras, were damaged by military activity during 1939–45, and in some places are tending to become overgrown by woodland scrub.

This last feature has caused concern in very many British dune systems since the advent of myxomatosis, whose effects were among the features at Newborough Warren described in a classic sequence of papers by the late Derek Ranwell (1958; 1959; 1960a; 1960b). Scrub eradication is now a necessary part of conservation practice in many European dune systems. Often it involves invasive species such as the bird-dispersed *Hippophae rhamnoides* which even when eradicated leaves a soil so enriched in nitrogen as to encourage high RGR ruderals including *Chamerion angusti-folium*. Gibraltar Point (N. Lincolnshire), Tenby (Pembrokeshire) and Cefn Sidan (Carmarthenshire) are among the systems where this shrub has caused difficulties. Rooney (1998) describes the use of specially adapted heavy machinery to clear scrub, much of it sea buckthorn, from the Ainsdale and Birkdale dunes on the Sefton Coast where substantial areas of open, unshaded terrestrial habitat are essential to the natter-jack toad (*Bufo calamita*). In this instance eutrophication was avoided by burning the cut scrub at a few fire sites, which were later buried and covered with clean sand.

The position at Berrow, North Somerset, is interesting as here spread of *Hippophae rhamnoides* has led to modification of the saltmarsh vegetation, much of which is now protected from the waves by the reinforcement which the shrub affords to the seaward dunes. *H. rhamnoides* was introduced to the area in the 1890s, has greatly extended its distribution on the dunes and has invaded the margins of the marsh; it will have raised local soil nitrogen levels. The seaward edge of most of the marsh is now bounded by a well-developed dune line which is much reducing tidal inflow; the enclosed marsh may soon become a freshwater *Phragmites* swamp (Willis, 1990; Packham and Willis, 1997). *P. australis* has invaded the Berrow Marsh extensively since 1963; this vigorous rhizomatous grass outcompetes many smaller species and often causes the over-stabi-lization of interesting damp dune slacks, as at Nicholaston Burrows on the Gower Peninsula, S. Wales.

Invasive plants are indeed a problem in all coastal communities based on granular substrates, where plant growth form and substrate properties are very relevant to establishment (Raunkiaer, 1934; Packham and Willis, 1995). *Spartina anglica* has greatly modified saltmarsh vegetation such as that at Cefni, Anglesey (Packham and Liddle, 1970), while *Rosa rugosa* has established on a shingle system as young as that at Black Rock, Brighton (Packham et al., 1995). The European *Ammophila arenaria* acts as an invasive alien in the Oregon Dunes National Recreation Area, USA (Packham and Willis, 1997; Wiedemann, 1984), and many other parts of the world.

Mowing and controlled grazing by domestic animals (Hewett, 1985) have proved to be important management tools in the maintenance of biodiversity at Newborough Warren, and also at Oxwich Dunes where *Pteridium aquilinum* invasion is a particular problem. Oxwich Dunes are vulnerable to heavy pressure by visitors from the Swansea area, whose barbecues have also damaged the vegetation.

The Aberffraw Dunes, Anglesey, were the setting for much detailed work on tracks, paths and trampling (Liddle, 1975a; 1975b; Liddle and Greig-Smith, 1975a; 1975b; Liddle and Moore, 1974). Despite the best efforts of the Meyrick Estate, damage here has greatly increased in recent years. The repeated passage of cars has in many places destroyed the surface grassland, and the sand beneath has subsequently been wind-blown, leaving tracks more than 1m lower than the adjacent turf.

Much dune management on the Sefton Coast (mowing, grazing, scrub eradication) is essentially recurring as it is in many of the sites discussed above. The removal of some of the long established conifer plantations (Sturgess, 1993), however, is essentially restoration management which needs to be carried out sympathetically, as does the restoration of some of the Dutch dunes involving the removal of artificial waterways.

Research directly concerned with biodiversity and the decline, for example, of rare plant species and community types is of especial value. That done at Kenfig and other sand dune systems in South Wales (Jones and Etherington, 1989; Jones et al., 1995) is a case in point. Interestingly, the dune system at Crymlin Burrows in South Wales, between Swansea Docks and the River Neath, though immediately adjacent to extensive industrial sites and subject to disturbance (Gilham, 1982), retains high biodiversity with *Matthiola sinuata* conspicuous in many seasons. Ranwell (1981) discussed introduced coastal plants and rare species in Britain.

Conclusions

Management problems, many involving increased growth of woody plants or possible climatic change, will continue to beset coastal dune systems. Increasing knowledge of the biology, ecology and dynamics of such systems will provide more refined management tools, but this must go hand in hand with the development of clearly defined objectives, such as those which have been defined by English Nature for Braunton Burrows. The conclusions reached by Breeds and Rogers (1998) and Oates et al. (1998) that mechanical control fails to provide the habitat quality of grazing, and that grazing needs to involve animals that can tackle coarse vegetation (cattle, ponies) as well as those which create a short sward (e.g. ponies, sheep) are particularly important.

Table 1. Monitoring the vegetation of sand dune systems

1. It is essential to maintain adequate **control areas** when experimenting with such interventions as mowing, grazing or mineral nutrient application. Records should also be kept of the abiotic environment, e.g. rainfall, wind patterns, topographic changes.

2. **Fixed quadrats** (monitored non-destructively) are well suited for studies of vegetational change. Where random samples are used within the areas being compared, it is essential to take enough for statistical validity.

3. The form of observations must be appropriate to the aims of the monitoring.
    a. **Presence.** It is useful to have complete species lists for the whole site, relevant parts of it, and for the experimental areas (where mean species number per unit area may also change over the duration of the experiment). The presence of rare species should be regularly monitored.
    b. **Cover** assessments for species present, either as percentages or as Domin values, are useful in assessing changes in the relative abundance of the species present.
    c. **Height** records are particularly important in monitoring regressive changes following the cessation of experimental or 'natural' grazing. For graminoids it may be worth recording mean vegetative and flowering heights.
    d. Measurements of **biomass** are time-consuming but often useful, particularly where nutrient applications are involved, and the sample can be retained for chemical analysis. Dry weight per unit area is a better measure than fresh weight; the use of relative bulk of above-ground biomass is an informative measure where fixed quadrats are to be assessed repeatedly (Willis et al., 1959a).

4. **Repeated fixed-point photography**, preferably using the same lens, affords a cost-effective and valuable long-term record. Changes in the proportion of bare sand or woodland and scrub are easily monitored in this way. Aerial photography, remote sensing and GIS are rapidly increasing in importance.

5. Although it may be helpful to make additional observations, it is important to maintain the **original recording pattern** throughout a long-term experiment.

6. It is important to consider the way in which the whole ecosystem is developing from time to time, and also to monitor fungi, lichens and animal groups (not only the grazers and their predators which have a primary impact on the vegetation). Undesirable invasive species, e.g. *Hippophae rhamnoides*, need careful monitoring (and are best controlled at an early stage).

In all dune systems continuous monitoring is an essential component of the efficient management of an environment in which great storms can cause major blowouts, while too many expensive experiments have failed either to establish effective baselines or to record adequate control plots throughout their duration. A number of points particularly relevant to the monitoring of coastal dunes are summarized in Table 1.

Dune management has inevitably to balance the creation of stability, which prevents excessive erosion, and the natural dynamism which allows areas of bare sand and encourages the growth and spread of less competitive and often rare species, such as *Liparis loeselii* (Jones, 1998). The landscape will be more interesting, and the biota remain more diverse, if future policies tolerate more sandblow and greater instability in suitable areas.

Loss of dune area to such features as roads and housing, formerly a major concern on the Sefton Coast and still a major concern on resort coasts elsewhere in Europe, is the greatest threat of all.

Acknowledgments

The authors are most grateful to Mr N.J. Musgrove for help with the diagrams, to Mr J.M. Breeds for measurements of the water table and rainfall, and to Mr P.C. Robinson for information on the composition of the vegetation related to grazing and mowing at Braunton Burrows in recent years. It is also a pleasure to acknowledge the co-operation afforded by the Christie Devon Estates Trust, English Nature (especially discussion with Dr R. Wolton) and the Ministry of Defence.

References

Anthony, J.C. (1998), 'An investigation into the ecological changes on Braunton Burrows Dune system, North Devon', BSc thesis, Department of Geography, University of Nottingham.

Boorman, L.A. (1989a), 'The grazing of British sand dune vegetation', *Proceedings of the Royal Society of Edinburgh*, **96B**, 75–88.

Boorman, L.A. (1989b), 'The influence of grazing on British sand dunes', in F. van der Meulen, P.D. Jungerius and J. Visser (eds.), *Perspectives in Coastal Dune Management*, SPB Academic Publishing, The Hague, 121–24.

Boorman, L.A. and Fuller, R.M. (1977), 'Studies on the impact of paths on the dune vegetation at Winterton, Norfolk, England', *Biological Conservation*, **12**, 203–16.

Boorman, L.A. and Fuller, R.M. (1982), 'Effects of added nutrients on dune swards grazed by rabbits', *Journal of Ecology*, **70**, 345–55.

Breeds, J. and Rogers, D. (1998), 'Dune management without grazing – a cautionary tale', *Enact: Management for Wildlife*, **6**, 19–22.

Burden, R.J. (1997), 'A hydrological investigation of three Devon sand dune systems: Braunton Burrows, Northam Burrows and Dawlish Warren', PhD thesis, University of Plymouth.

Connor, E.F. and McCoy, E.D. (1979), 'The statistics and biology of the species–area relationship', *American Naturalist*, **113**, 791–833.

Dony, J.G. (1977), 'Species–area relationships in an area of intermediate size', *Journal of Ecology*,

65, 475–84.

Gilham, M.E. (1982), *Swansea Bay's Green Mantle*, D. Brown and Sons, Cowbridge, South Wales.

Hewett, D.G. (1985), 'Grazing and mowing as management tools on dunes', *Vegetatio*, **62**, 441–47.

Hope-Simpson, J.F. (1997) 'Dynamic plant ecology of Braunton Burrows, Southwestern England', in E. van der Maarel (ed.), *Dry Coastal Ecosystems, General Aspects: Ecosystems of the World 2C*, 437–52.

Hope-Simpson, J.F. and Yemm, E. W. (1979), 'Braunton Burrows: developing vegetation in dune slacks', in R.L. Jefferies and A.J. Davy (eds.), *Ecological Processes in Coastal Environments*, Blackwell Scientific Publications, Oxford, 113–27.

Houston, J.A. (1997), 'Conservation management practice on British dune systems', *British Wildlife*, **8**, 297–307.

Jones, P.S. (1998), 'Aspects of the population biology of *Liparis loeselii* (L.) Rich. var. *ovata* Ridd. ex Godfery (Orchidaceae) in the dune slacks of South Wales, UK', *Botanical Journal of the Linnean Society*, **126**, 123–39.

Jones, P.S. and Etherington, J.R. (1989), 'Ecological and physiological studies of sand dune slack vegetation, Kenfig Pool and Dunes Local Nature Reserve, Mid-Glamorgan, Wales, U.K.', in F. van der Meulen, P.D. Jungerius and J. Visser (eds.), *Perspectives in Coastal Dune Management*, SPB Academic Publishing, The Hague, 297–303.

Jones, P.S., Kay, Q.O.N., and Jones, A. (1995), 'The decline of rare plant species and community types in the sand dune systems of South Wales', in M.G. Healy and J.P. Doody (eds.), *Directions in European Coastal Management*, Samara Publishing, Cardigan, 547–55.

Liddle, M.J. (1975a), 'A selective review of the ecological effects of human trampling on natural ecosystems', *Biological Conservation*, **7**, 17–36.

Liddle, M.J. (1975b), 'A theoretical relationship between the primary productivity of vegetation and its ability to tolerate trampling', *Biological Conservation*, **8**, 251–55.

Liddle, M.J. and Greig-Smith, P. (1975a), 'A survey of tracks and paths in a sand dune ecosystem. I. Soils', *Journal of Applied Ecology*, **12**, 893–908.

Liddle, M.J. and Greig-Smith, P. (1975b), 'A survey of tracks and paths in a sand dune ecosystem. II. Vegetation', *Journal of Applied Ecology*, **12**, 909–30.

Liddle, M.J. and Moore, K.G. (1974), 'The microclimate of sand dune tracks: the relative contribution of vegetation removal and soil compression', *Journal of Applied Ecology*, **11**, 1057–68.

Ministry of Defence, Plymouth (1998), *Braunton Burrows Management Plan*.

Oates, M., Harvey, H.J. and Glendell, M. (1998), *Grazing Sea Cliffs and Dunes for Nature Conservation*, The National Trust Estates Department, Cirencester.

Packham, J.R., Harmes, P.A. and Spiers, A. (1995), 'Development of a shingle community related to a specific sea defence structure', in M.G. Healy and J.P. Doody (eds.), *Directions in European Coastal Management*, Samara Publishing, Cardigan, 369–71.

Packham, J.R. and Liddle, M.J. (1970), 'The Cefni salt marsh, Anglesey, and its recent development', *Field Studies*, **3**, 331–56.

Packham, J.R. and Willis, A.J. (1995), 'Plant growth form, substrate properties and sediment stabilization in coastal ecosystems', in M.G. Healy and J.P. Doody (eds.), *Directions in European Coastal Management*, Samara Publishing, Cardigan, pp. 351–59.

Packham, J.R. and Willis, A.J. (1997), *Ecology of Dunes, Salt Marsh and Shingle*, Chapman and Hall, London.

Ranwell, D.S. (1958) 'Movement of vegetated sand dunes at Newborough Warren, Anglesey', *Journal of Ecology*, **46**, 83–100.

Ranwell, D.S. (1959), 'Newborough Warren, Anglesey. I. The dune system and dune slack habitat', *Journal of Ecology*, **47**, 571–601.

Ranwell, D.S. (1960a), 'Newborough Warren, Anglesey. II. Plant associes and succession cycles of the sand dune and dune slack vegetation', *Journal of Ecology*, **48**, 117–41.

Ranwell, D.S. (1960b), 'Newborough Warren, Anglesey. III. Changes in the vegetation on parts of the dune system after loss of rabbits by myxomatosis', *Journal of Ecology*, **48**, 385–95.

Ranwell, D.S. (1981), 'Introduced coastal plants and rare species in Britain', in H. Synge (ed.), *The Biological Aspects of Rare Plant Conservation*, John Wiley, Chichester, 413–19.

Ranwell, D.S. and Boar, R. (1986) *Coast Dune Management Guide*, Institute of Terrestrial Ecology, Huntingdon.

Raunkiaer, C. (1934) *The Life Forms of Plants and Statistical Plant Geography*, Clarendon Press, Oxford.

Read, D.J. (1989), 'Mycorrhizas and nutrient cycling in sand dune ecosystems', *Proceedings of the Royal Society of Edinburgh*, **96B**, 89–110.

Rooney, P. (1998) 'A thorny problem', *Enact: Management for Wildlife*, **6**, 12–13.

Sturgess, P.W. (1993), 'Clear-felling dune plantations: studies in vegetation recovery on the Sefton Coast', in D. Atkinson and J.A. Houston (eds.), *The Sand Dunes of the Sefton Coast*, National Museums and Galleries on Merseyside, Liverpool.

Wiedemann, A.M. (1984) 'The ecology of Pacific Northwest coastal sand dunes; a community profile', *U.S. Fisheries and Wildlife Service, FWS/OBS-84/02*, 130.

Willis, A.J. (1963), 'Braunton Burrows: the effects on the vegetation of the addition of mineral nutrients to the dune soils', *Journal of Ecology*, **51**, 353–74.

Willis, A.J. (1965), 'The influence of mineral nutrients on the growth of *Ammophila arenaria*', *Journal of Ecology*, **53**, 735–45.

Willis, A.J. (1985a), 'Plant diversity and change in a species-rich dune system', *Transactions of the Botanical Society of Edinburgh*, **44**, 291–308.

Willis, A.J. (1985b), 'Dune water and nutrient regimes – their ecological relevance', in J.P. Doody (ed.), *Sand Dunes and their Management*, Nature Conservancy Council, Peterborough, 159–74.

Willis, A.J. (1989), 'Coastal sand dunes as biological systems', *Proceedings of the Royal Society of Edinburgh*, **96B**, 17–36.

Willis, A.J. (1990), 'The development and vegetational history of Berrow salt marsh', *Proceedings of the Bristol Naturalists' Society*, **50**, 57–73.

Willis, A.J., Folkes, B.F., Hope-Simpson, J.F. and Yemm, E.W. (1959a; 1959b), 'Braunton Burrows: the dune system and its vegetation. Parts I and II', *Journal of Ecology*, **47**, 1–24 and **47**, 249–88.

Willis, A.J. and Jefferies, R.L. (1963), 'Investigations on the water relations of sand-dune plants under natural conditions', in A.J. Rutter and F.H. Whitehead (eds.), *The Water Relations of Plants*, Blackwell Scientific Publications, Oxford, 168–89.

Willis, A.J. and Yemm, E.W. (1961), 'Braunton Burrows: mineral nutrient status of the dune soils', *Journal of Ecology*, **49**, 377–90.

Wilson, K. (1960), 'The time factor in the development of dune soils at South Haven Peninsula, Dorset', *Journal of Ecology*, **48**, 341–59.

GRAZING AS A MANAGEMENT TOOL AND THE CONSTRAINTS OF THE AGRICULTURAL SYSTEM: A CASE STUDY OF GRAZING ON SANDSCALE HAWS NATURE RESERVE, CUMBRIA, NORTHWEST ENGLAND

PETE BURTON

Sandscale Haws Nature Reserve, England

ABSTRACT

Conservation benefits of stock grazing on a sand dune nature reserve in north-west England are reported. The site has a long history of sheep and cattle grazing that has maintained a healthy dune system, rich in specialist and rare plant and animal species. The grazing regime, however, relies on modern agricultural systems that supply the stock. Problems arising from this relationship include grazing preferences of modern breeds (compared to traditional breeds), the requirement for supplementary feeding, the BSE crisis, reduced value of fleece from the sheep because of the entrapped sand and reluctance of farmers to allow their stock to graze poor quality herbage. Rabbits have also been important grazers and negative impacts of enhanced vegetation succession have followed outbreaks of myxomatosis and Rabbit Viral Haemorrhagic Disease (RHVD).

Introduction

Grazing for nature conservation on sand dunes using domestic livestock is now viewed by most dune managers as the best prescription against the problems of past over-stabilization resulting from over-protection. By slowing down successional processes, and in some cases reversing them by rejuvenation of aeolian activity, a diverse and dynamic dune landscape can be achieved. Many species, from plants to invertebrates, benefit immensely from a suitable grazing regime and many rare or scarce species are almost confined to stock damaged areas.

The National Trust's dune nature reserve of Sandscale Haws is fortunate in having a very long history (at least 800 years) of grazing management. Until relatively recently, however, nature conservation was not the objective of the grazing regime. The highly valued dune landscape habitat had been maintained by default because of traditional agricultural practice. Unfortunately modern agricultural objectives may conflict with dune conservation.

Since the 1980s the National Trust has established a highly successful grazing regime on Sandscale Haws, although this has not been without problems. As with many other grazing schemes for nature conservation, the Trust must rely on the livestock of a local farmer. This reliance on the agricultural system therefore makes the ideal objectives for nature conservation much more difficult to achieve.

Experience and observations of the grazing at Sandscale Haws are reported here and the implications of the relationship of grazing with modern agricultural systems are discussed.

Site Description

Sandscale Haws is a 282ha foreland dune system on the eastern shore of the Duddon estuary only a few kilometres south of the Lake District National Park.

It is a Site of Special Scientific Interest (SSSI) in its own right and forms part of the much larger Duddon Sands SSSI, Special Protection Area and Ramsar Site. The reserve supports approximately 15 per cent of the British natterjack toad (*Bufo calamita*) population, a rich flora (over 500 species) including the Red Data Book British endemic dune helleborine (*Epipactis leptochila* var. *dunensis*), the largest population in England of coralroot orchid (*Corallorhiza trifida*) and an important invertebrate fauna including a number of Red Data Book species. It is also one of the most important coastal mycological sites in Britain with a number of species on both the British and European Red Lists.

The dune sand exhibits a broad range of pH, from nine in the frontal dunes to four to the rear, producing a range of vegetation communities from calcareous grassland to dune heath and acid grey dune, some of which are priority habitat types on Annex I of the Habitats Directive. Other typical dune communities include embryonic shifting dunes, shifting dunes with *Ammophila arenaria*, extensive fixed dune grassland and a broad range of slack types many of which flood in winter (over 20ha of water in some years). A range of mire types, carr woodland and open hawthorn scrub also occur.

Grazing Management

Sandscale Haws is known to have been grazed since at least the twelfth century. Although there is documented evidence of sheep grazing from this time, there is no evidence for the use of cattle until the early part of this century. The early owners, the Cistercian monks of Furness Abbey, ran the dunes as a rabbit warren and the rabbit has been an important component of the grazing system for hundreds of years.

The seven kilometre site boundary is largely unfenced. Internally a single-strand electric fence separates the central dunes from the outer. This fence was erected in about 1972 to prevent cattle wandering off the site. A further stock fence separates the southern 36ha, a Countryside Stewardship lowland heath regeneration scheme, from the rest of the reserve.

The types of livestock utilized on the site each fulfil differing management functions dependent on their breed, age, temperament and also the time of year. The types of livestock and how they are utilized on Sandscale are outlined below.

Cattle

Approximately 60–70 cattle, mainly modern beef out of Friesian and Hereford, are grazed throughout the year and are contained within the central dune area by the electric fence. During the summer months they are also allowed access to the heath regeneration area.

Cattle play an important role in opening up the sward by removing some of the taller coarse growth, thus allowing sheep to graze. These modern breeds however seem reluctant or unable to make any major impact on very coarse growth such as *Calamagrostis epigejos* in dune slacks. Where large stands of this grass have been able to develop the cattle will use it for lying up at night. *C. epigejos* has the potential to become a problem species should grazing be relaxed.

The effects of trampling assist the rejuvenation of aeolian activity by the development of small-scale internal blowouts. These are extremely important for many specialized invertebrates, such as solitary bees and wasps, a number of nationally scarce plants, and in some cases may provide new breeding habitat for the rare natterjack toad *Bufo calamita*. As an example of what can be achieved in a short period, in 1994 an overgrown area of dune previously not grazed by cattle was opened up for the winter period. By the following spring, grazing and trampling by the cattle had created a short open sward with bare sand which was colonized by hundreds of the mining bee *Andrena barbilabris*.

It is interesting to note that in the 1960s conservationists campaigned (luckily without success) to have the livestock removed from the reserve because of the perceived damage to the dunes due to trampling. A series of grazing exclosures set up in 1967 clearly demonstrated the effects of little or no grazing and are now dominated by 3–4m high *Betula*, *Salix* and *Rubus* spp. associated with *Chamerion angustifolium*.

Regular use of scrub areas by livestock for shelter can quickly lead to nutrient enrichment and the development of a rank, weedy vegetation. However, continued high pressure can actually lead to death of the scrub and subsequently renewed sand blow.

The modern beef breeds being used for grazing lack the hardiness of the traditional breeds and can quickly lose condition on the low productivity herbage of the dune system. As a consequence the farmer provides supplementary feed in winter in the form of big bale hay and silage to achieve his objective of obtaining the best possible return from his stock. The suggestion that winter feeding leads to enrichment and overgrazing of the dune habitat has not been borne out at Sandscale Haws. Here, enrichment is confined to feed sites and serious under-grazing occurs over the rest of the dune area due to feed site fidelity. This fidelity to feed sites can remain imprinted in the stock long after feeding has ceased and it can be many weeks or even months before the cattle start to wander further afield. With supplementary feeding, therefore, cattle are doing little or no grazing of the dunes and arguably should not be there.

In recent years Bovine Spongiform Encephalopathy (BSE) has had a major impact on beef farming in the UK and there are serious implications for conservation grazing if farms continue to go out of business. At Sandscale, as at other sites in the UK, the farmer is now reluctant to allow his cattle onto the lower productivity herbage of the heath as they are unable to gain slaughter weight before the government 30-month ruling (imposed because of the BSE crisis). This situation is mirrored in many other sites in the UK.

Sheep

Two very different sheep flocks are used on the reserve at different times of the year, each performing a different management role with differing constraints.

Winter grazing

The winter flock consists of approximately 200 Herdwick hogs (a hog is a first-year animal) along with approximately 100 Mule hogs (Mule is a Swaledale x Hexam Leicester cross). The winter flock is allowed free range over the dunes except for the dune heath to protect the heather from damage by the hill sheep. The Herdwick is a very hardy hill breed that will graze everything from the foredunes to the landward edge of the reserve. It is probably one of the best grazing tools at our disposal, ideal for restoration grazing for conservation. Herdwicks will browse scrub and will graze on very coarse vegetation even in quite wet situations.

Being quite a small breed, the Herdwick seems more susceptible to worrying by dogs and quite a number are lost or suffer injury each year. Also, being hefted (loyal to their birthplace) in the uplands, they have a habit of attempting to return to the Lake District mountains if over-pressurized by dogs.

The supply of upland sheep for dune grazing is still fairly good with many hill farmers willing to move their stock long distances to acquire suitable grazing. However, the dune manager cannot ignore the events occurring in the less favoured areas of the uplands, where many farms could go out of business in the near future. From a conservation perspective there is the dilemma that, while there is a clear need to reduce the massive over-grazing by sheep in the uplands, this will inevitably lead to a reduction in the availability of upland sheep breeds for dune grazing schemes.

Summer grazing

The summer flock of Swaledale Mule sheep (approximately 150 ewes plus lambs) are fairly hardy and rather selective feeders. This is a lamb production breed and though an important maintenance tool for plagioclimax communities, it does not make a great impact where succession has allowed the development of coarse growth.

This breed produces small-scale deflation hollows in warm weather by digging in order to find shade. This activity can provide ideal conditions for many rare dune invertebrates and plants. The resulting excess sand in the fleece, however, causes diffi-culties in shearing and a reduction in the fleece value. The farmer will frequently remove the stock if this is occurring.

Rabbit Grazing

Rabbits have played an important role in shaping the dunes and their vegetation for many hundreds of years. Their extensive burrows are frequently collapsed by cattle hoof pressure, resulting in deflation hollows which are an important locus for some nationally scarce plant species such as *Vulpia fasciculata* and *Erodium lebelii,* and the RDB early mining bee *Colletes cunicularius*. The introduction of myxomatosis caused massive mortality in the rabbit populations, which at Sandscale has resulted in steady reduction in bare sand habitats since 1957. The effects of myxomatosis can be a major problem in setting stock grazing levels due to the consequent fluctuating population size of the rabbits.

Since 1996 the rabbit population on the reserve has been reduced by approximately 90 per cent by a suspected outbreak of RVHD with no signs of population recovery. The effects are already apparent with increased grass growth and development of an increased litter layer, particularly in slacks.

The Effects of Combined Sheep and Cattle Grazing on Vegetation

The combined effect of both sheep and cattle on the dune vegetation is to create a tight, closed, species-rich sward. The species composition is diverse with an average of 26–30 species per 2m². Apart from a few small areas adjacent to scrub and supple-mentary feed sites there are few signs of enrichment despite the long grazing history; in fact many of the species which are known to respond to increased nutrients remain as dwarfed individuals in the sward.

Slack vegetation presents a more difficult challenge to grazing management than the dunes. Cattle show a preference for grazing on the dunes and the slightly improved grasslands to the rear of the dunes. It is not until later in the season that they move on to the coarser herbage of the dune slacks. The sheep also prefer the dune grass-lands and graze less frequently in the slacks, particularly when tall growth is present or the slacks are wet. The result is an increase in standing dead biomass leading to an increase in organic matter in slack soils. This is exacerbated by the fact that cattle appear preferentially to dung in taller slack vegetation.

Observations indicate that cattle grazing severely overgrown slacks tends to increase the ratio of grass to herb. Experimental grazing areas at Sandscale indicate that grazing alone cannot recreate the important early successional stages in dune slacks. How best to maintain or recreate these important pioneer stages, which often have rare specialist species, must be a priority in future work.

Concluding Remarks

Grazing with domestic livestock is without doubt one of the best management tools for maintaining the conservation interest of the dune habitat. However, if we are to maintain the biodiversity of the dunes in the future we must be aware of changes and developments in modern agricultural practice. Conservation will require a viable agriculture better tuned to conservation requirements with more flexible agri-environment support schemes that allow for the use of the most appropriate livestock in conservation grazing schemes.

RABBIT GRAZING AND RABBIT COUNTING

MARIJKE DREES and HAN OLFF

Wageningen Agricultural University, Tropical Nature Conservation and Vertebrate Ecology Group, The Netherlands

ABSTRACT

Since 1990 rabbit numbers in The Netherlands have declined. The reasons for the decline have been studied by analysing data on rabbit counts along transects, sampled from 1985 to 1997. Cold winters, disease and change in vegetation structure were considered. The conclusion is drawn that the main factor responsible for the continuing decline is a new rabbit disease, Rabbit Viral Haemorrhagic Disease (RVHD).

Introduction

Rabbits and society

The appreciation of rabbits, and the appreciation of the influence of rabbits on vegetation, has changed significantly through history. In medieval times rabbits were considered an asset. The dunes were the hunting ground of nobility, and the farmers on the landward side of the dunes were allowed to take measures to keep rabbits off their lands, but could not interfere with what happened in the dunes. As rabbits were not considered big game, the catching of rabbits was often leased to a *meijer* (warrener) who made his living out of it, and tried to increase the number of rabbits.

Modern times (starting at around 1790) have seen a different perception of wastelands like the dunes. Schemes were implemented to convert the land to productive use by stabilizing the blowing sand and initiating forestry and farming activities. Farming has been marginal, but afforestation, mainly with exotic pine trees, continued well into the 1900s, mainly on the central and landward dunes. The sea dyke has been carefully maintained with planted marram to protect the hinterland from the sea. For all these activities rabbits were seen as a nuisance; in particular they seemed to dig and eat at night what had been planted during the day.

Myxomatosis

Nowadays the rabbit is again considered to be an asset to the dunes. The obvious negative effects of myxomatosis have convinced nature managers that rabbits are beneficial in maintaining the biodiversity of dunes.

Myxomatosis arrived in the coastal dunes of The Netherlands in September 1953. The disease probably spread mainly from the north of France where it was introduced deliberately in 1952. People who moved infected rabbits to new areas for the purpose of rabbit control frequently assisted spread of the disease. The mortality at first was almost 100 per cent and after this dramatic decline it took a considerable time for the rabbit population to recover. At present, weaker strains of myxomatosis are permanently present in the population but the mortality rate is not high.

Following the outbreak of myxomatosis in the 1950s, rabbit numbers never recovered their former levels. The reasons for this have been subject to much debate; we hypothesize that when rabbit numbers were low, the resulting grass and scrub encroachment led to a decrease in high-quality food.

Rabbits, biodiversity and vegetation structure

Traditionally, people were convinced that rabbits had a negative effect on species diversity of natural vegetation. It is easy to get this impression in areas of intensive rabbit grazing where the very short vegetation does not flower and rabbits show a preference for rare plants. However, experiments with small exclosures (permanent quadrats where rabbits are kept out) have shown that the 'damage' is only temporary. In the first season without rabbit grazing the vegetation flowers; evidently many species were present that did not get the chance to produce flowers under heavy rabbit grazing. In general, therefore, the influence of rabbit grazing on biodiversity is positive. The effect of the virtual eradication of rabbits by myxomatosis was the same as that demonstrated by small exclosures: initially the dune vegetation flowered and rare plants spread, but this was soon followed by the successional development of coarse grasses and scrub.

Typical rabbit habitat in dune grassland areas exhibits a small-scale mosaic of short and taller vegetation patches. Rabbit grazing can inhibit scrub encroachment and the development of undergrowth in woodland. If the vegetation develops to the next successional stage, the food quality becomes too low, and the rabbits leave.

Vegetation as food of the rabbit

Food preference

An investigation of rabbit food preferences demonstrates why sand dunes, with the associated sparse, patchy vegetation, are such a good rabbit habitat. Attempts were made to make a preference list of plant species but the results were very variable. The main conclusion was that rabbits seemed to like a varied diet. Comparison of the contents of the stomach and the gut of rabbits shot in a patchy environment showed that they even vary their meals over the course of the day, by visiting different vegetation types (Wallage-Drees et al., 1986).

The general experience is that rabbits do not like vegetation that is wet or dominated by tall grasses. They show a preference for dry places with sparse, short vegetation. They are able to eat almost anything, even plants such as *Senecio jacobaea* that are considered poisonous to ruminants. Since the rabbit is not a ruminant, it is not so sensitive to high alkaloid content or other substances in a plant that might be harmful to the flora of the gut.

Food digestibility is the key to understanding rabbit grazing behaviour; they select on the basis of a high energy content, and can only get this from plants with a low fibre content. This means a preference for dicotyledons, and, when they eat grasses, for young, actively growing plants. Rabbits therefore prefer short vegetation that is kept short by their own constant nibbling.

The introduction of larger grazers

Since the 1970s the increase in atmospheric nitrogen deposition, which indiscriminately enriches agricultural and nature areas alike, has led to an increased rate of vegetation succession resulting in the development of monotonous areas of *Calamagrostis epigejos* and scrub encroachment. Rabbits do not seem able to push back these vegetation types and they just disappear from those areas.

To reverse this trend nature managers have introduced large grazers such as cattle. The resulting short vegetation should attract rabbits. There is much positive experience of the effect of larger grazers on vegetation. There are, however, strong indications that the central dune *Taraxacum* landscape (Doing, 1989) with *Viola rupestris* is damaged by these grazers.

Population Dynamics

Population dynamics have been studied in two ways:

1. Wallage-Drees (1988) studied a marked population from 1977 to 1981 and studied food choice by analysing the gut contents of the game bag in one area.
2. Olff and Boersma (1998) analysed data collected on a much larger scale and over a longer time-span. Data were collected by counting rabbits along set routes in many areas of the mainland dunes from 1985 to 1997, studying 150 separate locations, covered by 250,000 separate counts.

The circumstances of the two studies differ in three ways:

1. In 1967 the fox appeared in the Dutch dunes. During the first study fox numbers were still low but during the second they seemed stabilized at a high level.
2. In 1990 a new disease appeared, RVHD (Rabbit Viral Haemorrhagic Disease). Because of the presence of foxes that eat or bury the carcasses it was not possible to obtain reliable data on the impact of the disease.
3. During the first study there was still rabbit control by game wardens in many areas, which has now has stopped.

Observations of a marked population

In this study the following conclusions were drawn:

Habitat

The close proximity of dry burrowing areas and feeding grounds makes stabilized dunes a good rabbit habitat. At high population densities rabbits keep their habitat suitable by grazing and digging. At (temporary) low population densities vegetation succession takes place, which reduces the area of suitable feeding habitat.

Predation and disease

The co-evolution of the virulence of the Myxoma virus and of resistance in the rabbit should result in outbreaks of myxomatosis continuing to occur at irregular intervals in western European rabbit populations. In the coastal dunes myxomatosis has its main impact in spring and autumn, mostly on the new generation of juveniles. In addition, rabbits were killed by stoats and hunters.

Data from both the game bag and the field study show that the increase in fox numbers in the Noord-Holland Dune Reserve since 1970 has not caused a decrease in rabbit numbers in autumn and winter. The impact of total predation might even be less than before because of the decrease in stoat numbers (Mulder, 1990).

Food in winter

The winter diet in the coastal dunes, assessed from quality of stomach contents, had a sufficient proportion of basic protein, but a level of digestibility below that required for body maintenance.

The length of the breeding season was modified by an interaction between population density and food supply. The timing of the breeding season seemed to be adapted to the changes in quality of the vegetation. The first young were born when the proportion of dicotyledons in the vegetation increased. High population densities kept the breeding season short and litter frequency low.

The main conclusion therefore is that, due to the low quality of winter food and the absence of sufficient regulating factors, food supply is the major factor limiting the density of the breeding population of rabbits in the dunes. There is, however, some density dependence in the rate of increase or decrease.

Longear: rabbit counting

To further investigate the fluctuations in rabbit populations, nature managers along the Dutch mainland set up a long-term monitoring programme involving counting rabbits along set routes (the transect method). In a project called 'Longear', all data collected in this way were analysed. The project had two aims:

1. to advise the managers on continuation of the monitoring in the most efficient way; and
2. to determine the major factors affecting rabbit density.

Method

The variation in the level of above-ground activity needs to be taken into account when assessing population size by sight counts. Emergence and appearance times do not vary over the year but the percentage of the population above ground at any one time does. When comparing rabbit counts, therefore, one should compare one season with the same season in the next year.

The method used was to count all rabbits in the headlights of a slow-moving car along a transect for eight evenings in autumn and another eight evenings in spring. From the eight counts the highest was chosen, as the low numbers and the mean are too much influenced by incidental occurrences such as a passing jogger. Each transect was divided into sections and the structure of the vegetation described for each section. This was done in a number of areas on the mainland and on the islands. Neither Natuurmonumenten nor the State Forestry was willing to support our study, so we did not interpret the data collected on the islands.

We consider the mainland rabbits to be divided into two metapopulations: north and south of the North Sea Canal (crossing the coastal dune ridge near Amsterdam). It is not possible for these two groups to have any contact nor is it possible for rabbits to migrate from one group to the other.

Results

Figure 1 shows the results of changes in the counted number of animals over the years in four areas. In two representative areas above the North Sea Canal (Zwanenwater and Duin en Kruidberg) the population either strongly fluctuated or increased over the first decade. The synchronicity of the fluctuations is shown in Figure 2, for which the transects were divided into sections and the frequency of sections with decrease or increase was counted. This synchronicity was strong before 1990, while it disappeared after this date. After 1990, we observed a downward trend, especially in the sites north of the North Sea Canal. Introducing large grazers did not reverse this trend (see Fig. 1).

The per capita population change in the rabbits since 1990 was strongly density dependent (Fig. 3). The decline was most marked in the areas with initially (in 1990) high densities. The rate of decrease of the curve is significantly larger in the northern area, where even in areas with initially low densities the population exhibited a negative per capita change (more deaths than births).

Next, *temporal autocorrelations* between consecutive counts (between autumn and the following spring, autumn and the following autumn and so on) were studied. The only significant relationships were those with one lag: *there is a dominant long-period trend*. When comparing these autocorrelations between areas, an interesting pattern emerges. Taking the northern and southern populations separately, we found that the strongest correlations occurred between areas situated at a short distance from each other.

Both of these results indicate that, apart from dominant factors such as extreme

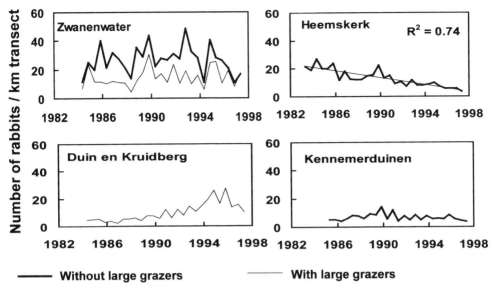

Figure 1. *Changes in the average density of rabbits per km of counting route at four different locations. For each year both the spring and autumn density is presented. The average density for each season was calculated by taking for each 1km section of transect the maximum of 6–8 censuses in that period, and averaging these maxima over all sections of a site. The results are presented separately for sections grazed by large herbivores (thin lines) and without large herbivores (thick lines). Regression lines with R^2 are presented only if the regression was significant (P< 0.05).*

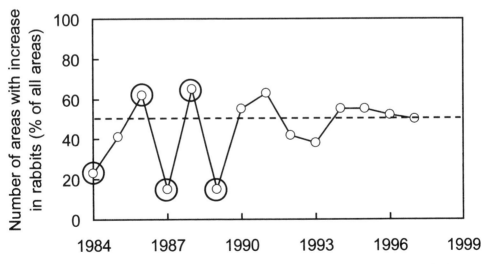

Figure 2. *Changes with time in the number of 1km transect sections in which the number of rabbits counted is decreasing over two subsequent spring censuses. This provides a measure of synchronicity independent of the actual number observed. A high synchronicity implies that the observed number either decreases in all sections or increases in all sections. Particularly high or low values are circled.*

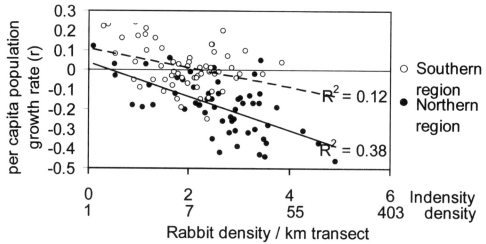

Figure 3. Per capita density changes since 1990 in rabbit counts, as a function of the initial density in 1990. This density dependence of the relative rate of change of the population density is explored separately for the northern part (above the North Sea Canal, filled circles) and the southern part (below the North Sea Canal) of the Dutch coastal dunes. Each point represents a 1km section of a counting transect. Per capita density changes are calculated as slope of the regression of the ln-transformed densities against time. For example, a per capita change of -0.20 implies that during the observed period the density in each year was on average 20% lower than the density in the previous year. Only sections without large grazers were used in this analysis.

winters, affecting all populations, there must be another factor that is transported between rabbits.

Vegetation structure
Aerial photographs showed that the area of scrub has increased significantly over the last 10 years, and also that the increase has been greater north of the North Sea Canal. Comparing trends in rabbit numbers and the development of the vegetation it can be seen that the increase in scrub parallels the decline in rabbit numbers.

Fluctuations in rabbit numbers
Fluctuations in rabbit numbers were highest in open landscapes. We also found that woodland with some small open patches, or large patches of scrub (*Hippophae rham-noides*) were the optimal habitats with the most stable rabbit numbers. One might hypothesize that in these landscapes the distances between burrows and foraging areas are shortest so the rabbits are less likely to come into contact with the disease; this is, however, very speculative.

Discussion

The autocorrelations of rabbit numbers show that adjacent areas influence each other, a pattern that could be explained by a disease transported by migrating rabbits. Combining this pattern with the timing of the start of the decline (1990, the year the first rabbits with RVHD were found) indicates RVHD as the major cause of decline. This has led us to conclude that the influence of the disease is much larger than is apparent from the occasional finds of dead rabbits killed by RVHD.

Many questions still remain. The most important is perhaps that of the interaction between the larger grazers (sheep, cows, horses) and rabbits. In the studies reported here we could not find the expected facilitation of rabbits by the larger grazers.

In the years of the field study myxomatosis had already become much less virulent, and it was concluded that neither predation nor disease could keep rabbit numbers so low that they would not reach the level set by the food availability. They are food-limited. When rabbit numbers are food-limited they are expected to fluctuate with food availability in winter, i.e. spring numbers should vary according to the cold and wetness of the preceding winter. This cannot explain the steady decrease since 1990.

What is new since 1990 is the appearance of RVHD. We conclude that this effect is so large as to keep rabbit numbers below the level set by available food. This hypothesis would explain why we did not see the expected facilitation by the introduction of large grazers.

Rabbit Cycles

Following analysis of data on rabbit populations and vegetation change we hypothesize the existence of cycles of rabbit populations of about 30 years. Since areas with much scrub are a very good habitat, the mobility of the rabbits is low, the impact of disease is less, and so rabbit populations increase. The high rabbit density results in less scrub, a higher rabbit mobility and greater impact of disease. Following on from this hypothesis, the introduction of larger grazers which control scrub has two effects: it increases the food availability, but also increases rabbit mobility.

In the early years of rabbit counting reported here it was difficult to detect any clear patterns in the data; numbers seemed to fluctuate erratically. Persistence in continuing the data collection and the addition of a new approach from a second scientist has enabled a clearer understanding of the factors affecting the fluctuations in rabbit numbers.

References
Doing, H. (1989), *Landscape Ecology of the Dutch Coast*, European Union for Coastal Conservation, Leiden.
Mulder, J.L. (1990), 'The stoat *Mustela erminea* in the Dutch dune region, its local extinction,

and a possible cause: the arrival of the fox *Vulpes vulpes*', *Lutra*, **33(1)**, 1–22.

Olff, H. and Boersma, S.F. (1998), *Lange-termijn veranderingen in de konijnenstand van Nederlandse duingebieden. Oorzaken, en gevolgen voor de vegetatie*, Report, Nature Conservation and Plant Ecology Group, Wageningen Agricultural University.

Wallage-Drees, J.M., Immink, H.J., de Bruyn, G.J. and Slim, P.A. (1986), 'The use of fragment-identification to demonstrate short-term changes in the diet of rabbits', *Acta Theriologica*, **31(22)**, 293–301.

Wallage-Drees, J.M. (1988), 'Rabbits in the coastal sand dunes; weighed and counted', doctoral thesis, Rijksuniversiteit Leiden.

DONKEY DIET IN A FLEMISH COASTAL DUNE AREA IN THE FIRST YEAR OF GRAZING

MAURICE HOFFMANN, ERIC COSYNS
University of Ghent, Belgium
MIEKE DECONINCK, INDRA LAMOOT and ARNOUT ZWAENEPOEL
Institute of Nature Conservation, Belgium

Plant species nomenclature according to Lambinon et al. (1998)

ABSTRACT

The composition of the diet of a small herd of donkeys in an 80ha nature reserve in the coastal dunes of Flanders (Belgium) was investigated in the first year after its introduction. The animals predominantly ate grasses with a clear preference for *Festuca juncifolia, Elymus repens* and *Koeleria albescens* and the sedge *Carex arenaria. Calamagrostis epigejos* and *Ammophila arenaria* appeared as only slightly desirable species. Apart from *Hieracium umbellatum*, characteristic herbaceous species of dune grassland (e.g. *Galium verum*) seem to be avoided. Browsing of woody species (which were dominant in much of the study area) was rather limited, although some non-native species (*Syringa vulgaris, Fallopia aubertii*) were eaten fairly frequently. Leaves are the preferred plant parts, while dead and shrivelled plant material forms a small minority of the diet as compared to fresh green plant material. Some seasonal trends in diet composition could be detected.

Introduction

During the nineteenth and the beginning of the twentieth century grazing by domesticated livestock was a common practice in the Western Flemish coastal dunes (De Smet, 1961). The area was used relatively intensively by small-scale farming fishermen and labourers of the large prosperous farms on the neighbouring polders. Small fields were created for the cultivation of potatoes, rye and vegetables. Livestock consisting of sheep, cattle, donkeys and horses grazed natural vegetation. For example, in 1828 the dune area of the Western Flemish coast (approximately 2500ha) was grazed by 450 sheep, 240 cows, 112 donkeys and 51 horses. The remaining scrub was regularly cut down and used as firewood. Dune vegetation largely consisted of grey dunes, moist slack vegetation and dune grassland (Massart, 1908b). Scrub was a rare phenomenon, clearly illustrated by the fact that until the 1950s hardly any epiphytes were known from the area (Massart 1904; 1908a; Barkman, 1990; Hoffmann, 1993). After this type of agricultural land use was abandoned the entire dune area showed a strong tendency towards scrub development, dominated by species like *Hippophae rhamnoides, Ligustrum vulgare*, and *Salix* div. spp., which largely replaced the species-rich *Cladonio-Koelerietalia* vegetation. Today scrub occupies approximately one third of the remaining dune area which has not been subject to afforestation, urban development or agriculture. This trend in vegetation succession decreases the area of open dune vegetation and is believed to threaten its relatively large number of specialist dune species (among which are a large proportion of red-list species).

Following the development of an ecosystem approach for the entire coastal dune area (Provoost and Hoffmann, 1996) the Flemish Community decided to reintroduce large domestic herbivores in a number of their nature reserves. Monitoring the effects of this management initiative is one of our major activities (Bonte et al., 1998). Investigating behaviour of the grazers is one approach to this monitoring. Here we report the immediate effect of grazing by donkeys, which was investigated by direct observation of donkey diet at Houtsaegerduinen nature reserve.

Materials and Methods

Houtsaegerduinen is one of the Flemish nature reserves where domesticated grazers were introduced. It is an 80ha dune area, mainly occupied by *Hippophae rhamnoides* scrub, with relatively small patches of dune grassland and grey dune (*Cladonio-Koelerietalia*). Old, deteriorating *Hippophae* scrub is generally replaced by *Calamagrostis epigejos* dominated, species-poor grassland. Part of the area was planted with *Alnus glutinosa* and several non-native tree species (*Populus* div. spp.). Several other exotic plant species were introduced as hedge plants along small fields; others escaped from neighbouring gardens (e.g. *Syringa vulgaris, Fallopia aubertii*). The area is almost entirely enclosed by residential development, camping areas and roads and is therefore strongly influenced by human activities (Hoys et al., 1996).

In April 1997 a small herd of six donkeys (1 male, 5 female) of Romanian origin was introduced for year-round grazing over the entire area (1 animal per 13ha). Donkeys were frequent grazers in the Flemish dunes during the last century and were therefore considered a reasonable choice for introduction. Although principally a more southern animal of dry or semi-desert environments, the first documented observations of the presence of domesticated donkeys in Belgium date back to the Roman period (Verbeeck et al., 1991). They appear to cope quite well with the mild maritime climate along the Flemish coast. They were further chosen as a comparison to the Shetland ponies introduced to the neighbouring nature reserve (the Westhoek), thus enabling evaluation of the differences in grazing and browsing activities of both herbivore species.

Very little is known of the diet of donkeys in natural West European vegetation although there are some data on donkey diet from a Flemish inland nature reserve (Van Assche, 1993). Apart from grasses the donkeys appeared to eat thistles and nettles frequently and during winter periods they browsed quite a large number of shrub and tree species (*Ulmus* sp., *Fraxinus excelsior, Prunus serotina, Salix* div. spp., *Alnus glutinosa*). Therefore we expected the donkeys to eat mainly grasses during the entire year, while they were expected to browse shrubs and trees, especially in winter. This would coincide with the management aim to suppress spread of *Calamagrostis epigejos* and to stop further scrub encroachment. Finally, donkeys are known to be grazers of food plants of low nutritional value. This is a very suitable characteristic for grazing dune vegetation.

To determine the exact diet composition we followed the animals intensively during summer (5 August–16 September 1997), autumn (14 October–10 November 1997) and winter (15 February–15 March 1998) (Deconinck, 1998). After a habituation period of 14 days the animals were followed closely (within a maximum range of 2 metres) during separate daytime sessions of 6 hours. The behaviour of the male animal was clearly influenced by the observation method and was therefore excluded from the analysis. The female behaviour was not noticeably influenced by the observation method and all five females were included in the analysis. Every session started with the random choice of one female that was followed for 15 minutes. For every bite we recorded the plant species and plant part (stem, rhizome, leaf, inflorescence, etc.) which was consumed and the condition of the plant material (fresh green or dead shrivelled). Mixed bites were registered as different bites of each of the plant species. Every 15 minutes another randomly chosen female animal was followed. From these data we can determine what species compose the diet of the donkeys during their first year of presence in the nature reserve. In addition we know what plant parts are preferred, whether they prefer fresh green plant material to dead shrivelled material and whether there are seasonal diet trends.

To estimate plant species preferences, necessary to compare different herbivore species, we used two different methods. The mean number of bites per species in a 15-minute observation period gives a first indication of preference. To calculate the mean we only used those observation periods in which the plant species was actually bitten. We assume that non-biting means that the animal did not meet the plant in that specific 15-minute period.

A second, presumably more accurate approach is the diet-availability ratio (Colebrook et al., 1987), discussed by Stuth (1991):

$$(\% \ Diet - \% \ Availability) \ / \ (\% \ Diet + \% \ Availability)$$

As a measure of diet we used the number of bites in summer. As a measure of availability we used the above-ground biomass of every species in summer in the vegetation patches visited by the animals during the observation sessions. Above-ground biomass, expressed in dry weight, was measured by clipping several 0.25m² plots within the most frequented vegetation types. After separation of the different species, the plant material was air-dried for at least one week and oven-dried for at least 24 hours at 75°C until no further weight loss was observed. We estimated overall species availability by calculating the above-ground biomass of every species per vegetation type and multiplying it by the total area of that vegetation type.

Results

General

During 3930 observation minutes in summer, 1785 in autumn and 1845 in winter 74,895 bites were registered with a mean of 8.53 bites per minute in summer, 7.72 in autumn and 14.96 in winter. Out of approximately 340 species present in the area 112 were seen bitten. Grasses and sedges represent 81.5 per cent of the donkey diet, herbs 9.7 per cent, woody species 6.6 per cent, non-identifiable, fallen litter 1.2 per cent and mosses and lichens 1.0 per cent. Table 1 illustrates the seasonality in general diet composition. Herbaceous plants are more frequently bitten in summer and autumn than in winter, while unidentifiable, fallen litter and mosses and lichens are slightly more often bitten in winter. Browsing does not show significant seasonal differences.

Diet at the plant species level

Carex arenaria is the most frequently bitten plant species (Table 2). Eight of the ten most bitten species are grasses (*Elymus repens, Arrhenatherum elatius, Festuca juncifolia, Calamagrostis epigejos, Avenula pubescens, Ammophila arenaria, Elymus athericus* and *Koeleria albescens*). *Hieracium umbellatum, Diplotaxis tenuifolia* and *Achillea millefolium* are the most frequently bitten herbaceous species. Most frequently bitten woody species are *Rubus caesius, Ligustrum vulgare* and *Fallopia aubertii. Hippophae rhamnoides* is rarely bitten, even though it is a dominant species in much of the area. Surprisingly enough some poisonous, extremely hairy or prickly plant species were bitten fairly frequently, for example *Ligustrum vulgare, Sambucus nigra, Prunus serotina, Chaerophyllum temulum, Cynoglossum officinale* and *Robinia pseudoacacia*.

Table 1. General diet composition (percentage of total number of bites) of five female donkeys in the study area in the first year after their introduction

Plant group	Summer	Autumn	Winter	Total
Grasses/sedges	78.2	77.9	87.3	81.5
Herbaceous plants	14.4	13.0	2.3	9.7
Woody species (browse)	7.1	8.9	5.1	6.6
Fallen litter	0.2	< 0.1	2.9	1.2
Mosses and lichens	0.1	0.1	2.4	1.0

Table 2. Total number of bites, percentage of the total number of bites and bites per minute per plant species of the most frequently bitten species, the most frequently bitten herbaceous and woody species and Hippophae rhamnoides during summer 1997, autumn 1997 and winter 1997–98 by five female donkeys in the study area

Species	Number of bites				Percentage of bites				Bites per minute			
	Sum.	Aut.	Win.	Total	Sum.	Aut.	Win.	Total	Sum.	Aut.	Win.	Total
Carex arenaria	7684	2221	3486	13391	22.93	16.11	12.63	17.88	1.96	1.24	1.89	1.77
Elymus repens	3245	1156	7867	12268	9.68	8.38	28.50	16.38	0.83	0.65	4.26	1.62
Arrhenatherum elatius	2841	1562	4339	8742	8.48	11.33	15.72	11.67	0.72	0.88	2.35	1.16
Festuca juncifolia	2988	149	4331	7468	8.92	1.08	15.69	9.97	0.76	0.08	2.35	0.99
Calamagrostis epigejos	2166	4088	502	6756	6.46	29.65	1.82	9.02	0.55	2.29	0.27	0.89
Avenula pubescens	1175	108	1602	2885	3.51	0.78	5.80	3.85	0.30	0.06	0.87	0.38
Ammophila arenaria	1208	133	1217	2558	3.61	0.96	4.41	3.42	0.31	0.07	0.66	0.34
Elymus athericus	548	1265	0	1813	1.64	9.18	0.00	2.42	0.14	0.71	0.00	0.24
Hieracium umbellatum	1643	1	0	1644	4.90	0.01	0.00	2.20	0.42	0.01	0.00	0.22
Koeleria albescens	1347	13	131	1491	4.02	0.09	0.47	1.99	0.34	0.01	0.07	0.20
Diplotaxis tenuifolia	988	471	4	1463	2.95	3.42	0.01	1.95	0.25	0.26	0.01	0.19
Festuca rubra	1410	0	0	1410	4.21	0.00	0.00	1.88	0.36	0.00	0.00	0.19
Achillea millefolium	18	573	0	591	0.05	4.16	0.00	0.79	0.01	0.32	0.00	0.08
Rubus caesius	397	207	404	1008	1.18	1.50	1.46	1.35	0.10	0.12	0.22	0.13
Ligustrum vulgare	24	2	753	779	0.07	0.01	2.73	1.04	0.01	0.01	0.41	0.10
Fallopia aubertii	155	178	161	494	0.46	1.29	0.58	0.66	0.04	0.10	0.09	0.07
Hippophae rhamnoides	54	175	0	229	0.16	1.27	0.00	0.31	0.01	0.10	0.00	0.03
Syringa vulgaris	91	42	86	219	0.27	0.30	0.31	0.29	0.02	0.02	0.05	0.03
Other species (incl. litter)	5524	1443	2719	9686	16.49	10.47	9.85	12.93	1.41	0.81	1.47	1.28
Fresh green	31807	12781	23775	68363	94.90	92.70	86.10	91.30	8.09	7.16	12.89	9.04
Dead and shrivelled	1701	1006	3827	6532	5.10	7.30	13.90	8.70	0.43	0.56	2.07	0.86
Total	33506	13787	27602	74895	100.00	100.00	100.00	100.00	8.53	7.72	14.96	9.91

Table 3. Diet composition at the plant part level (percentage of total number of bites) of five female donkeys in the study area in the first year after introduction to the area. Overall distribution and a selection of plant species with distinctive distributions are given

Species	Leaves	Stems	Inflorescence	Fruits	Juveniles
Overall	89.78	3.85	3.42	2.43	0.52
Carex arenaria	100.00	–	–	–	–
Elymus athericus	94.56	–	5.44	–	–
Diplotaxis tenuifolia	62.44	13.80	12.73	11.02	–
Rubus caesius	50.59	43.60	2.27	3.54	–
Chelidonium majus	50.41	5.69	0.27	–	43.63
Hieracium umbellatum	47.81	1.76	47.84	2.59	–
Ligustrum vulgare	33.99	66.01	–	–	–
Rosa pimpinellifolia	0.14	0.14	–	99.72	–

For some plant species the donkeys show strong seasonal differences (Table 2). *Carex arenaria* is bitten equally frequently in all three seasons, but *Calamagrostis epigejos*, for example, is far more frequently bitten in autumn than in summer or winter. *Elymus repens* on the other hand is bitten more frequently in winter, while most herbaceous species (e.g. *Hieracium umbellatum*, *Achillea millefolium* and *Diplotaxis tenuifolia*) are hardly bitten in winter. Among the more important woody species there is a trend towards more frequent use in autumn and winter than in summer.

Diet at the plant part level

For the majority of species leaves are by far the most preferred plant parts (Table 3). Leaf preference is certainly true for all grass and grass-like species, but several herbaceous and woody species show a different distribution. Of *Rosa pimpinellifolia* only fruits were seen bitten, while for several herbaceous species the inflorescence is an important diet component, particularly for *Hieracium umbellatum*. Of the woody species, stems are often eaten, e.g. *Ligustrum vulgare* and *Rubus caesius*. True peeling of tree or shrub bark has not actually been observed, although some poplar trunks in the area clearly show signs of peeling.

Diet at the plant state level

Dead, shrivelled plant material is far less frequently consumed than fresh green plant material, although there is a slight increase in dead plant material consumption towards the winter (Table 2). *Urtica dioica* was only consumed as dead plant material.

Plant species preferences expressed by the mean number of bites per observation period

For statistical purposes only those species that show a high frequency in the area, are

Table 4: Mean number of bites per min during 15 minute observation periods. Zero-bite observations per species are left out of the calculation of the mean, assuming that during these zero-bite periods the plant species was not met with by the donkey. Significant differences of the mean (Welch's approximate t-test, Sokal and Rohlf, 1995, 404) between the 14 most frequently bitten plant species, available in all three seasons and frequently present in at least one of the visited vegetation types, are indicated

***: $p<0.001$; **: $p<0.01$; *: $p<0.05$; ns: $p \geq 0.05$

Species	Mean per min	2	3	4	5	6	7	8	9	10	11	12	13	14
1 Festuca juncifolia	5.86	ns	ns	ns	**	**	***	***	***	***	***	***	***	***
2 Avenula pubescens	5.83	–	ns	ns	**	**	***	***	***	***	***	***	***	***
3 Carex arenaria	3.82		–	ns	ns	ns	ns	*	*	ns	*	**	**	**
4 Festuca rubra	3.36			–	ns	ns	ns	*	*	*	*	ns	**	***
5 Koeleria albescens	2.42				–	ns	ns	ns	ns	ns	ns	**	**	***
6 Elymus repens	2.16					–	ns	ns	*	ns	*	***	***	***
7 Arrhenatherum elatius	1.92						–	ns	ns	ns	ns	***	***	***
8 Phragmites australis	1.57							–	ns	ns	ns	**	**	**
9 Ammophila arenaria	1.56								–	ns	ns	***	**	***
10 Fraxinus excelsior	1.55									–	ns	*	*	*
11 Ligustrum vulgare	1.33										–	*	*	**
12 Diplotaxis tenuifolia	0.72											–	ns	ns
13 Calamagrostis epigejos	0.65												–	ns
14 Rubus caesius	0.52													–

present in all three seasons and have a relatively high bite frequency (number of 15-minute periods in which the plant is bitten n > 20) are taken into account. Difference of the mean is tested with Welch's approximate t-test, because of different sample sizes and differences in variance (Sokal and Rohlf, 1995).

The mean number of bites per 15-minute observation period (Table 4) indicates a clear preference for *Festuca juncifolia, Avenula pubescens* and *Carex arenaria* (>50 bites per 15 min). These species are significantly more often bitten per 15-minute period than most other species. *Festuca rubra* seems to be a preferred species as well, but the mean number of bites is hardly ever significantly different from the other species.

A second group of species (between 10 and 50 bites per 15 min: *Elymus repens, Arrhenatherum elatius, Phragmites australis, Ammophila arenaria, Fraxinus excelsior* and *Ligustrum vulgare*) shows an intermediate position. They are significantly less bitten per 15-minute period than the first group but are significantly more often bitten than the third group.

Table 5. *Area covered by the more important vegetation types in the study area and the percentage grazing time in each of those vegetation types observed by Lamoot (1998). Above-ground biomass production and fallen litter quantity per vegetation type (tonnes per hectare) is also given (n = number of 0.25m² plots used for the biomass calculation)*

Vegetation type	% Area	% Grazing time	Biomass	Litter	n
Blond dune with very sparse vegetation	1.59	7.21	–	–	–
Ammophila arenaria dominated dunes	1.83	14.73	7.16	3.61	4
Grey dunes	5.93	8.74	0.52	0.00	8
Dune grassland	1.18	12.90	2.93	0.48	5
Nitrophytic grassland	11.96	22.85	5.51	2.21	4
Rosa pimpinellifolia dominated vegetation	3.80	18.70	5.71	2.45	6
Dune slack	0.37	0.42	1.02	0.37	5
Reed beds	0.29	1.11	–	–	–
Calamagrostis epigejos dominated grassland	2.26	2.32	5.79	4.76	4
Salix repens dwarf scrub	1.37	1.16	–	–	–
Vital *Hippophae rhamnoides* scrub	24.69	4.93	–	–	–
Dying *Hippophae rhamnoides* scrub	16.70	1.80	–	–	–
Ligustrum vulgare scrub	12.84	0.58	–	–	–
Other scrubs	15.18	2.55	–	–	–

Table 6. *Overall above-ground biomass production in summer 1998 (in kg) of some of the more frequently bitten plant species in the entire study area, against the observed number of bites in summer 1997. Biomass data are used as availability indication; bite data are used as diet indication*

Species	Biomass (summer 98)	Bites (summer 97)
Arrhenatherum elatius	29264	2841
Calamagrostis epigejos	10103	2166
Ammophila arenaria	6038	1208
Avenula pubescens	4361	1175
Galium verum	2651	9
Achillea millefolium	2468	18
Festuca rubra	2273	1410
Carex arenaria	1616	7684
Koeleria albescens	261	1347
Festuca juncifolia	167	2988
Hieracium umbellatum	154	1643
Elymus repens	496	3245
Melandrium album	41	322

The third group of species (≤ 10 bites per 15 min: *Diplotaxis tenuifolia, Calamagrostis epigejos* and *Rubus caesius*) is not preferred by the donkeys. They are significantly less frequently bitten per 15-minute period than species of the first two groups.

Plant species preferences expressed by the diet-availability (D:A) ratio

In Table 5 the area of the different vegetation types in the Houtsaegerduinen is given (after Hoys et al., 1996) and the percentage grazing time observed by Lamoot (1998) in those vegetation types. From these figures we can conclude that scrub vegetation is hardly ever grazed. We therefore omitted them from calculations of availability of the more frequently bitten plant species. Table 5 gives the production estimates in summer of the more frequently visited vegetation types. Table 6 shows summer production figures of the most frequently bitten plant species in the entire study area. These figures are used to calculate the D:A ratio.

According to Stuth (1991) three preference classes can be identified:

$$D:A > 0.35 \quad \text{Preferred species}$$
$$-0.35 < D:A \le 0.35 \quad \text{Desirable species}$$
$$D:A \le -0.35 \quad \text{Undesirable, avoided or forced species}$$

Using this classification some of the frequently bitten plant species can be classified as follows (Table 7).

Table 7. Plant species preferences classified by D:A (diet-availability) ratio

Species	D:A ratio
Preferred	
Festuca juncifolia	0.96
Hieracium umbellatum	0.94
Melandrium album	0.91
Elymus repens	0.90
Koeleria albescens	0.87
Carex arenaria	0.86
Desirable	
Festuca rubra	0.28
Avenula pubescens	-0.13
Calamagrostis epigejos	-0.24
Ammophila arenaria	-0.28
Undesirable, forced, avoided	
Arrhenatherum elatius	-0.57
Achillea millefolium	-0.96
Galium verum	-0.98

Discussion

Only one year after the animals were introduced general conclusions about the diet of a small herd of donkeys in the Flemish Nature Reserve, the Houtsaegerduinen, still remain preliminary. Earlier research in other areas has already shown that their diet and their preferences change considerably over the years (Van Assche, 1993). In addition, we do not have diet data for the spring period.

There is a strong seasonality in the mean number of bites. In winter the number of bites doubles compared to summer and autumn. This indicates that the nutritional value of a bite in winter is much lower than in summer and autumn and that the donkeys have to spend much more energy on feeding in winter than in both other seasons. In addition tourists who feed the animals were less active during the winter months than during summer and autumn; this might be implicated in the lower number of bites observed in these seasons.

During their first year of dune experience the donkeys clearly performed like most other horses (New Forest ponies – Putman et al., 1987; Exmoor ponies – Gates, 1982; free ranging horses in the Camargue – Mayes and Duncan, 1986; feral horses in Alberta

– Salter and Hudson, 1979), which is as grazers, i.e. grass-eating animals. Woody species that form the dominant vegetation types are rather rarely eaten, although some exotic species are eaten quite frequently (*Syringa vulgaris* and *Fallopia aubertii*).

A study of the location selection (Lamoot, 1998) clearly shows that the donkey distribution is restricted to a minor part of the fenced area of 80ha. Only about 24ha (i.e. the non-scrubby vegetation) is actually visited and grazed by the animals. Thus the actual effective herd density is increased in this non-scrubby vegetation from one donkey per 13ha to one donkey per 4ha. Even so, primary production of grass and grass-like species in the area appears to be much higher than can be consumed at the present herd density.

Both preference estimates (mean number of bites per 15-minute observation period and D:A ratio) indicate that the grass species *Elymus repens, Festuca juncifolia* and *Koeleria albescens* and the sedge *Carex arenaria* are preferred species. The animals do not particularly favour the abundant grass species *Calamagrostis epigejos* and *Ammophila arenaria*, although, in absolute figures, both are eaten quite frequently; *Arrhenatherum elatius* must be considered an undesirable species. Hardly any of the characteristic herbaceous plant species of dune grassland are eaten frequently. All of these species, with the exception of *Hieracium umbellatum*, seem to be avoided by the animals.

The estimate of diet quantity, which we merely based on the number of bites, has a major disadvantage in that we have no estimate of the quantity of plant material consumed with every bite (de Vries, 1994). One bite of a tussock-forming species, such as *Calamagrostis epigejos* or *Ammophila arenaria*, is generally considerably larger than a bite of a species with long rhizomes and scattered shoots, such as *Festuca juncifolia*, *Elymus repens* and *Carex arenaria*. Further research has to be done therefore on the quantity of plant material consumed per bite of a particular species. This should lead to a more accurate quantification of the donkey diet.

Preference estimates are important in the comparison between different herbivore species. At present only donkey preferences were studied and an evaluation of the different introduced large herbivore species in coastal dune nature reserves in Flanders (donkeys, Konik horses, Shetland ponies and Highland cattle) cannot yet be made. Data on all herbivore species have to be collected before sound decisions can be made on future selection of herbivores for grazing.

Despite some shortcomings (limited length of observation time and methodology), we can already conclude that the management aims within the Houtsaegerduinen (decrease of *Calamagrostis* and *Hippophae* dominance) can only be reached with a higher donkey density in the area. Increase of the herd size is therefore recommended. Change of the herd composition, however, is not recommended since the evaluation time is still too short to draw general conclusions on the effects of donkey grazing and data on other herbivore species are not yet available for the Flemish coastal dunes. Further observations on herbivore behaviour as such and on the behavioural and diet change under changing vegetation conditions and changing herd size are necessary for decisions on herbivore choice for specific management aims.

Acknowledgements

We wish to thank the Ministry of the Flemish Community (Nature Division) for permitting this research project in their nature reserve. The project contributes to the Life-Nature Project 'Integral Coastal Conservation Initiative', co-financed by the European Community and the Ministry of the Flemish Community, in which the effects of several nature management techniques are being monitored.

References

Barkman, J.J. (1990), 'The epiphytic flora and vegetation along the Belgian and Northern French coast in the fifties', *Mém. Soc. Roy. Bot. Belg.*, **12**, 11–19.

Bonte, D., Ampe, C., Hoffmann, C.M., Langohr, R., Provoost, S. and Herrier, J.L. (1998), 'Monitoring research in the Flemish dunes: from a descriptive to an integrated approach', *Proc. European Seminar on Coastal Dunes, Management, Protection and Research*, Skagen, Denmark.

Colebrook, W.F., Black, J.L. and Kenney, P.A. (1987), 'A study of factors influencing diet selection by sheep', in M. Rose (ed.), *Herbivore Nutrition Research*, Aust. Soc. Anim. Prod., Brisbane, 85–86.

Deconinck, M. (1998) 'Soortspecifieke begrazing door ezels in het Vlaams natuurreservaat de Houtsaegerduinen (De Panne, West-Vlaanderen)', unpublished dissertation, University of Ghent.

De Smet, J. (1961), 'Onze duinen in 1828', *Biekorf (Westvlaams Archief voor geschiedenis, oudheidkunde en folklore)*, **62(9)**, 257–66.

Gates, S.A. (1982), 'The Exmoor pony – a wild animal?', *Nature in Devon (Journal of the Devon Trust for Nature Conservation)*, **2**, 7–30.

Hoffmann, M. (1993), 'Verspreiding, fytosociologie en ecologie van epifyten en epifytengemeenschappen in Oost-En West-Vlaanderen', unpublished PhD thesis, University of Ghent.

Hoys, M., Leten, M. and Hoffmann, **M.** (1996), *Ontwerpbeheersplan voor het staatsnatuurreservaat De Houtsaegerduinen te De Panne (West-Vlaanderen)*, University of Ghent, Lab. Botany, under the authority of the Ministry of the Flemish Community (Nature Division).

Lambinon, J., De Langhe, J.E., Delvosalle, L. and Duvigneaud, J.(1998), *Flora van België, het Groothertogdom Luxemburg, Noord-Frankrijk en de aangrenzende gebieden. Pteridofyten en Spermatofyten*, 3rd edn, Nat. Bot. Garden, Meise.

Lamoot, I. (1998), 'Ezels als beheerders van het Vlaams natuurreservaat De Houtsaegerduinen (De Panne): staat hun gedrag in functie van de vegetatiesamenstelling en –structuur?', unpublished dissertation, University of Ghent.

Massart, J. (1904), 'Les muscinées du littoral Belge', *Bull. Soc. Roy. Bot. Belg.*, **42**, 141–62.

Massart, J. (1908a), *Essai de géographie botanique des districts littoraux et alluviaux de la Belgique*, Brussels, Lamertin.

Massart, J. (1908b), 'Les districts littoraux et alluviaux', in Ch. Bommer and J. Massart (eds.), *Les aspects de la végétation en Belgique*, Nat. Bot. Garden, Meise.

Mayes, E. and Duncan, P. (1986), 'Temporal patterns of feeding behaviour in free ranging horses', *Behaviour*, **96**, 105–29.

Provoost, S. and Hoffmann, M. (eds.) (1996), *Ecosysteemvisie voor de Vlaamse Kust*, University of Ghent and Institute of Nature Conservation (Brussels), under the authority of the Ministry of the Flemish Community (Nature Division), 3 volumes.

Putman, R.J., Pratt, R.M., Ekins, J.R. and Edwards, P.J. (1987), 'Food and feeding behaviour of cattle and ponies in the New Forest', *J. Appl. Ecol.*, **24**, 369–80.

Salter, R.E. and Hudson, R.J. (1979), 'Feeding ecology of feral horses in western Alberta', *Journal of Range Management*, **32**, 221–25.

Sokal, R.R. and Rohlf, F.J. (1995), *Biometry*, Freeman, New York, 3rd edn.

Stuth, J.W. (1991), 'Foraging behavior', in R. K. Heitschmith and J. W. Stuth (eds.), *Grazing Management: An Ecological Perspective*, Timber Press, Portland, Oregon, 65–84.

Van Assche, L. (1993), 'Ezels in de Pont. De impact van 3 jaar ezelbegrazing', *Euglena (Jeugdbond voor Natuurstudie en Milieubescherming)*, **12(4)**, 16–24.

Verbeeck, M., Lentacker, A., Van Neer, W. and Charlier, C. (1991), 'Première approche interdisciplinaire du site d'Erps-Kwerps (Brabant, Belgique): archéologie, archéozoologie et anthropologie', *Acta Archeologica Lovaniensia*, **30**, 21–39.

Vries, M.F.W. de (1994), 'Foraging in a landscape mosaic: diet selection and performance of free-ranging cattle in heathland and riverine grassland', unpublished PhD thesis, University of Agriculture, Wageningen, The Hague.

USING BEHAVIOURS TO IDENTIFY RABBIT IMPACTS ON DUNE VEGETATION AT ABERFFRAW, NORTH WALES

J.A. POTTER and C.A. HOSIE

Environment Research Group, University College Chester, England

ABSTRACT

The rabbit (*Oryctolagus cuniculus*) is a widespread, naturalized herbivore of many British dune systems. This paper presents preliminary results from Aberffraw, North Wales which illustrate how rabbit behavioural ecology should be integrated with the study of vegetation to elucidate the spatial and temporal impacts of this species on dune system dynamics. A pathway by which individual rabbit behaviours within dune systems may feed into community ecology and feed back to individual rabbit responses is presented. The potential for a behavioural approach to rabbit activities to explain more fully the mechanisms and patterns of rabbit influences across a range of scales is discussed.

Introduction

In primary successional systems, such as sand dunes, it might be expected that the impact of herbivores would be particularly acute because of the importance of facilitation in driving community changes (Connell and Slayter, 1977). The facilitation model proposes that the temporal change of successional flora is driven by plant-mediated alteration of abiotic factors, such as soil structure and chemistry and above-ground microclimate. Thus herbivory would be expected to decrease the rate of succession. It has been proposed to effect this on dune systems by specialist insect herbivores (Bach, 1994) and generalist grazers such as rabbits (Edmondson et al., 1993), and the species-specific influences of domestic livestock are used to manipulate dune vegetation.

Rabbits have a variety of impacts on the vegetation as a result of the many different activities or behaviours they exhibit. Each behaviour and its associated impact(s) show unique, although often related, patterns of distribution. Collectively, the behaviours modify plant community composition by impacts on target plants and changes to the physical environment over a variety of different spatial and temporal scales. A composite view of the influence of rabbit presence on community floristic composition may indicate that rabbit presence influences succession processes, but may hide important behaviour-specific influences on individual plant species.

The range of behaviours and impacts may increase both biotic environmental heterogeneity (e.g. warren-centred, selective grazing maintains localized short swards which support a diverse range of species) and abiotic environmental heterogeneity (e.g. excavating scrapes and burrows exposes bare sand in vegetation stands; the use of latrines results in topographic variation as mounds develop and causes localized changes in soil fertility). However, both will ultimately alter plant population dynamics by differentially influencing individual plant survival and reproduction. As the activities of rabbits change plant distribution, abundance, growth form and chemistry, these factors feed back up the trophic system as a determinant of animal–plant community structure (Hunter, 1994, and see Fig. 1). The pervasive influence of rabbits in controlling both plant and animal population dynamics has been well documented (see Sumption and Flowerdew, 1985) and implies that rabbits are in fact a keystone species in many British dune habitats.

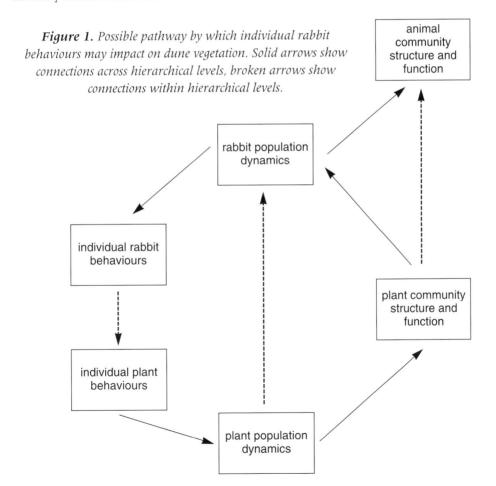

Figure 1. *Possible pathway by which individual rabbit behaviours may impact on dune vegetation. Solid arrows show connections across hierarchical levels, broken arrows show connections within hierarchical levels.*

This paper presents preliminary results from the heavily grazed dune system at Aberffraw, North Wales on the spatial distribution of rabbits and rabbit activities. Specifically, the work aims (i) to identify a role for behavioural studies in clarifying the relationships between rabbits and dune vegetation; (ii) to highlight the importance of sampling across a range of scales when drawing conclusions on the complex impacts rabbit have on their environment; and (iii) to emphasize the importance of acknowledging the interactive cycle of response which occurs between rabbits and the habitat in which they live.

The Aberffraw Dune System

The Aberffraw dunes form part of the Tywyn Aberffraw SSSI on the southwest coast of Anglesey, North Wales (between Ordnance Survey grid references SH358675 and SH377703). The dunes stretch more than 3km inland from the seaward dune ridge in a bay facing directly into the prevailing winds. The dunes are common land and are grazed by sheep and cattle (Ashall et al., 1995) as well as by a large population of rabbits.

The data presented here were collected during spring–summer 1998 from the southwest of the site where semi-fixed/fixed dune grasslands (National Vegetation Classification SD 7 and 8, Rodwell, 1999) intergrade on the leeward side of mobile dune (SD 6) ridges. The vegetation pattern is complicated by the presence of areas of slack vegetation (SD 16 and 17).

Individuals: Relationships between Rabbit Behaviour and Plant Responses

Rabbits respond to their environment interactively and their activities can be measured quantitatively. A wide range of rabbit behaviours leave tangible impacts on the environment and the responses of plants to such behaviours can be measured. However, the wide-ranging published literature on rabbits and dune vegetation contains very few studies of specific rabbit behaviours and their subsequent impacts on individual plants. Most plant responses to specific rabbit behaviours are measured at the population level (e.g. presence/absence of a species) or at the community level (e.g. species richness, species diversity). Such studies are unlikely to clarify the mechanisms of the rabbit–plant relationship they aim to investigate due to the mismatch of the appropriate scale of study between the rabbit behavioural impact and the vegetation response (see Fig. 1).

At Aberffraw, a simple investigation to elucidate the impact of rabbit grazing on *Salix repens* agg. within dune slacks was undertaken. Assuming that grazing intensity is inversely related to the distance from a burrow or warren (Wood et al., 1987), the impact of rabbit grazing on *Salix repens* agg. was measured by counting the number of live and dead terminal buds at 3m and 10m from active burrows in April 1998. In

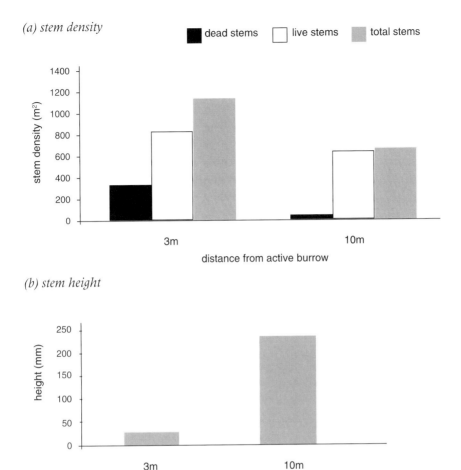

Figure 2. Mean stem densities and stem heights of Salix repens *agg.*

dune slack vegetation, burrows were sparsely and singly distributed. We therefore assume that one burrow indicates the presence of approximately one rabbit (Myers et al., 1994). Total stem densities of *S. repens* were significantly higher (Z = -2.1, p<0.05, n = 30) closer to burrows and this was attributed to a significantly greater number of grazed stems (Z = -4.7, p<0.05, n = 30). In addition plant height was significantly lower closer to the active burrows (Z = -2.5, p<0.001, n = 30, see Fig. 2). This simple experiment highlights how rabbit grazing affects plant growth form, one of a number of factors which may feed into population-level responses or community system-responses as proposed by Hunter (1994).

Populations: Relationships between Rabbit and Plant Populations

Research on plant population-level responses to rabbit behaviours has concentrated on the impacts of variations in rabbit population density, measured directly by counts or indirectly by surrogate measures of activity traces. Rabbit density can be spatially manipulated by building exclosures, while studies over longer time periods have focused on the changes in vegetation before and after rabbit population crashes associated with the myxomatosis epidemic. These have given much valuable information; however, the underlying assumption of a linear relationship between rabbit numbers and the magnitude of the plant response is not proven. Furthermore, it is unlikely given the complexity of behaviours exhibited by rabbit populations and the mutability of individual rabbit behaviours in response to both external environmental factors and changes in rabbit population density and structure (see for example Myers et al., 1994).

At Aberffraw, the response of *Pilosella officinarum* agg. (previously *Hieracium pilosella* agg.) to rabbit population density was investigated in swards and the presence or absence of the species was measured in response to specific rabbit behaviours in semi-fixed/fixed dune vegetation (Table 1). Population density was on average 45 per cent lower ($Z = -2.0$, $p<0.05$, n = 5) in swards near active rabbit burrows than in swards beyond the limits of the warren from which the active burrow was sampled. Rosette density varied more than sixfold between the quadrat pairs. The species was not found

Table 1. *Effects of rabbit population and behaviours on the distribution and abundance of* Pilosella officinarum *agg. in semi-fixed/fixed dune vegetation. (a) Rosette density and (b) per cent occurrence. Numbers in parentheses indicate sample size. A dash (–) identifies a parameter for which no data were collected*

(a) rosette density (m²)

	Distance from an active burrow	
	3m	10m
mean	60.2 ± 20.3 (5)	105.2 ± 22.9 (5)
minimum	23	58
maximum	139	187

(b) occurrence (%)

	Distance from behavioural trace		
	0m	1m	1.5m
latrine	0 (25)	–	5 (25)
scrape	0 (25)	8 (25)	–
grass sward	25 (20)	–	–

in either latrine sites or scrapes. *Pilosella* rosettes are not thought to be eaten by rabbits (Bishop and Davy, 1984) and the results imply that population density changes are likely to be due to the physical disturbance of the soil caused by the presence of rabbits, irrespective of whether the disturbance accelerates or decelerates succession processes. The large variation in *Pilosella* population densities between quadrat pairs seemed to reflect differences in the sizes of the warrens sampled (personal observation), further supporting the argument that the mechanism of rabbit impacts on this species may be mediated by physical disturbance.

Landscape: Relationships between Rabbit and Plant Communities

Management of plant community structure and diversity is often the driving force initiating experimental work on the relationship between rabbits and dune vegetation. Many studies relate direct or indirect measures of rabbit population density to plant community structure. However, rabbit population density alone is unlikely to be a good indicator of rabbit impacts (Boorman, 1989). For example, estimates of rabbit populations within vegetation community types at Aberffraw (Table 2) would fail to identify the potentially important role of rabbits in the transition zone between semi-fixed/fixed dune communities and dune slack. Examination of the distribution and characteristics of rabbit warrens along a 120m transect in this transition zone was undertaken (Figure 3a). All warrens were situated within the semi-fixed/fixed dune area but most were less than 10m from *Salix repens* agg., an indicative species of dune slack at Aberffraw. It is therefore likely that the impacts of rabbits on dune slack vege-

Table 2. *Extent of dune vegetation communities at Aberffraw, North Wales, community species richness and maximum density estimates for rabbits. Mean values are presented ± SE.* [a] *identifies data modified from Ashall et al. (1995),* [b] *maximum rabbit densities were calculated using burrow-centred 20m × 20m quadrats and assuming 1.3 rabbits per active burrow (Myers et al., 1994)*

Dune vegetation by National Vegetation Classification	Total area (ha)[a]	Species richness[a]	Maximum rabbit density[b]
SD6: mobile dune	28.7	6.3 ± 1.5 (n = 12)	12.6 ± 0.9 (n = 3)
SD 7/8: semi-fixed and fixed dune	102.8	17.3 ± 0.8 (n = 46)	39 ± 6.0 (n = 3)
SD16/17: dune slack	39.7	18.0 ± 0.9 (n = 34)	1.0 ± 0.3 (n = 3)

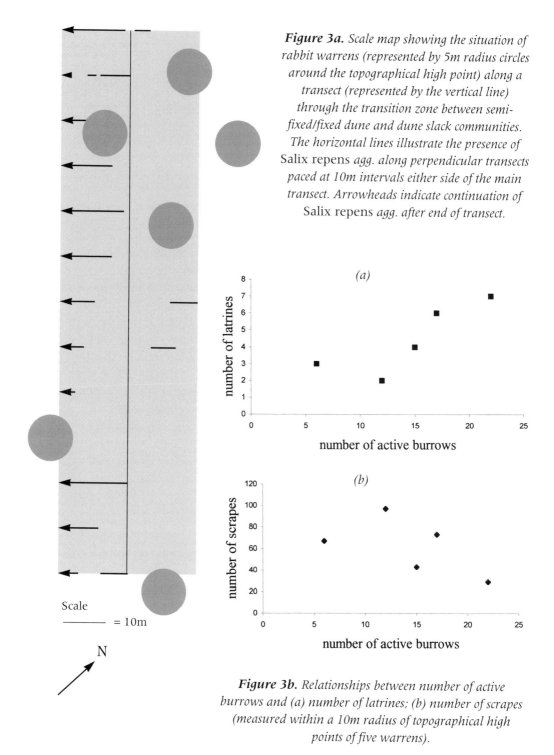

Figure 3a. Scale map showing the situation of rabbit warrens (represented by 5m radius circles around the topographical high point) along a transect (represented by the vertical line) through the transition zone between semi-fixed/fixed dune and dune slack communities. The horizontal lines illustrate the presence of Salix repens *agg.* along perpendicular transects paced at 10m intervals either side of the main transect. Arrowheads indicate continuation of Salix repens *agg.* after end of transect.

Scale
———— = 10m

N

Figure 3b. Relationships between number of active burrows and (a) number of latrines; (b) number of scrapes (measured within a 10m radius of topographical high points of five warrens).

tation are underestimated by simply calculating maximum rabbit densities. This reinforces the need to look not only at rabbit numbers within vegetation patches but also at rabbit distributions between vegetation patches, particularly when neighbouring vegetation patches are of very different resource quality. Detailed mapping of the characteristics of warrens reveals further detail of the relationships between individual rabbit behaviours (Fig. 3b). Two individual behaviours (latrine and scrape formation) which impact differently on the direction of succession show opposite directional trends across a range of warren sizes, measured as the number of active burrows.

Conclusions

Despite a wealth of published information on the impacts of rabbits on dune systems, most studies have failed to provide mechanistic explanations for the relationships they identify. This is due, in large part, to mismatched spatial scales for the measurement of rabbit impacts and vegetation responses. When some links in the cycle shown in Figure 1 are bypassed, problems also arise as a result of the mismatch in the temporal scales at which individuals, populations or communities respond. The few data presented above highlight the need to concentrate on collecting and integrating data from the range of temporally and spatially matched scales to allow a full understanding of the interactions between rabbits and dune habitats. This could be achieved by concentrating on behaviour and the associated impacts of rabbits. Behaviour is a flexible response to the environment which synthesizes the reaction of an individual rabbit to both rabbit and plant dynamics across a range of spatial and associated temporal scales. In dune systems such studies are facilitated as many rabbit behaviours leave clear traces which can be easily observed and often dated. As rabbit numbers increase in Britain, it is timely to clarify the detail of their relationships with dune habitats.

References

Ashall, J., Duckworth, J., Holder, C. and Smart, S. (1995), *No. 45. Sand Dune Survey of Great Britain. Site Report No. 123. Aberffraw, Ynys Mon, Wales, 1991*, Joint Nature Conservation Committee, Peterborough.

Bach, C.E. (1994), 'Effects of a specialist herbivore (*Altica subplicata*) on *Salix cordata* and sand dune succession', *Ecological Monographs*, **64**, 423–45.

Bishop, G.F. and Davy, J. (1984), 'Significance of rabbits for the population regulation of *Hieracium pilosella* in Breckland', *Journal of Ecology*, **72**, 273–84.

Boorman, L.A. (1989), 'The grazing of British sand dune vegetation', *Proceedings of the Royal Society of Edinburgh*, **98B,** 75–88.

Connell, J.H. and Slayter, R.O. (1977), 'Mechanisms of succession in natural communities and their role in community stability and organization', *American Naturalist*, **111,** 1119–144.

Edmondson, S.E., Gateley, P.S., Rooney, P.J. and Sturgess, P.W. (1993), 'Plant communities and succession', in D. Atkinson and J. Houston (eds.), *The Sand Dunes of the Sefton Coast*, National Museums and Galleries on Merseyside, 65–84.

Hunter, M.D. (1994), 'Interactions within herbivore communities mediated by the host plant: the keystone herbivore concept', in M.D. Hunter, T. Ohgushi and P.W. Price (eds.), *Effects of Resource Distribution on Animal–Plant Interactions,* Academic Press, San Diego, 278–325.

Myers, K., Parer, I., Wood, D. and Cooke, B.D (1994), 'The rabbit in Australia', in H.V. Thompson and C.M. King (eds.) *The European Rabbit,* Oxford Scientific Press, Oxford, 108–57.

Rodwell, J.S. (ed.) (1999) *British Plant Communities, Volume 5. Maritime Communities and Vegetation of Open Habitats,* Cambridge University Press, Cambridge.

Sumption, K.J. and Flowerdew, J.R. (1985), 'The ecological effects of the decline in rabbits (*Oryctolagus cuniculus*) due to myxomatosis', *Mammal Review,* **15,** 151–86.

Wood, D.H., Leigh, J.H. and Foran, B.D. (1987), 'The ecological and production costs of rabbit grazing', cited in Myers et al., 1994.

THE MANAGEMENT OF COASTAL SAND DUNE WOODLAND FOR RED SQUIRRELS (*SCIURUS VULGARIS* L.).

C.M. SHUTTLEWORTH
Ynys Mon (Anglesey), Wales
and J. GURNELL
School of Biological Sciences, Queen Mary and Westfield College, England

Introduction

Woodland plantations and scrub encroachment have often contributed to reductions in the area of sand dune. The maintenance or recovery of these dune habitats frequently involves woodland clearance (Wheeler et al., 1993). This will consequently affect woodland species such as the red squirrel (*Sciurus vulgaris*) and may lead to conflict between different conservation objectives. For example, red squirrels are vulnerable to extinction in the UK, and are listed under Appendix III of the Bern Convention and protected in the UK under the Wildlife and Countryside Act (1981) and the Wildlife (Northern Ireland) Order (1995). Their long-term survival depends on managing habitats in appropriate ways (see Gurnell and Lurz, 1997).

This paper explores the impact of scrub and woodland clearance on resident red squirrel populations by reviewing the ecology of red squirrels in coastal pine plantations based on a study at Ainsdale between 1994 and 1997 (Shuttleworth, 1997). We then suggest ways of promoting sand dune management that is sympathetic to the needs of red squirrels in relation to Ainsdale and coastal areas elsewhere. We also discuss ways of improving retained areas of coastal woodland for red squirrels.

Coastal Woodland Habitat at Ainsdale Sand Dunes National Nature Reserve

Red squirrels were studied between 1994 and 1997 in a 30ha area within the 90ha of rear woodlands at Ainsdale NNR (see Fig. 1). These plantations were dominated by mature Corsican pine (*Pinus nigra*) interspersed with a few scattered sycamore (*Acer pseudoplatanus*), lodgepole pine (*Pinus contorta*) and maritime pine (*Pinus pinaster*) trees. In 1997, the studies were extended to a 15ha area of frontal woodland, close to the coastline. This frontal habitat was more diverse, with both Corsican and lodgepole

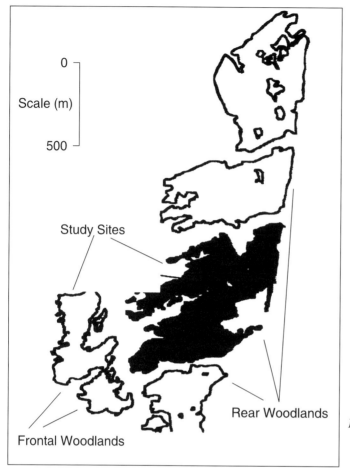

Figure 1. Rear and frontal woodland study sites at Ainsdale NNR.

pine present along with small amounts of sycamore, Scots (*Pinus sylvestris*) and maritime pine. On the seaward edge, scrub pine was interspersed with a mixture of sea buckthorn (*Hippophae rhamnoides*), birch (*Betula*), hawthorn (*Crataegus monogyna*), willow (*Salix*), and poplar (*Populus alba* and *candicans*). A more detailed description of these habitats can be found in Edmondson et al. (1993) and Shuttleworth (1997).

Squirrel Population Densities

Population densities within the rear woodland fluctuated between 0.5 and 4.2 squirrels ha^{-1} over a four-year period (Fig. 2). There was a reasonable correlation between population density and the amount of feeding on pine seeds (r = 0.6, n = 24, p <0.01). In contrast, the densities in the frontal woodland were lower and varied between 0.9 and 1.2 squirrels ha^{-1}, although these animals were only studied for one year. Densities in the spring of 1997 were 1.7 and 0.9 squirrels ha^{-1} for the rear and frontal woodlands respectively.

Breeding

In 1997, the frontal woodland was more productive with 70 per cent of adult females (n = 10) producing litters relative to 18 per cent (n = 28) in the rear. This difference probably resulted from heavy thinning operations undertaken in the rear woodland during the previous winter (Shuttleworth, 1997). In the previous three years the proportion of lactating females in the rear woodland was 79 per cent, 2 per cent and 50 per cent. Only two litters were actually found at Ainsdale so an accurate mean litter size could not be calculated. However, litters tend to range between 2.7 and 3.2 (Pulliainen, 1982; Tonkin, 1984). This suggests that the 1996 reproductive rate of 50 per cent from a spring population of 0.7 animals per hectare would not account for the observed dramatic population rise in the summer of 1996 (see Fig. 2). Approximately 50 per cent of 'new captures' were juveniles and sub-adults and immigration rates are likely to have been high. Continuity of woodland along the coast with woodlands inland is necessary to facilitate such dispersal movements.

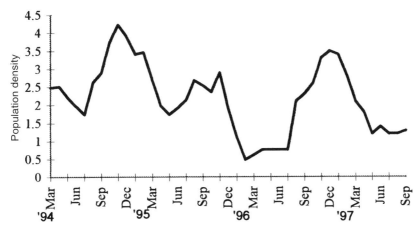

Figure 2. *Long-term population densities per hectare in rear woodland stands of Ainsdale Sand Dunes National Nature Reserve.*

Food Availability and Utilization

The level of cone crop production varied between years in the Corsican pine stands within the rear study area (see Fig. 3). The squirrels began feeding upon green cones in June when they were not fully ripe. The level of exploitation reflected changes in cone abundance, consumption peaking in the late summer, and declining through the winter into late spring when any cones remaining in the canopy opened and shed their seed.

In the frontal woodland, squirrels were most frequently found foraging in Corsican or lodgepole pine. During the late winter and spring animals also exploited sea

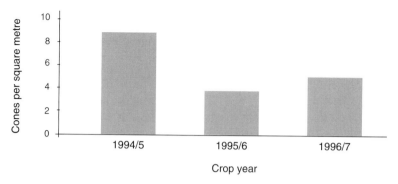

Figure 3. *The estimated size of the annual Corsican pine cone crop in the rear woodland study area (number of cones m^{-2}). Red squirrels fed or cached 91%, 100% and 98% of the cones each year respectively.*

buckthorn (buds, flowers) and the few scattered maritime pines (shoots, flowers and seed) (see Table 1). The buds and flowers of other broadleaves including sycamore, willow and balsam poplar were also taken. Pine seed was becoming increasingly scarce at that time and the animals appear to have responded by using 'secondary' food items. In addition, squirrels were also observed exploiting the seed in scattered Scots pine trees within an area of birch scrub. Although individuals were occasionally found in birch or white poplar they were never observed feeding but rather appeared to be using these trees as routes between other woodland blocks or tree species.

In the rear woodland Corsican pine stands, animals were occasionally recorded feeding on fungi, and the seed, buds, shoots, and flowers of a variety of other tree

Table 1. *Red squirrel habitat selection in the rear woodland at Ainsdale National Nature Reserve. The species selection ratio (w_i) was derived from the percentage of the total number of radiotracking fixes recorded in each tree species divided by the proportion of the total area covered by that species. A value of $w_i = 1$ shows no selection, > 1 indicates preference, < 1 indicates avoidance*

		Spring		Summer		Autumn		Winter	
Tree species	Area (ha)	% of fixes	w_i	% of fixes	w_i	% of fixes	w_i	% of fixes	w_i
Corsican pine	10.0	44.8	0.67	62.4	0.94	70.0	1.05	64.1	0.96
Scots pine	0.2	10.2	7.65	3.1	2.35	0.0	0.00	7.1	5.33
Maritime pine	0.3	16.1	8.05	1.0	0.50	0.0	0.00	4.3	2.15
Lodgepole pine	3.6	14.0	0.58	29.3	1.22	28.7	1.20	21.3	0.89
Sea buckthorn	0.3	9.0	4.50	1.7	0.85	0.0	0.00	1.5	0.75
Other broadleaves	0.6	5.9	1.48	2.5	0.63	1.3	0.33	1.7	0.45

Table 2. Mean monthly minimum convex polygon range areas (hectares) using radiotracking data ± standard deviation, n = number of ranges recorded. 100% ranges included all fixes, 70% ranges included the 70% of the fixes closest to the range centre and are considered to be core areas (see Wauters and Dhondt, 1985)

Study site	Male			Female		
	70% Ranges	100% Ranges	n	70% Ranges	100% Ranges	n
Front	1.6 ± 0.17	3.5 ± 0.15	32	1.4 ± 0.29	3.3 ± 0.66	32
Rear	0.6 ± 0.10	1.5 ± 0.05	45	0.8 ± 0.09	1.9 ± 0.20	83

species. However, these foods were limited, and squirrels were therefore heavily reliant on Corsican pine buds and flowers when pine seed was scarce. The late winter exhaustion of the pine cone crop resulted in a dramatic population decline in early 1996 (see Fig. 2). Only 7 per cent of radio-fixes were in tree species other than Corsican pine during this time period. Of thirteen adults studied using radiotracking techniques during January–April 1996, five (38 per cent) were found dead, and four (31 per cent) went missing. In addition, three other animals were found dead; all the dead squirrels were emaciated.

Animals in the rear woodland had smaller range areas than those in the frontal woodland study site (see Table 2). All these ranges were smaller than those reported by Lurz (1995) in northern spruce-dominated plantations where the mean size for males was 16ha and that for females 9ha, but similar in size to the 100 per cent ranges of 1.9–2.7ha recorded in Belgian pine woodlands by Wauters and Dhondt (1985).

It was found that the squirrels adjusted their ranging behaviour in response to changes in the spatial and temporal availability of different food items. In the rear woodlands, both the 70 per cent and 100 per cent ranges were negatively related to the amount of conifer seed which animals were consuming. Although no such relationship was found from the short study in the frontal woodlands (see Table 3), there was evidence that the animals moved centres of foraging activity rather than expanding range size.

Coastal Woodlands as Red Squirrel Habitats

Bryce et al. (1997) estimated a density of 4 red squirrels ha^{-1} in 30-year-old coastal Scots pine woodland on sand dunes in Elie, Fife, Scotland. Scots pine of 20–40 years of age is a particularly good habitat for red squirrels, providing both good food and cover (Gurnell, 1996). The rear woodland stands at Ainsdale also contained high spring densities (1.7–2.5) in three out of four study years. Such densities are considerably greater than those recorded in similar but inland habitats elsewhere (e.g. densities in

Table 3. Correlation coefficients of monthly mean adult range size, population density and the number of cones eaten per squirrel per ha. n = the number of months

Study Site		Males			Females		
		100%	70%	n	100%	70%	n
Rear	Density	-0.83**	-0.74**	23	-0.58**	-0.51*	20
	Cones	-0.68**	-0.72**	23	-0.56**	-0.53*	20
Front	Density	-0.18	-0.25	9	-0.21	-0.22	9
	Cones	-0.02	0.02	9	0.18	0.34	9

* $P < 0.05$, ** $P < 0.01$

pine-dominated habitats are generally between 0.3 and 1.4 squirrels ha^{-1}; see Moller, 1986; Gurnell, 1983; 1987; Wauters and Dhondt, 1990). In contrast, the frontal woodland was shown to support a population density more typical of the species (0.9 squirrels ha^{-1}). The study also illustrated that when primary food sources are exhausted the availability of alternative foods is of critical importance. Within the sand dune environment, individuals preferentially exploited the seed, flowers, shoots and buds of other coniferous species such as sea buckthorn, sycamore and hawthorn. Woodland area requirements for red squirrels are considered below.

Seasonal and annual fluctuations in population density in the rear woodland (see Fig. 2) are attributable to annual variation in Corsican pine cone production. The inclusion of additional tree species within the plantation would reduce the chance of simultaneous seed failure across the habitat (Gurnell and Pepper, 1993; Lurz et al., 1995) and help to sustain animals during years when the main tree seed crop was small, as observed at Ainsdale in 1995–96.

A critical requirement of red squirrel habitats is the absence of the grey squirrel, *S. carolinensis* (Gurnell and Pepper, 1993). Grey squirrels were not present in Ainsdale NNR during the studies, but there have been over 80 records within the Sefton Coast since 1993 (Shuttleworth, unpublished data).

Coastal Woodland Management

The studies at Ainsdale and elsewhere (e.g. Lurz and Garson, 1997; Gurnell et al., 1997; Gurnell and Pepper, 1993) indicate how coastal woodlands can be managed to favour red squirrel populations. On sand dune systems, woodland clearance operations can be carried out sympathetically to the needs of the remaining squirrel populations (see Table 4). In addition, management should be considered for stands that are to be retained.

Table 4. Scrub and woodland management sympathetic to the requirements of the red squirrel

Scrub and woodland clearance to facilitate recovery of dune habitats

1. *Scrub removal operations* adjacent to coniferous plantations should seek to retain small pockets at intervals at c. 200m. Preference should be given to areas dominated by sea buckthorn, hawthorn, and/or balsam poplar at the expense of birch, willow or white poplar.

2. *Clearfelling operations* should maintain woodland corridors between fragments which would otherwise be left isolated. Corridors need not necessarily consist of tree species which are major food sources or be of a continuous nature. Small patches of woodland at 50–100m distances might work but would not be ideal.

Management of woodlands which are to be retained on sand dune systems

1. *Size:* woods should consist of at least 200ha of connected woodland, and ideally should be much larger than this.

2. *Buffer zones:* between 1km and 3km of unsuitable squirrel habitat should be left between red squirrel woodland and grey squirrel habitats.

3. *Tree species diversity:* should be encouraged, although this should not be at the expense of significant reductions in the area covered by particularly beneficial species such as Scots, Corsican or lodgepole pine. On suitable soils, the inclusion of small areas of large seeding broadleaves such as beech and oak would be particularly beneficial, if there is no risk of grey squirrel encroachment.

4. *Scrub:* small pockets of sea buckthorn- or hawthorn-dominated scrub should be retained.

5. *Corridors:* the simultaneous clearfelling of large areas of woodland should be avoided where possible. Overmature stands should be removed gradually and corridors left to connect woodland patches.

6. *Thinning regimes:* should seek to retain corridors to facilitate movement between seed-producing areas within blocks. The thinning of maturing stands should be at intervals sufficient to produce a full and continuous canopy.

7. *Woodland structure:* the management of coniferous plantations should aim to develop a forest structure within which 50–60% of trees are old enough to produce seed. A mosaic of stands of different ages is preferable to large continuous blocks of coeval woodland. In commercial woodlands, a suggested age-class structure is 30% young trees (i.e. pre-cone production, 0–15 years in Scots pine or 0–30 years in Corsican pine), 30% middle-age trees (16–30 years in Scots pine, 30–60 years at Ainsdale), and 40% older trees.

Clearfell operations

Red squirrels tend to avoid areas of clearfell and heavy thinning where large distances are left between trees (Cartmell, 1997). In the short term, individuals from clearfelled areas move into neighbouring stands (Wolff, 1975), but their long-term fate is unknown. Squirrels will forage within the brash of clearfelled compartments but this is generally restricted to within 30m of the edge of adjacent woodland (Shuttleworth, personal observation). To enhance the dune vegetation and improve the habitat for sand lizards (*Lacerta agilis*) and natterjack toads (*Bufo calamita*), 11ha of frontal woodland at Ainsdale (including 4.5ha of pine) were removed in 1992 and 14ha of Corsican pine in 1996; proposed operations will encompass the remaining 15ha. This represents an absolute loss of habitat for red squirrels and a reduction in carrying capacity.

Scrub removal

At Ainsdale, the flowers and buds of sea buckthorn, hawthorn, and sycamore were heavily used by red squirrels in the frontal woodlands during the late winter and spring when pine seeds were scarce. The retention of small patches of deciduous scrub close to the woodland edge would clearly benefit red squirrels (cf. Wheeler et al., 1993). Studies on squirrel movement at Ainsdale suggest that leaving patches of scrub at intervals of about 200m would not restrict access to these food sources. Where possible patches should contain a mixture of tree species with as little birch, willow and white poplar as possible because these species have a minimal food value for squirrels, although they may provide structure in corridors between fragments of more suitable habitat.

Areas of naturally regenerated pine scrub do not represent valuable habitat until they reach an age when seed production begins. For example, red squirrels avoid Corsican pine stands less than 20 years of age (Gurnell et al., 1997; Cartmell, 1997). The removal of this type of scrub would not represent a significant loss of food source in the short term, but would be adverse if the stands provided a habitat corridor between adjacent blocks of more mature woodland. However, in the long term, such regeneration may provide an important source of seed as it matures. As a consequence, the removal of this habitat type should be viewed against long-term development of the habitats within the entire woodland.

Management of remnant woodland

Size of woodland

Red squirrel habitat quality is a function of species diversity, stand age structure, and the abundance of tree species which are particularly valuable food sources (Gurnell and Pepper, 1988; Gurnell et al., 1997; Lurz and Garson, 1997). The area and habitat quality of woodland fragments will determine the number of animals they can support (Verboom and van Apeldoorn, 1990). The minimum viable population size (MVP) for red squirrels is not known, but it has been suggested that it should consist of at least

200 individuals (Gurnell, 1996). Using 1997 densities at Ainsdale, this suggests that there should be about 100ha of connected rear woodland. However, although the densities in the rear woodland were relatively high, in poor seed crop years numbers dropped to low levels (e.g. 60 squirrels in spring 1996), when the population may not be viable (Gurnell, 1983). It is advisable that conifer woodlands should be at least 200ha in area, and where possible considerably larger than this (Patterson and Pepper, 1998). To achieve this, the 300ha (Wheeler et al., 1993) of Sefton Coast woodland should be managed as a single unit and links between isolated woodland blocks established. The defence of the area against incursions of grey squirrels should also form an important part of habitat management (see Gurnell and Pepper, 1993).

Thinning operations
Thinning operations to increase crown depth or to encourage natural regeneration should seek to maintain canopy integrity to as large an extent as possible. Open canopies force squirrels to move more frequently across the ground, so reducing foraging efficiency and elevating predation risks. In some areas line thinning may be better than selective thinning regimes (Gurnell and Pepper, 1993) depending on site characteristics. Thinning across rather than along planted rows can also minimize the size of gaps in the canopy (Gurnell, personal observation). Where possible, drey trees should not be removed and thinning operations should be confined to October to February to minimize disturbance to breeding females. In single-species-dominated stands, areas of diversity should be preferentially retained. Finally, heavy thinning operations were implicated in the population decline and reduction in reproduction at Ainsdale (Shuttleworth, 1997) and should be avoided.

Maintenance of corridors
Woodland corridors are an important feature of red squirrel conservation, particularly when the distances between habitat fragments exceed several hundred metres (Wauters, 1997). In this and other studies (Gurnell et al., 1997), radiotracked animals were frequently observed foraging between stands separated by distances of 50–100m. However, while travelling across open ground the animals face elevated predation risks (Bak and Lagendijk, 1995). For example, red squirrels are taken by foxes (*Vulpes vulpes*), and such predation accounted for 18 per cent (3 out of 17) of the Ainsdale frontal woodland population in December 1996 (Shuttleworth, 1997). Consequently, it would be ideal if corridors were continuous in nature so that animals do not have to venture to the ground.

Woodland tree species and age structure
Where replanting or underplanting of thinned stands is considered, the selection of species will be restricted to those which tolerate soils with a high sand content. Such conditions principally favour Corsican, lodgepole or Scots pine. Scots pine is the optimal tree species for the red squirrel as it produces regular cone crops and seed

production begins earlier than other species of pine at 15 years of age (Gurnell, 1983; Gurnell et al., 1997).

In selected parts of the woodland, such as ride or woodland edges, mixtures of tree species, including pines and hawthorn, would provide a more stable seed supply from one year to the next (Gurnell and Pepper, 1988; Patterson and Pepper, 1998). Design plans to maximize seed production should take advantage of south-facing slopes and ride edges, with the provision of islands of good seed trees in clearfelled areas, and irregularly shaped compartments to maximize woodland edges. In the UK, planting large-seeded broadleaved species such as beech and oak should be avoided as they promote immigration and survival of grey squirrels.

In commercial conifer plantations, an age structure of 30 per cent young trees (i.e. pre-cone production, 0–15 years in Scots pine), 30 per cent middle-age trees (16–30 years), and 40 per cent older trees (over 30 years of age) has been proposed (Gurnell and Pepper, 1993). The age classes can be extended in woodlands where nature conservation is the primary objective and stands are retained beyond their economic optimal age. A suggested age structure within the Ainsdale pine plantations would be 30 per cent 0–30 years, 30 per cent 31–60 years, and 40 per cent over 60 years. However, the current age structure of these and other Sefton Coast woodlands is not ideal, with many of the plantations being between 55 and 90 years of age. The long-term management plan should therefore consider establishing felling and regeneration rotations with the age and size of felling coups selected to achieve a better age structure. Management should seek to maintain woodland corridors and the overall carrying capacity of the coastal woods as a whole.

Acknowledgments

The research at Ainsdale was funded by the People's Trust for Endangered Species and English Nature as part of its programme of research into nature conservation. We are grateful to Harry Pepper and Joanne Lello for their comments on an earlier draft of this manuscript.

References

Bak, A. and Lagendijk, A. (1995), *Ruimtegebruik van de rode eekhoorn, Sciurus vulgaris L., in een gefragmenteerd habitat*, Doctoraalverslag, Universiteit Utrecht.

Bryce, J., Pritchard, J.S., Waran, N.K. and Young, R.J. (1997), 'Comparison of methods for obtaining population estimates for red squirrels in relation to damage due to bark stripping', *Mammal Review*, **27**, 165–70.

Cartmell, S. (1997), 'A study of red and grey squirrels in Clocaenog forest, North Wales: A preliminary study', in Gurnell and Lurz (eds.), 1997, 89–95.

Edmondson, S.E., Gateley, P.S., Rooney, P.J. and Sturgess, P.W. (1993) 'Plant communities and succession', in D. Atkinson and J. Houston, *The Sand Dunes of the Sefton Coast*, National Museums and Galleries on Merseyside/Sefton Metropolitan Borough Council, 65–84.

Gurnell, J. (1983), 'Squirrel numbers and the abundance of tree seeds', *Mammal Review*, **13**,

133–48.

Gurnell, J. (1987), *The Natural History of Squirrels*, Christopher Helm, London.

Gurnell, J. (1996), 'Conserving the red squirrel', in P. Ratcliffe and J. Claridge (eds.), *Thetford Forest Park: The Ecology of a Pine Forest*, Forestry Commission, Edinburgh.

Gurnell, J. and Pepper, H. (1988), 'Perspectives on the management of red and grey squirrels', in D.C. Jardine (ed.), *Wildlife Management in Forests*, CF, Edinburgh, 92–109.

Gurnell, J. and Pepper, H. (1993), 'A critical look at conserving the British Red Squirrel *Sciurus vulgaris*', *Mammal Review*, **23**, 127–37.

Gurnell, J. and Lurz, P.W.W. (eds.) (1997), *The Conservation of Red Squirrels, Sciurus Vulgaris*, The People's Trust for Endangered Species, London.

Gurnell, J., Clark, M.J., and Feaver, J. (1997), 'Using geographic information systems for red squirrel conservation management', in Gurnell and Lurz, 1997, 153–59.

Lurz, P.W.W. (1995), 'The ecology and conservation of the red squirrel (Sciurus vulgaris L.) in upland conifer plantations', PhD thesis, Newcastle University.

Lurz, P.W.W., Garson, P.J., and Rushton, S.P. (1995), 'The ecology of squirrels in spruce dominated plantations: implications for forest management', *Forest Ecology and Management*, **79**, 79–90.

Lurz, P.W.W. and Garson, P.J. (1997), 'Forest management for red squirrels in conifer woodlands: a northern perspective', in Gurnell and Lurz, 1997, 145–51.

Mollar, H. (1986), 'Red squirrels (*Sciurus vulgaris*) in a Scots pine plantation in Scotland', *Journal of Zoology*, **209**, 61–84.

Patterson, G. and Pepper, H. (1998), *The Conservation of Red Squirrels*, Forestry Practice Note 5, Forestry Commission, Edinburgh.

Pulliainen, E. (1982), 'Some characteristics of an exceptionally dense population of the red squirrel, *Sciurus vulgaris* L. on the southern coast of Finland', *Aquilo Series Zoologica*, **21**, 9–12.

Shuttleworth, C.M. (1997), *The Impact of Woodland Management upon the Red Squirrel (Sciurus Vulgaris) Population in Ainsdale National Nature Reserve*, Report to English Nature, Research Contract GL73, Peterborough.

Tonkin, J.M. (1984), 'Red squirrel ecology', PhD thesis, University of Bradford.

Verboom, B. and van Apeldoorn, R. (1990), 'Effects of habitat fragmentation on the red squirrel, *Sciurus vulgaris* L.', *Landscape Ecology*, **4**, 171–76.

Wauters, L. (1997), 'The ecology of red squirrels (*Sciurus vulgaris*) in fragmented habitats: a review', in Gurnell and Lurz, 1997, 5–12.

Wauters, L.A. and Dhondt, A.A. (1985), 'Population dynamics and social behaviour of red squirrel populations in different habitats', *VII Congress Int. Union Game Biologists, Brussels*, 311–18.

Wauters, L.A. and Dhondt, A.A. (1990), 'Red squirrel (*Sciurus vulgaris* L. 1758) population dynamics in different habitats', *Zeitschrift für Saugetierkunde*, **55**, 161–75.

Wheeler, D.J., Simpson, D.E., and Houston, J.A. (1993), 'Dune use and management', in D. Atkinson and J. Houston (eds.), *The Sand Dunes of the Sefton Coast*, National Museums and Galleries on Merseyside/Sefton Metropolitan Borough Council, 129–49.

Wolff, J.O. (1975), 'Red squirrel response to clearcut and shelterwood systems in interior Alaska', *USDA Forest Service Research Note PNW-255*, Portland, Oregon.

A GIS STUDY OF BREEDING BIRD HABITATS IN THE FLEMISH COASTAL DUNES AND ITS IMPLICATIONS FOR NATURE MANAGEMENT

DRIES BONTE

Laboratories of Animal Ecology and Botany, Department of Biology, University of Ghent, Belgium
and MAURICE HOFFMANN
Laboratory of Botany, University of Ghent, and Institute of Nature Conservation, Belgium

ABSTRACT

In 1997–98 an area of approximately 1300ha of Flemish coastal dunes was censused for breeding birds. Territory maps were drawn and compared with maps on vegetation, hydrology and recreation pressure by the overlay method in a GIS. Twinspan recognized seven breeding bird communities; a DCA-ordination revealed the dominant importance of vegetation succession and the degree of urbanization for breeding bird community variance. Breeding habitats are characterized for all territorial songbirds. The comparison of the distribution of potential habitat and actually occupied habitat stresses the importance of increase of the groundwater level and recreation pressure for specific species. At the landscape level a positive relationship between area of shrub and grey dunes and species number was detected; an opposite relationship exists between area and total territory density.

Introduction

In recent decades the Flemish coastal dune area has suffered significant degradation from urbanization, mass recreation, water collection and scrub encroachment (Provoost and Hoffmann, 1996). To conserve the specific flora and fauna of mesophytic dune grassland and wet dune slacks, the Flemish Community invested in scrub removal and introduction of domesticated grazers. Simultaneously, a monitoring programme was started to investigate the effects of these nature management measures on flora, vegetation, fauna and soil (Bonte et al., 1998). One of the major aims is to study the effects of landscape changes due to human impact (nature management, recreation, urbanization) on the breeding bird fauna. The application of a Geographical Information System (GIS) enables us to model specific habitat or landscape structure characteristics of breeding bird species. Once modelled, the GIS approach enables us to make predictions of changes in the breeding bird composition in relation to the above-mentioned landscape changes.

Materials and Methods

Breeding birds were censused in the Flemish coastal dunes in 1997 and 1998. The dune areas involved are the Dunes du Perroquet (Bray-Dunes, France), the Westhoek and Houtsaegerduinen (De Panne, Belgium), the Noordduinen, the Doornpanne and Ter Yde (Koksijde, Belgium) and the Ijzermonding (Nieuwpoort, Belgium). We used the territory mapping method described by Hustings et al. (1985) as it is the only effective way to obtain detailed information about breeding bird numbers and territory distribution in the area. The only disadvantage of this method is its time-consuming nature during the inventory period.

During each breeding season (March to July), we visited the dune areas about 10 times. In two years we censused an approximate area of 1300ha of coastal dunes in detail. The interpretation criteria of Hustings et al. (1985) were used to construct the specific territory maps. These maps were digitized with the GIS software package Genamap 6.2. A 1ha grid was used to analyse breeding bird community composition. All grid cells not entirely covered by a censused area (predominantly border cells) were excluded from the analysis. All of the remaining 965ha were included and for these squares the number of territories per breeding bird species was calculated. Overlays were made with maps of present-day vegetation, hydrology and recreation pressure. Twinspan (Hill, 1979a) and Detrended Correspondence Analysis (DCA) (Hill, 1979b) were used for analysis of community composition and ecology.

Results

Breeding bird communities
Seventy-two breeding bird species were recorded in the surveyed area of coastal dunes (see Table 1). Twinspan analysis of the 700 1ha samples revealed seven main bird communities. We named them after the most characteristic bird species: the dunnock–collared dove type, the woodpigeon–blackcap type, the nightingale–turtle dove type, the whitethroat–willow warbler type, the house sparrow–black redstart type, the linnet–partridge type and the little ringed plover–crested lark type. Mean density per hectare of the species per community is listed in Table 1. DCA-ordination of the breeding bird communities is given in Figure 1, the individual species in Figure 2. The first DCA-axis has an eigenvalue of 0.571 and represents a vegetation density gradient from open, bare dune towards dune woodland. The second axis (eigenvalue 0.412) is significantly correlated with the degree of urbanization of the 1ha grid samples.

Table 1. Species list with densities per hectare grid square in the Twinspan end groups (e = excluded from the analysis; ? = no absolute density data). Red List status: 1 = critically endangered; 2 = endangered; 3 = vulnerable; Z = rare; D = declining; N = not on Red List; ? = insufficiently known.

Key to Twinspan end groups: 000 dunnock–collared dove type, 0010 woodpigeon–blackcap type, 0011 nightingale–turtle dove type, 010 whitethroat–willow warbler type, 011 house sparrow–black redstart type, 10 linnet–partridge type, 11 little ringed plover–crested lark type.

Species	English name	000	0010	0011	010	011	10	11	Red List
				Maximum value for each species is emboldened					
Ardea cinerea	Grey heron	0,000	**0,275**	0,000	0,000	0,000	0,000	0,000	N
Tadorna tadorna	Shelduck	0,000	0,025	0,012	0,043	0,052	**0,072**	0,000	N
Anas platyrhynchos	Mallard	0,000	**0,066**	0,006	0,007	0,000	0,028	0,000	N
Accipter nisus	Sparrow hawk	0,000	**0,025**	0,003	0,000	0,000	0,000	0,000	N
Buteo buteo	Buzzard	0,000	**0,008**	0,000	0,000	0,000	0,000	0,000	N
Falco tinnunculus	Kestrel	0,000	**0,016**	0,000	0,000	0,000	0,000	0,000	N
Falco subbuteo	Hobby	0,000	**0,016**	0,000	0,000	0,000	0,000	0,000	N
Perdix perdix	Partridge	0,000	0,000	0,003	0,102	0,102	**0,347**	0,000	3
Phasianus colchicus	Pheasant	0,000	0,216	**0,332**	0,284	0,105	0,072	0,000	N
Rallus aquaticus	Water rail	0,000	0,000	**0,006**	0,000	0,000	0,000	0,000	?
Gallinula chloropus	Moorhen	0,000	**0,075**	0,000	0,000	0,000	0,000	0,000	N
Haematopus ostralegus	Oystercatcher	0,000	0,000	0,000	0,000	0,000	**0,028**	0,000	N
Charadrius dubius	Little ringed plover	0,000	0,000	0,000	0,017	0,000	0,050	**0,482**	N
Charadrius alexandrinus	Kentish plover	0,000	0,000	0,000	0,000	0,000	0,000	**0,003**	1
Vanellus vanellus	Lapwing	0,000	0,000	0,000	0,000	0,000	**0,028**	0,017	N
Columba oenas	Stock dove	0,000	0,050	0,003	0,007	**0,157**	0,014	0,000	N
Columba palumbus	Woodpigeon	1,235	**1,266**	0,963	0,169	0,000	0,014	0,000	N
Streptopelia decaocto	Collared dove	**1,352**	0,225	0,018	0,015	0,210	0,014	0,000	N
Streptopelia turtur	Turtle dove	0,000	0,083	**0,470**	0,128	0,052	0,014	0,000	D
Merops apiaster	Bee-eater	e	e	e	e	e	e	e	?
Cuculus canorus	Cuckoo	0,058	0,225	**0,381**	0,343	0,052	0,246	0,068	N
Athene noctua	Little owl	0,000	0,000	0,000	**0,002**	0,000	0,000	0,000	N
Asio otus	Long-eared owl	?	?	?	?	?	?	?	N
Picus viridis	Green woodpecker	0,058	0,241	**0,480**	0,302	0,000	0,043	0,034	N
Dendrocopus major	Great spotted woodpecker	0,000	**0,300**	0,107	0,002	0,000	0,000	0,000	N
Dendrocopus minor	Lesser spotted woodpecker	0,000	**0,016**	0,000	0,000	0,000	0,000	0,000	N
Galerida cristata	Crested lark	0,000	0,000	0,000	0,017	0,315	0,115	**0,655**	1
Alauda arvensis	Skylark	0,000	0,000	0,000	0,015	0,000	**0,087**	0,000	D
Anthus pratensis	Meadow pipit	0,000	0,008	0,006	0,146	0,052	**1,275**	0,586	D
Troglodytes troglodytes	Wren	0,411	**1,216**	0,572	0,058	0,000	0,028	0,000	N
Prunella modularis	Dunnock	**1,117**	0,241	0,418	0,584	0,526	0,144	0,000	N
Erithacus rubecula	Robin	0,000	**0,375**	0,055	0,002	0,000	0,000	0,000	N
Luscinia megarhynchos	Nightingale	0,058	0,141	**0,895**	0,317	0,000	0,000	0,000	3

Species	English name	000	0010	0011	010	011	10	11	Red List
Phoenicurus ochruros	Black redstart	0,647	0,030	0,015	0,023	**0,789**	0,028	0,137	N
Phoenicurus phoenicurus	Redstart	**0,176**	0,141	0,073	0,007	0,000	0,000	0,000	3
Saxicola torquata	Stonechat	0,000	0,008	0,030	0,246	0,157	**0,318**	0,068	2
Oenanthe oenanthe	Wheatear	0,000	0,000	0,000	0,033	0,052	**0,449**	0,000	1
Turdus merula	Blackbird	**1,235**	0,833	0,689	0,335	0,421	0,028	0,068	N
Turdus philomelos	Song thrush	**0,235**	0,191	0,049	0,038	0,000	0,000	0,000	N
Turdus viscivorus	Mistle thrush	**0,176**	0,083	0,080	0,100	0,052	0,000	0,000	N
Cettia cetti	Cetti's warbler	0,000	0,016	**0,043**	0,010	0,000	0,000	0,000	Z
Locustella naevia	Grasshopper warbler	0,000	0,000	0,135	**0,171**	0,000	0,000	0,000	3
Acrocephalus palustris	Marsh warbler	0,000	0,000	0,033	**0,035**	0,000	0,014	0,000	N
Hippolais icterna	Icterine warbler	0,000	0,033	**0,523**	0,015	0,000	0,000	0,000	N
Hippolais polyglotta	Melodious warbler	0,000	0,000	**0,003**	0,000	0,000	0,000	0,000	N
Sylvia curruca	Lesser whitethroat	0,058	0,075	**0,292**	0,158	0,000	0,057	0,000	N
Sylvia communis	Whitethroat	0,000	0,100	0,870	**0,928**	0,000	0,057	0,034	N
Sylvia borin	Garden warbler	0,000	0,075	**0,283**	0,061	0,000	0,000	0,000	N
Sylvia atricapilla	Blackcap	0,000	**1,091**	0,630	0,030	0,000	0,000	0,000	N
Phylloscopus collybita	Chiffchaff	0,294	**0,925**	0,790	0,069	0,000	0,000	0,000	N
Phylloscopus trochilus	Willow warbler	0,000	0,033	**0,883**	0,697	0,105	0,000	0,000	N
Regulus regulus	Goldcrest	0,000	**0,033**	0,006	0,000	0,000	0,000	0,000	N
Aegithalos caudatus	Long-tailed tit	0,059	0,058	**0,172**	0,033	0,000	0,000	0,000	N
Parus cristatus	Crested tit	0,000	0,000	**0,006**	0,000	0,000	0,000	0,000	N
Parus caeruleus	Blue tit	**0,235**	0,183	0,224	0,020	0,000	0,000	0,000	N
Parus major	Great tit	**0,529**	0,416	0,486	0,048	0,000	0,000	0,000	N
Sitta europaea	Nuthatch	0,000	**0,016**	0,000	0,000	0,000	0,000	0,000	N
Certhia brachydactyla	Short-toed treecreeper	0,000	**0,283**	0,036	0,000	0,000	0,000	0,000	3
Oriolus oriolus	Golden oriole	0,000	**0,291**	0,056	0,002	0,000	0,000	0,000	N
Garrulus glandarius	Jay	0,000	0,091	**0,095**	0,020	0,000	0,000	0,000	N
Pica pica	Magpie	**0,235**	0,208	0,169	0,053	0,052	0,014	0,000	N
Corvus monedula	Jackdaw	0,000	0,008	0,000	0,005	**0,210**	0,000	0,000	N
Corvus corone	Carrion crow	0,000	**0,183**	0,055	0,013	0,000	0,000	0,000	N
Sturnus vulgaris	Starling	0,000	0,000	0,003	0,000	**0,105**	0,014	0,000	D
Passer domesticus	House sparrow	1,823	0,108	0,021	0,020	1,421	0,072	0,000	D
Passer montanus	Tree sparrow	0,000	0,000	**0,015**	0,005	0,000	0,000	0,000	D
Fringilla coelebs	Chaffinch	0,000	**0,066**	0,012	0,000	0,000	0,000	0,000	N
Serinus serinus	Serin	**0,059**	0,000	0,000	0,000	0,000	0,000	0,000	?
Chloris chloris	Greenfinch	e	e	e	e	e	e	e	N
Carduelis carduelis	Goldfinch	e	e	e	e	e	e	e	N
Carduelis spinus	Siskin	0,000	**0,025**	0,005	0,000	0,000	0,000	0,000	Z
Carduelis cannabina	Linnet	0,058	0,025	0,375	0,054	0,315	**0,739**	0,137	N
Carduelis flammea	Redpoll	0,000	0,000	**0,040**	0,000	0,000	0,000	0,000	Z
Pyrrhula pyrrhula	Bullfinch	0,000	**0,025**	0,012	0,000	0,000	0,000	0,000	N
Emberiza schoeniclus	Reed bunting	0,000	0,000	0,012	**0,066**	0,000	0,000	0,000	D

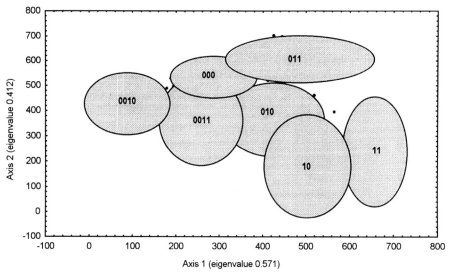

Figure 1. *DCA-plot of the breeding bird communities based on species composition in 1ha grid squares. Approximate areas of the Twinspan end groups on the DCA-plot are shown.*

Key to Twinspan end groups:

000 Dunnock–collared dove type	010 Whitethroat–willow warbler type
0010 Woodpigeon–blackcap type	011 House sparrow–black redstart type
0011 Nightingale–turtle dove type	10 Linnet–partridge type
	11 Little ringed plover–crested lark type

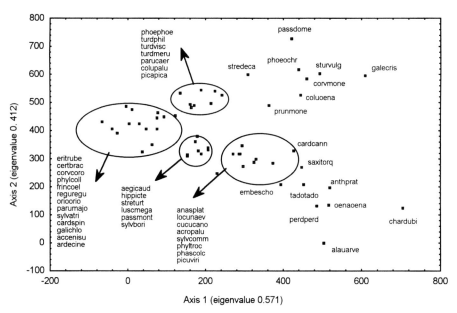

Figure 2. *DCA-plot of the breeding bird species based on species composition in 1ha grid squares. Bird species are labelled by an abbreviation of the species names listed in Table 1, four letters for the genus and four for the species.*

Effects of recreation and lowering of the water table

After characterization of the habitat of the several species via overlays of species territory with vegetation maps (Bonte and Hoffmann, in prep.), we were able to locate potentially suitable habitats in the study area adjacent to the actually occupied territories. Comparison of all suitable hydrologically unaltered (wet) habitats with dried-out dune areas enabled us to determine which species are vulnerable to lowering of the groundwater table (Welch's independent t-test, $p<0.05$). Marsh warbler and reed bunting are the only species which show a significant preference for wet scrub mosaics. The moorhen is also vulnerable to lowering of the water table. Only the whitethroat reached significantly higher densities in the dried-out dune areas.

The same method was applied to determine the effect of recreation pressure. Densities of shelduck, partridge, wheatear and stonechat (characteristic species of open, mosaic dunes) appeared to be significantly lower in the areas under recreation pressure (Welch's independent t-test or Mann-Whitney U-test, $p<0.05$). As expected, no species showed a significant preference for dunes with a certain degree of recreation pressure.

The relation between scrub size and breeding bird numbers and densities

Superimposition of the vegetation map with all territory maps allows the detection of landscape-ecological relationships between the size of homogeneous vegetation patches and territory density and total number of breeding birds.

The positive relationship between total breeding bird numbers and the size of mixed scrub vegetation patches is illustrated in Figure 3. An analogous, although less signif-

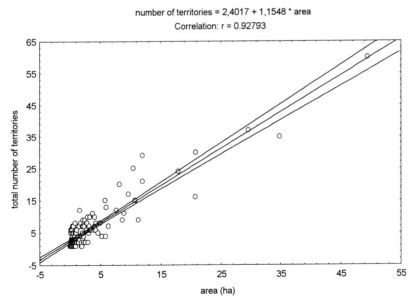

Figure 3. *The relationship between mixed scrub patch size (hectares) and the total number of bird territories.*

icant relationship exists for breeding birds of grey dunes and their characteristic dune habitat (see Fig. 4).

Total bird density–area relationships are characterized by a clear tendency towards density decrease with increasing size of scrub or grey dune vegetation patch (see Figs. 5 and 6). After a natural logarithm transformation of the data this significant negative relationship is clearly demonstrated (see Fig. 7). Despite this general negative relationship, specific species density–patch area relationships reach a maximum at differing, not necessarily small, patch sizes. Figures 8 and 9 illustrate this relationship for the turtle dove and the lesser whitethroat in mixed scrub.

Discussion and Implications for Flemish Dune Management

Although the Flemish coastal dunes have already been subject to a lot of ornithological research, specific habitat characteristics of breeding birds had not previously been studied. Despite much detailed research on breeding bird communities (Bibby et al., 1989; Fuller et al., 1989; Gillings et al., 1998; Smith et al., 1992 [woodlands]; Lysaght, 1989 [agricultural landscape] and Stillman and Brown, 1994 [upland vegetation]), no analogous studies are available of dune breeding birds. Only Lust et al. (1995) investigated the relationship between vegetation and total bird densities in the dune area of the eastern Flemish coast. For the western Flemish coast (Bonte, 1994) and for the dune area between Katwijk and Scheveningen (Netherlands) (Van der Meer, 1996) only general breeding bird inventories are known.

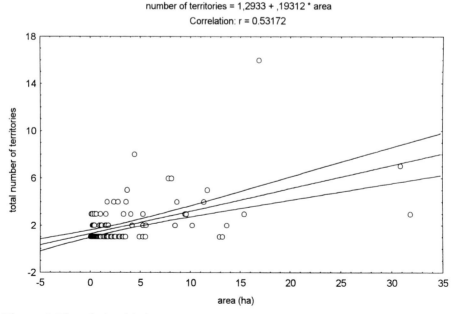

Figure 4. The relationship between grey dune patch size (hectares) and the total number of bird territories.

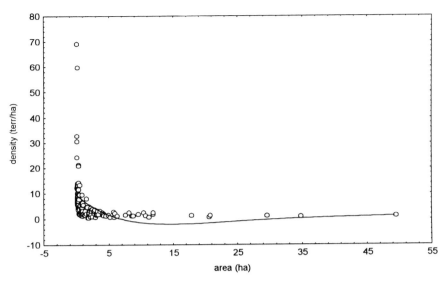

Figures 5–7. *The relationship between vegetation patch area (hectares) and density of breeding bird territories (number of territories per hectare).*
Figure 5. Hippophae-Sambucus *patch size and bird density.*

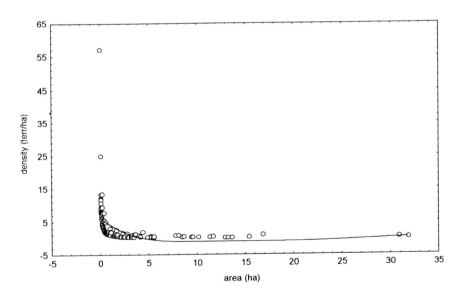

Figure 6. *Grey dune patch size and bird density.*

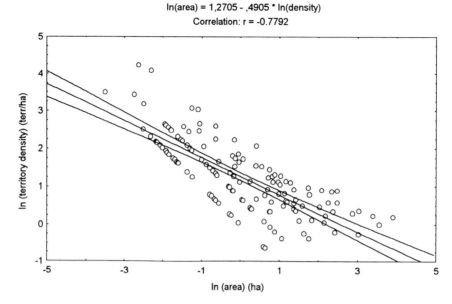

Figure 7. Scrub patch size and bird density natural logarithm transformation.

In general, the main determining parameters for bird communities are hydrology, scrub and woodland proportion or altitude. The breeding bird communities in the Flemish coastal dunes are determined by the same parameters, of which vegetation succession appears to be the most important. Flemish Red-List (Devos and Anselin, in prep.) species in the coastal dune area are characteristic of open, sandy or short, grazed vegetation, scrub mosaic and (less important) woody scrub vegetation and dune woodland. Species of grey dunes and short grazed grasslands in mosaic with low scrubs are especially threatened in Flanders. Taking into consideration the vulnerability of these specific species to recreation pressure, conservation of this habitat should be one of the main aims of the present management.

The area–species number relationship of scrub species emphasizes the importance of large vegetation patches for conservation and enhancement of the total number of breeding birds. Although total numbers increase with patch size, territory density is strongly negatively correlated with the patch size. This can be explained by the species-specific optimal scrub patch sizes: even typical scrub species do not increase their density with increasing scrub patch size, but reach their maximum density at relatively small patch sizes. As a consequence total territory density would increase with an increase of the area of scrub vegetation in mosaic with other vegetation types (for example short grassland, woodland).

Together with abiotic (altitude, climate) and biotic (competition, predation, nutrition) parameters, vegetation patch size is an important (but overlooked) characteristic that can explain superficially large differences (see Lust et al., 1995) in

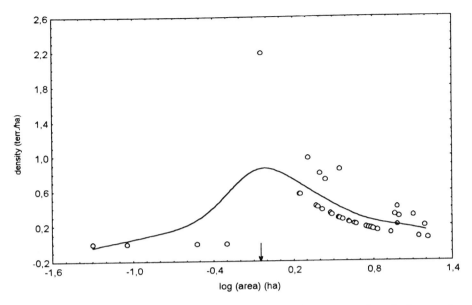

Figure 8. *Relationship between mixed scrub patch area (hectares) and turtle dove territory density (number of territories per hectare).* ↓ *indicates optimal patch size.*

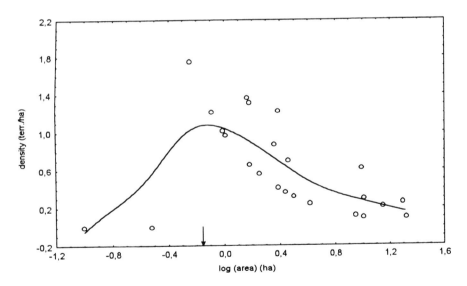

Figure 9. *Relationship between mixed scrub patch area (hectares) and lesser whitethroat territory density (number of territories per hectare).* ↓ *indicates optimal patch size.*

breeding bird densities between analogous scrub types in different study areas.

Since the main goal of contemporary nature management is the optimization of biological (including avian) diversity, grazing at low herd density should be encouraged in the Flemish coastal dunes. It is probably the only management tool that can effectively increase the area of short grassland, in fine- to large-scale mosaics with scrub and dune woodland. In this way a lot of extra suitable habitat can be generated for Red-List species, while common scrub species can reach higher densities in the remaining small scrub patches.

At present domesticated livestock (Highland cattle, donkeys, Shetland ponies and Konik horses) graze approximately 220ha of the dune area. Since annual monitoring of all breeding bird species in the managed dune areas would be too time consuming, a selection should be made of typical species from the identified breeding bird communities. Since all communities have typical Red-List species, detailed monitoring of these should reveal important data on the effect of landscape changes due to nature management or natural succession on individual breeding birds and on breeding bird communities.

Acknowledgments

We wish to thank the Ministry of the Flemish Community (Nature Division) for permitting this research project in their nature reserves. This project was funded by Life-Nature 'Integral Coastal Conservation Initiative', co-financed by the European Community and the Ministry of the Flemish Community, in which the effects of several nature management measures are monitored.

References

Bibby, C.J., Bian, C.G. and Burges, D.J. (1989), 'Bird communities of highland birchwood', *Bird Study*, **36,** 123–33.

Bonte, D., Ampe, C.,Hoffmann, M., Langohr, R., Provoost, S. and Herrier, J.-L. (1998), 'Monitoring research in the Flemish dunes: from a descriptive to an integrated approach', in C.H. Ovesen (ed.), *Proc. Eur. Seminar on Coastal Dunes, Management, Protection and Research*, Skagen, Denmark.

Bonte, D. (1994), 'Broedvogelinventarisatie van het duinencomplex Ter Yde-Groenendijk te Oostduinkerke', *Mergus*, **8(1)**, 25–43.

Bonte, D. and Hoffmann, M. (in prep.), Territory characteristics of breeding birds in the Flemish coastal dunes.

Devos, K. and Anselin, A. (in prep.), *Een gedocumenteerde Rode Lijst van de broedvogels van Vlaanderen*, Institute of Nature Conservation, Brussels.

Fuller, R.J., Stuttard, P. and Ray, C.M. (1989), 'The distribution of breeding songbirds within mixed coppice woodland in Kent, England, in relation to vegetation age and structure', *Ann. Zool. Fennici*, **26,** 265–75.

Gillings, S., Fuller, R.J. and Henderson, A.C.B. (1998), 'Avian community composition and patterns of bird distribution within birch-heath mosaics in northeast Scotland', *Ornis Fennica*, **75,** 27–37.

Hill, M.O. (1979a), *TWINSPAN: A FORTRAN Program for Averaging Multivariate Data in an Ordered Two-Way Table by Classification of the Individual and Attributes*, Section of Ecology and Systematics, Cornell University.

Hill, M.O. (1979b), *DECORANA: A FORTRAN Program for Detrended Correspondence Analysis and Reciprocal Averaging*, Section of Ecology and Systematics, Cornell University.

Hustings, M.F.H., Kwak, R.G.M., Opdam, P.F.M. and Reynen, M.J.S.M. (eds.) (1985), *Vogelinventarisatie, Achtergronden, richtlijnen en verslaggeving, Natuurbeheer in Nederland*, Pudoc, Wageningen.

Lust, P., De Scheemaeker, F. and Gilis, L. (1995). 'Broedvogelinventarisatie van enkele duingebieden aan de Vlaamse Oostkust (Zwinbosjes te Knokke tot de Vosseslag te De Haan) in 1993. Deel1: teksten en tabellen', *Mergus,* **9,** 149–450.

Lysaght, L.S. (1989), 'Breeding bird populations of farmland in mid-west Ireland in 1987', *Bird Study,* **36,** 91–98.

Provoost, S. and Hoffmann, M. (eds.) (1996), *Ecosysteemvisie voor de Vlaamse Kust*, University of Ghent and Institute for Nature Conservation (Brussels), under the authority of the Ministry of the Flemish Community (Nature Division), 3 volumes.

Smith, K.W., Burges, D.J. and Parks, R.A. (1992), 'Breeding bird communities of broad-leaved plantation and ancient pasture woodlands of the New Forest', *Bird Study,* **39,** 132–41.

Stillmann, R.A. and Brown, A.F. (1994), 'Population sizes and habitat associations of upland breeding birds in the south Pennines, England', *Biological Conservation,* **69,** 307–14.

Van der Meer, H.P. (1996), *Atlas van broedvogels tussen Katwijk en Scheveningen,* Duinwaterbedrijf Zuid-Holland.

PREDATION OF THE SAND LIZARD *LACERTA AGILIS* BY THE DOMESTIC CAT *FELIS CATUS* ON THE SEFTON COAST

CARL T. LARSEN and RUTH E. HENSHAW
Liverpool Hope University College, England

ABSTRACT

The predation of the sand lizard *Lacerta agilis* by the domestic cat *Felis catus* is considered. Evidence to support the predation of the former by the latter was obtained from behavioural literature on the domestic cat; data derived from tracking radio-tagged cats and anecdotal information from questionnaires of cat owners. Radio-tagged cats were observed hunting in *L. agilis* colonies and potential habitat. Tracking showed that more than one cat hunts at the same *L. agilis* colony and that cats travelled significantly faster when moving outside colonies than within them and made more stops within colonies than outside them. Tagged cats also return to precise locations on their hunting routes. The degree of compactness of a *L. agilis* colony was considered in relation to the level of cat predation at that site, and showed that cats were more likely (p<0.05) to visit colonies with a compactness ratio of <0.675 (where 1 is a perfect circle), implying that lizards within such colonies were more susceptible to predation. Questionnaire data shows that 5.7 per cent of the cats surveyed at Hesketh are known to have caught lizards within the last year and that 67 per cent of the cats from the questionnaire were regular visitors to Hesketh golf course. There is a high proportion of cats of 'suitable' hunting age per number of households that live in close proximity to a known *L. agilis* colony or potential habitat. Cat outdoor activity strongly correlates with *L. agilis* activity. There are cats in the Hesketh locality of feral background and which are un-neutered. Behavioural literature suggests that despite being fed by their owners a cat's urge to hunt is not suppressed. Literature also suggests that cats can become specialist hunters potentially capable of depleting an entire prey population. The maps denoting the lizard colonies and the routes of the radio-tagged cats are not included for reasons of confidentiality. Management proposals are offered. A comprehensive reference list is provided for future workers.

Introduction

L. agilis belongs to the order Squamata, sub-order Lacertilia, family Lacertidae and genus *Lacerta* (Smith, 1951). It has been reported as the most threatened native species of reptile in Britain (Corbett, 1989). As a result of population decline over the last thirty years it acquired protection through the Council of Europe's Convention on the Conservation of European Wildlife and Natural Habitats (the 'Bern Convention') of 1979. It is also protected by Schedule 5 of the Wildlife and Countryside Act 1981 and more, recently, through the European Union's Habitats and Species Directive of 1992, and the Conservation (Natural Habitats &c.) (Amendment) Regulations 1997. Continuing threats to *L. agilis* have warranted its inclusion in English Nature's Species Recovery Programme, which began in 1994.

Today the species only occurs naturally in scattered populations in Dorset, Surrey and Merseyside (English Nature, 1996). The principal reasons for *L. agilis* decline are documented as being habitat loss resulting from building developments and golf course construction (Stafford, 1989), habitat fragmentation through conifer afforestation with the increasing number of associated fires (Cooke, 1994) and a less favourable climate (Jackson, 1979). Consequently, today only an estimated 2100ha of dune survives of an area which once exceeded 3000ha (Doody, 1991). The Sefton Coast, 32km in length, lies between Southport and Seaforth, bordering the Irish Sea, and is home to the Merseyside *L. agilis* population in northwest England.

The Sefton Coast *L. agilis* population is at the northern limit of its natural range in Britain. It is of particular conservation value, partly due to its isolation by some 300km from the English south coast population. Consequently, this may have resulted in genotypic as well as phenotypic divergence (Beebee, 1978). Nicholson (1980) found that climatically the north west was no more suitable for *L. agilis* than other parts of England where it does not occur. This observation suggests that there are other factors influencing the survival of the northwest *L. agilis* population.

Prestt et al. (1974) discuss *L. agilis* populations in the north west before 1930; figures are estimated at 8000–10,000. Jackson and Yalden (1977) estimated that the population had declined to just a few hundred. More recent figures for *L. agilis* populations in northwest England are reported by Corbett (1994) as being 255 natural and 305 if including the reintroduced individuals. The Nature Conservancy Council (1986) state that 'In the 30 years up to 1974, 92 per cent of the known *L. agilis* colonies in the area have been lost' (NCC, 1986, 169).

Although Jackson (1979) acknowledges the human causes of the decline of this species she also points out that the species has disappeared from many areas still supporting a dune habitat. This suggests that there may be other factors responsible for decline, such as predation, potential inbreeding depression resulting in lower fecundity, eggshell thinning and proliferation of recessive alleles. Predation of *L. agilis* by ground beetles has also been suggested. There is also the issue of diseases and parasites, as mentioned by Smith (1951), which may be contributing to the decline of the species. In addition, predatory birds such as the kestrel *Falco tinnunculus* and the magpie *Pica pica* are potential predators of *L. agilis*. Simms (1966) talks of adult *L. agilis* being major predators on their young. The increasingly stable sand dune habitat is also an important constraint on sand lizard survival.

Losses also occur during hibernation. The Nature Conservancy Council (1986) report that figures for one Lancashire site show that for both Lacertid species, roughly one third of juveniles failed to survive hibernation, while another third of survivors had died by early summer.

However, in the light of the literature and the findings of this study, it appears that the potential threat of predation by domestic cats is very strong. Fox, a British predation biologist, reports (1997) that domestic cats slay 210 million birds and animals a year – almost 80 per cent of all wildlife killed – and maim a further 42 million. He also states that domestic cats are driven to catch prey, no matter how well fed they are. Llewellyn Smith (1997) notes that, on small islands, cats have wiped out entire species.

The relationship between the decline of *L. agilis* on the Sefton Coast and the intro-duction of *F. catus* has, until recently, been overlooked by the scientific community. Corbett (1994) in his 'Pilot study for sand lizard species recovery programme' was one of the first to acknowledge that cat predation may be a cause of decline. English Nature (1992) also mention cat predation from urban fringes but there is no evidence cited. McDonald (1997) conducted a preliminary investigation into the potential threat to *L. agilis* of predation by domestic cats, concluding that there was indeed a potential threat of predation. Our study aimed to determine further whether the accelerating decline of *L. agilis* is a consequence of predation by the domestic cat (*Felis catus*), by radio-tagging the cats to determine their foraging excursions, and to see whether these foraging excursions encroached on known and potential *L. agilis* colonies. Finally, an assessment of the demography of the cat populations adjacent to the *L. agilis* colonies was conducted using questionnaires, to make inferences about the scale of the potential threat the resident cats posed.

Leyhausen (1979), Laundre (1977) and Liberg (1980) have observed hunting grounds, which cats regularly use within their home ranges, implying a degree of permanency. Bradshaw (1992) states that home ranges of adult males are usually much larger than those of adult females.

Turner and Meister (1988) discuss how cats hunt in areas that are different from the surrounding habitat. These include clearings, for example, where there may be increased chances of finding prey. They mention that a cat may travel directly to such a place. Numerous studies have shown that domestic cats can become specialized predators (Heidemann and Vauk, 1970; Lups, 1972; Tabor, 1986). However, Turner and Meister (1988) write that they are unaware of any field data demonstrating the existence of prey specialization. Experimental studies on captive cats have indicated that early experiences with a particular prey type or diet influence later behaviour and preferences (Caro, 1980). Turner and Meister (1988) believe that the prey brought to kittens by their mothers may influence prey specialism. They state that cats have an excellent memory for locality and will often return to the precise location of an earlier capture days or weeks later. This implies that a cat may potentially exhaust an entire prey population.

Inselman and Flynn (1973) found that folliculin increased the readiness of female cats to catch prey in the laboratory while other sex hormones inhibited it. Leyhausen (1979) reports that cats catch considerably more prey when they have young. Turner and Meister (1988) briefly discuss the effects of neutering on the hunting behaviour of both male and female cats. They report that neutered cats do catch, eat and bring home prey, but point out that as far as they are aware, no field study has compared hunting activity of cats of either sex before and after neutering.

Predatory behaviour in domestic cats is reported by Bradshaw (1992) and Leyhausen (1979) to be disconnected from hunger. Turner and Meister (1988), however, report that well-fed house cats may typically hunt for up to a quarter of each day, while feral (un-fed) cats can spend twelve out of every 24 hours searching for food. Similarly, Biben (1979) showed that although cats engage in some predatory behaviour whether hungry or not, the tendency to kill increases with hunger. Although the intensity of hunting may vary, therefore, there is no doubt that cats will engage in at least some hunting irrespective of hunger.

Coleman et al. (1997) report that extensive studies of the feeding habits of free-ranging domestic cats over 50 years and four continents indicate that small mammals comprise approximately 70 per cent of cats' prey while birds make up 20 per cent. They also found that the diets of free-ranging cat populations reflect the food locally available. Brooker (1977) recorded 78 of the small dragon lizard *Tympanocryptis lineata* and 24 other reptiles in just 9 cats from the Nullarbor Plain in Australia. Brooker also recorded a larger proportion of reptiles than mammals in the guts of cats. Clearly domestic cats are predators to lizard species in areas of the world such as Australia and North America. The literature reviewed so far implies that the domestic cat is a likely predator to the Sefton Coast *L. agilis*.

Method

In order to investigate the Hesketh cat population, it was necessary to distribute questionnaires to those houses surrounding the golf course. The design of the questionnaire is explained in Table 1.

Radiotracking

As Kenward (1993) states, 'biologists use animal radio tags for two main purposes: to locate study animals in the field, and to transmit information about the physiology or behaviour of wild or captive animals'. Custom-built, collar-mounted radio tags were fitted to suitable domestic cats, i.e. those living near to suitable *L. agilis* habitat. The tag comprises a single-stage transmitter, powered by a TW3.5 AA battery (3.5 volts),

Table 1. Explanation of questionnaire design

Required information	Questions to be used
Whether the household owns a cat or not	Do you own a cat? If so, how many?
The sex of the cat (as research has indicated that females are more avid hunters)	Is it male or female?
The age of the cat (research suggests that younger cats are more active predators than older cats)	What is its age?
Whether the cat is likely to be breeding	Is it neutered?
Approximate times when the cat will not be hunting, and whether the cat is used to a routine	What time(s) of day is your cat fed?
Times when the cat is most active, and whether this time coincides with expected *L. agilis* activity	How much time does your cat spend outdoors?
Whether the cat is a specialist hunter or not	Has your cat ever brought lizards back? If so, how often? Are catches concentrated within one season?
The likely foraging route of the cat: does this encroach potential or known *L. agilis* habitat?	Do you know where your cat goes when out of the boundaries of your home?
Whether the owner is willing for their cat to be radiotracked	Would you be prepared to allow your cat to be radiotracked?

with a magnetic on/off switch supplied by Biotrack Ltd. The tag transmits pulsed signals at a frequency of 150 MHz, which are detected by a portable M-57 (Marinar radar) receiver, attached to a 1m, three-element Yagi antenna.

Tracking took place between May and the end of September to cover the period of greatest activity of *L. agilis*. Tracking was conducted on warm dry days as both cats and *L. agilis* are known to be less active during rainy, cold periods, and also in unusually hot conditions. Tracking in the dune habitat proved difficult because of signal reflection and diffraction and typical triangulation methods for identifying the tagged cats' positions were not always applicable. Non-triangulation methods proved more successful and involved obtaining positional data by following the transmitted signal's increasing strength until the instrumented cat was actually observed. The only disadvantage of this 'homing' technique was its time-consuming nature. The cat's location was then marked on a 1:2,500 map of the area. To overcome problems of locating the animal's position on the map a Garam hand-held Global Positioning System was used. Locational readings were taken every five minutes. Caution was required when following the cat as it was desirable that the cat was unaware that it was being observed, and thus behaved naturally.

Big Balls Hill at Ainsdale was the focus of tracking between May and mid-August. Hesketh golf course cats were tracked in September. Tracking was attempted 3 times per week over a period of 24 weeks. This amounted to 72 tracking attempts. Of these, 22 attempts were abandoned due to unsuitable weather conditions, e.g. rain or intense sunshine, and 11 due to the fact that the tagged cats were locked in their homes. In addition, the data from a further 8 tracking attempts were omitted because the tagged cat's signal was lost for two or more consecutive readings (or a period of ten minutes). Other circumstances accounted for the failure of a further 6 tracking sessions. On these bases, of the 72 attempts, only 25 were successful, a success rate of 34.72 per cent.

These data were digitized using MapInfo Professional. The maps created combined aerial photographs that depicted individual cat-owning houses with *L. agilis* colony information and the radiotracking data. It was then possible to calculate cat journey distance and approximate time spent within each *L. agilis* site.

The area and perimeter of the *L. agilis* colonies/potential habitats were measured to provide patch characteristics. Forman (1995) shows that an elongated patch is less effective in conserving its internal resources than a compact patch. This is because an elongated boundary provides a greater probability per unit area of movements across the boundary, in one or both directions. Further, in an elongated patch the ratio of the perimeter to the area is greater. Therefore, the edge is proximately closer to the 'centre' and so individuals in the 'centre' are more vulnerable to predation. Patch shape was investigated by examining the compactness ratio (c-ratio) using the software IDRISI program (after Eastman, 1992). The formula for this is:

Where C = the compactness ratio (*C*-ratio)

Ap is the area of the colony

$$C = \sqrt{\left(\frac{Ap}{Ac}\right)}$$

Ac is the area of a circle having the same perimeter as the polygon under consideration

Ac is derived using πr^2

A value of 1.0 indicates perfect compactness – a circle

The lower the value the greater the ratio of edge to area

The compactness of a specific colony is often more informative than measuring its area and is simply a ratio of a polygon's area and perimeter. Indeed, due to extreme topographical variability within and between colonies it would be erroneous to make comparisons between colonies based solely on their respective areas, without incorporating some measure of their topographical attributes.

Results

Questionnaire

Of the 120 households surveyed in Hesketh, 27 (22.5 per cent) were cat owners. It is important to notice that there were 36 cats to 120 houses. These cats are all housed adjacent to Hesketh golf course, and thus present a likely threat to the *L. agilis* population.

Based on information from conversations with cats' owners it was assumed that a domestic cat is likely to be hunting most actively between the ages of 2 and 12 years old. In accordance with this, the 33.3 per cent of the known Hesketh cat population which fell into the age category 3–10 years were targeted for radiotracking.

With regard to the daily activity of the sampled cats, 55.5 per cent were active outdoors in the morning, and 27 per cent were active outdoors in the afternoon. These findings coincide with *L. agilis* activity as the NCC (1983) have pointed out that *L. agilis* bask on warm sunny days, mainly in the mornings after emergence and during late afternoons. Furthermore, literature suggests that *L. agilis* become more active at higher temperatures, yet avoid intense heat (Langton, 1989; Stafford, 1989). Interestingly, tagged cats stayed indoors during periods of hot sunshine. Bradshaw (1992) mentions that cats can be flexible in timing their hunting, usually to coincide with the main activity periods of the most readily available prey. The implications from these similarities in behaviour indicate the potential threat to *L. agilis* by domestic cats.

Approximately 60 per cent of surveyed cats were fed early in the morning, even though, as mentioned above, 55.5 per cent were also active outdoors. This reiterates Bradshaw's (1992) findings that cats will start out on a hunting expedition immediately after consuming a meal provided by their owner. Therefore the regular feeding of domestic cats is no insurance against their predatory behaviour, emphasizing that *L. agilis* are just as much at threat despite cats being fed by their owners. Fewer cats are fed during the afternoon.

The percentage of cats known to visit Hesketh golf course regularly is approximately 67 per cent, and although data were not available for 23 per cent of the cat population in the survey, the close proximity of their homes to the golf course suggests that they too may utilize the site.

5.7 per cent of the surveyed cats have been known to catch lizards. It should not be assumed that the caught lizards were *L. agilis*. However, the cats' close proximity to the known *L. agilis* colony at the golf course suggests that there is a strong likelihood of the lizards being *L. agilis*, as opposed to common lizards, *L. vivipara*.

Data from interviews showed that cats at Hesketh frequently brought lizards back to the house. Cats were observed eating the lizards head-first and then omitting to eat the lizards' stomach and tail. Questionnaire data from Big Balls Hill, compiled by McDonald (1997), showed one particular cat to be a confirmed predator of *L. agilis*.

Behavioural observations of this cat support this view. It is most active outdoors during the morning and early evening (4–7 pm); it regularly catches lizards, identified as *L. agilis* (approximately 2–3 per week), and finally, its home location is adjacent to a *L. agilis* colony. Considering that cats can become specialized predators, as noted above, the potential threat to this colony is clear.

Radiotracking

The sensitive nature of the mapped data containing the *L. agilis* colonies and potential habitat, and the foraging routes of the radio-tagged cats, prohibits their inclusion in this book. However, Table 2 summarizes statistics from 25 tracking sessions, conducted on 5 cats.

The mean values are probably the most useful with regard to predicting the foraging behaviour of the tagged cats. During a typical foraging excursion, cats will cover a mean distance of 307.6m but only travel to a mean maximum distance from their home of 113.5m. This suggests that they actually take a sinuous outward or homeward route and thus cover a larger area. However, it may also indicate that *L. agilis* colonies further than the mean maximum distance from a cat's home will experience a lower rate of visitation. Surprisingly, during a mean foraging journey of 98 minutes, cats only stop 4.16 times.

One tagged cat was recorded travelling 760.6m during a foraging excursion over a 135-minute duration. The cat revisited the same location (at the boundary of a *L. agilis* colony) twice during this session and behaved in a manner typical of a hunting cat, i.e. taking cover in the grass, remaining motionless and then pouncing. The return of a cat to an exact location was also found by Turner and Meister (1988) and implies that the cat may have made a successful catch here in the past. As the cat owner has observed many lizards being brought home by the cat, this return to an exact locality suggests that the cat regularly hunts *L. agilis*.

Table 2. Summary statistics of foraging route characteristics for radio-tagged cats

	Distance of journey (metres)	Duration of journey (minutes)	Maximum distance to furthest point (metres)	Number of times when stationary
Total	7689.30	2300.0	2837.7	104.00
Minimum	79.50	25.0	40.0	0.00
Maximum	832.50	165.0	305.7	12.00
Mean	307.60	92.0	113.5	4.16
S.D.	228.70	52.3	79.4	4.70
Median	347.25	137.5	134.2	3.00

Table 3. The mean relative travel velocities of cats foraging within and outside L. agilis *colonies and suitable habitat*

	Within colony/ suitable habitat	Outside colony/ suitable habitat
Distance travelled (metres)	3025	5657
Duration (minutes)	1301 (<22 hours)	1149 (>19 hours)
Mean relative velocity (metres/hour)	135.5	295.4

Based on the measure of distance covered per unit time, the mean relative travel velocities of foraging cats were significantly higher when recorded moving outside than within *L. agilis* colonies/potential habitat (Table 3). Turner and Meister (1988) observed cats moving directly to known food sources, thus supporting this.

However, this may be an arbitrary measure when considering factors such as the degree of topographical undulation. It is apparent that the terrain outside the colonies is considerably flatter, comprising road networks, railway tracks, footpaths etc. Furthermore, not only do such features tend to be less undulating than habitat within the colonies, they are also often surfaced and are thus easier to negotiate.

Characteristics of *L. agilis* colonies and cat visitation

There are a number of attributes that each colony possesses which may make them more susceptible to predation. These include their proximity to local predators, their area, their perimeter and finally their compactness. The level of compactness (or circularity) of a sand lizard colony was investigated to see if it influenced the likelihood of visitation by a foraging cat. Assuming substrate and habitat homogeneity, a compact patch should theoretically be less susceptible to predation, due to the proximity of its perimeter to the 'centre'.

At Hesketh, the patches exemplifying least compactness include *L. agilis* sites 2, 3 and 7 (Table 4). Site 3 experienced particularly high levels of visitation. It is important to notice that the known *L. agilis* colony at site 4 has a higher compactness ratio ($c = 0.721$) than the least compact colonies at Big Balls Hill. Based on our knowledge of the hunting preference of cats for uncompact territories it may therefore be less susceptible to predation. The elongated nature of *L. agilis* sites 2, 3 and 7 implies that a strong 'edge effect' may be taking place. Raptors, cats, canines and other predators often focus their foraging on edges (Forman, 1995) and one would predict high levels of predation.

It was found that *L. agilis* sites 2–4 and 7–12 were the most compact at Big Balls Hill (Table 5). Therefore it could be suggested that *L. agilis* at these sites have a greater chance of survival than those at the less compact *L. agilis* sites in the area. For example, as *L. agilis* site 12 is the most compact of the sites at Big Balls Hill ($c = 0.832$), the *L. agilis* survival rate would be expected to be higher than at *L. agilis* site 6 ($c = 0.447$). Despite its small area ($256.9m^2$), it is 236.7m from the nearest cat-owning house and

Table 4. L. agilis *colony/potential habitat characteristics and cat visitation. Hesketh*

Colony number	Area (m²) Ap	Perimeter (metres)	Compactness ratio $C = \sqrt{\left(\frac{Ap}{Ac}\right)}$	Distance to centre of colony from nearest cat house (metres)	Cumulative residence time of cats within each area (minutes)
1	1141	127.6	0.776	48.02	125
2	2741	268.1	0.705	72.52	45
3	3517	357.5	0.675	126.90	183
4	1998	230.1	0.721	127.40	116
5	4286	256.5	0.715	80.77	51
6	1895	184.6	0.742	232.50	10
7	2531	244.6	0.710	84.45	137

Table 5. L. agilis *colony/potential habitat characteristics and cat visitation. Big Balls Hill*

Colony number	Area (m²) Ap	Perimeter (metres)	Compactness ratio $C = \sqrt{\left(\frac{Ap}{Ac}\right)}$	Distance to centre of colony from nearest cat house (metres)	Cumulative residence time of cats within each area (minutes)
1	12,280.0	549.10	0.664	67.25	189
2	808.9	129.80	0.794	169.50	25
3	339.0	72.94	0.822	207.10	0
4	1193.0	171.30	0.776	223.20	26
5	24,300.0	977.10	0.532	115.20	172
6	42,820.0	1584.00	0.447	230.60	156
7	8015.0	107.20	0.806	191.30	18
8	326.3	71.25	0.827	225.20	0
9	354.7	75.59	0.822	99.99	3
10	548.6	90.84	0.815	166.60	0
11	599.9	93.88	0.817	233.50	0
12	256.9	64.53	0.832	236.70	0

so survival rate at this colony may also be further enhanced. Interestingly, sites 3, 8 and 10–12 experienced no visitation by cats despite their proximity to cat-owning houses. Possibly, due to their relatively small areas of <600m² they may remain unde-tected by foraging cats. The least compact colonies include L. agilis sites 1, 5 and 6. The radiotracked cat hunted in all of these colonies. Their respective compactness ratios of c = 0.664, c = 0.532 and c = 0.447 imply that there may be a relationship between the degree of compactness of patch shape and cat visitation.

Indeed, analysis showed that cats spent significantly longer ($U = 4$; $p<0.05$) within the least compact colonies and close to their boundaries (Tables 3 and 4). However, potential L. agilis site 1 is a particularly compact site (c = 0.776), and radiotracked cats did visit this site almost as frequently as they did the others. Although this seems to contradict the previous finding, the close proximity to the cat's home offers an expla-nation for the cat focusing on the site, rather than the compactness of the site being the main influence. Furthermore, this site was also visited en route to other colonies. It may suggest that this potential habitat does indeed support a L. agilis population on which the cats are preying.

In summary, analysis of the length of time foraging cats hunted along the edges of potential colonies of varying compactness showed a significant preference for hunting along the least compact colonies. This may have important implications for the long-term management of these fragmented colonies, where relocation may prove to be the only option for the conservation of the species. However, cats also hunted at known colony sites with relatively high compactness ratios. This was a factor of the proximity of the colonies to the cats' respective homes. The focus of cat foraging at certain potential habitat sites may indeed indicate the presence of lizards at that site. Cats also made significantly more 'stops' within colonies than outside of them and travelled at significantly higher velocities outside than within colonies. Both observations suggest that the cats were foraging within the colonies.

Management proposals

Corbett and Moulton (1994–95) reported on management techniques to improve habitat for British L. agilis. They investigated the possibilities of linking the L. agilis colonies of Birkdale and Hillside, which has now been completed (Corbett and Moulton, 1995–96). Other possibilities include:

- The relocation of L. agilis from colonies that are most under threat of predation by cats. These threatened sites include Hesketh, Birkdale, and Ainsdale principally.
- Restrictions could be put on cats' freedom by keeping them indoors to eliminate unwanted reproduction, predation on wild animals and the spread of disease.
- De-clawing is an effective reducer of hunting success.
- Limitations on cat ownership could be introduced.
- L. agilis could be radio-tagged to monitor their daily activity, as well as to inves-tigate their fates.

Summary

It is evident from the findings of this study that domestic cats do hunt in known *L. agilis* colonies and also hunt at sites that have been classed as potential habitat for *L. agilis*. It is also evident that cats return to precise locations within the *L. agilis* colony, behaviour that may indicate that a previous capture has been made there. Although it is likely that previous prey catches have included *L. agilis*, particularly as specific cats have been seen eating lizards by their owners, there is still no documented evidence that these lizards were *L. agilis* despite wardens having identified some lizards killed by cats at another site as *L. agilis*. Both field data and anecdotal evidence have shown that the domestic cat is a predator of lizards and there are many reasons cited and proposed that suggest these lizards are indeed *L. agilis*. Furthermore, the almost simultaneous decline of the *L. agilis* population on the Sefton Coast in conjunction with the growth of the human population (accompanied by an increased cat population) also supports this conclusion. However, despite the likely threat of predation imposed by the resident cat population, surely it is the loss of the dune habitat on which the lizards depend, which has been caused by the increased human population, that will ultimately lead to their demise.

Acknowledgments

Our thanks go to the team at the Sefton Life Project, the cat owners who allowed their cats to take part in the study and who filled out the questionnaire, the owners of Hesketh golf course for allowing access and Liverpool Hope University College for funding the study.

References

A comprehensive reference list is provided for future workers.

Adaniec, R.E. (1976), 'The interaction of hunger and preying in the domestic cat (*Felis catus*): an adaptive hierarchy?', *Behavioural Biology*, **18**, 263–72, Abstract No. 5328.

Aebischer, N. and Robertson, P.A. (1993), 'Compositional analysis of habitat use from animal radio-tracking data', *Ecology*, **74(5)**, 1313–25.

Atkinson, D. and Houston, J. (eds.) (1993), *The Sand Dunes of the Sefton Coast*, National Museums and Galleries on Merseyside.

Beebee, T.C. (1978), 'An attempt to explain the distributions of the rare herptiles *Bufo calamita*, *Lacerta agilis* and *Coronella austriaca* in Britain', *British Journal of Herpetology*, **5,** 763–70.

Biben, M. (1979), 'Predation and predatory play behaviour of domestic cats', *Animal Behaviour*, **27**, 81–94.

Boddicker, M.L. (1983), 'House cats (feral)', in Institute of Agriculture and Natural Resources, *Prevention and Control of Wildlife Damage*, University of Nebraska, Lincoln, Nebraska.

Bradshaw, J.W. (1992), *The Behaviour of the Domestic Cat*, C.A.B International, Oxford.

Brooker, M.G. (1977), 'Some notes on the mammalian fauna of the western Nullarbor Plain, Western Australia', *Western Australian Naturalist*, **14**, 2–15.

Caro, T.M. (1980), 'The effects of experience on the predatory patterns of cats', *Behavioural and Neural Biology*, **29**, 1–28.

Carss, D.J. (1995) 'Prey brought home by two domestic cats *Felis catus* in northern Scotland', *Journal of Zoology, London*, **237**, 678–86.

Churcher, P.R. and Lawton, J.H. (1987), 'Predation by domestic cats in an English village', *Journal of Zoology, London*, **212**, 439–55.

Coleman, J.S., Temple, S.A. and Craven, S.R. (1997), 'Cats and wildlife: a conservation dilemma', http://www.wise.edu/wildlife/extension/catflyz.htin (25/7/97).

Cooke, A.S. (1991), 'The habitat of *Lacerta agilis* at Merseyside', Research Survey Number 41, Nature Conservancy Council.

Cooke, A.S. (1994), 'The habitat of the sand lizard *Lacerta agilis* on the Sefton Coast', in D. Atkinson and J. Houston (eds.), *The Sand Dunes of the Sefton Coast*, National Museums and Galleries on Merseyside, 123–26.

Corbett, K.F. (1981), 'Reptile conservation in Britain – its possible applications to Europe', *Proceedings of the European Herpetological Symposium*, Cotswold Wildlife Park, Oxford, 91–95.

Corbett, K.F. (1989), *Conservation of European Reptiles and Amphibians*, Christopher Helm, London.

Corbett, K.F. (1994), 'Pilot study for sand lizard species recovery programme', *English Nature Research Report No. 102*, English Nature, Peterborough.

Corbett, K.F. and Moulton, N.R. (1994–95), 'Sand lizard species recovery programme: first year (1994–95) report', *English Nature Research Reports. No 134*, English Nature, Peterborough.

Corbett, K.F. and Moulton, N.R. (1995–96), 'Sand lizard species recovery programme: second year (1995–96) report', *English Nature Research Reports. No 187*, English Nature, Peterborough.

Corbett, K.F. and Tamarind, D.L. (1979), 'Conservation of the sand lizard *Lacerta agilis* by habitat management', *British Journal of Herpetology*, **5**, 799–823.

Corbett, L.K. (1979), 'Feeding ecology and social organisation of wild cats (*Felis sylvestris*) and (*Felis catus*) in Scotland', PhD thesis, University of Aberdeen.

Coward, T.A. (1901), 'The sand lizard in the North of England', *Zoologist 4th Series*, **5**, 355–57.

Dent, S. and Spellerberg, F. (1987), 'Habitats of the lizards *Lacerta agilis* and *Lacerta vivipara* on forest ride verges in Britain', *Biological Conservation*, **42**, 273–86.

Doody, J.P. (1991), *Sand Dune Inventory of Europe*, Joint Nature Conservation Committee, Peterborough.

Doody, J.P. (1993), 'Foreword', in D. Atkinson and J. Houston (eds.), *The Sand Dunes of the Sefton Coast*, National Museums and Galleries on Merseyside.

Eastman, J.R. (1992), *Idrisi Technical Reference V4.0*, Clark University, Worcester, MA.

English Nature (1996), 'Reptile survey methods: proceedings of a seminar held on 7 November 1995 at the Zoological Society of London's meeting rooms, Regent's Park, London', *English Nature Science*, No 27, English Nature, Peterborough.

Environmental Advisory Unit (1992), *Sand Lizard Conservation Strategy for the Merseyside Coast*, Environmental Advisory Unit, University of Liverpool.

Eringe, S., Goransson, G., Hogstedt, G., Jansson, G., Liberg, O., Lornan, J., Nilson, I.N., Schantz, T.V. and Sylven, M. (1984), 'Can vertebrate predators regulate their prey?', *American Naturalist*, **123(1)**, 125–33.

Feldman, H. (1995), 'The behaviour of the domestic cat', *Animal Behaviour*, **6**, 563–64.

Ficarra, A. (1997), 'Feral cats: why they need to be controlled', http://www.ucsc.edu/ucsc/nat-reserves/papers/Ad-Ficarra-management.httl (25/7/97).

Fitzgerald, B.M. (1988), 'Diet of domestic cats and their impact on prey populations', in D.C. Turner and P. Bateson (eds.), *The Domestic Cat: The Biology of its Behaviour,* Cambridge University Press, Cambridge.

Forman, R.T. (1995), *Land Mosaics: The Ecology of Landscapes and Regions,* Cambridge University Press, Cambridge.

Fox, N. (1997), 'Domestic predator', *BBC Wildlife,* 70–73.

Gautestad, A.O. and Mysterud, I. (1993), 'Physical and biological mechanisms in animal movement processes', *Journal of Applied Ecology,* **30,** 523–35.

George, W.G. (1974), 'Domestic cats as predators and factors in winter shortage of raptor prey', *Wilson Bulletin,* **86,** 384–96.

Heidemann, G. and Vauk, G. (1970), 'Zur Nahrungsökologie wildernder Hauskatzen (*Felis sylvestris, F. catus* Linne, 1758)', *Zeitschrift für Saugeflerkunde,* **35,** 185–90.

Henriksen, P., Dietz, H. and Henriksen, S. (1994), 'Fatal toxoplasmosis in five cats', *Vet. Parasitol,* **55(1–2),** 15–20.

Henshaw, R.E. (1998), 'An investigation to determine if the domestic cat *Felis catus* is a predator of the Sefton Coast sand lizard *Lacerta agilis*', MSc dissertation, Liverpool Hope University College.

House, S.M. and Spellerberg, I.F. (1980) 'Ecological factors determining the selection of egg incubation sites by *Lacerta agilis* L. in southern England', in J. Coborn (ed.), *European Herpetological Symposium,* Cotswold Wildlife Park Ltd, 41–54.

House, S.M. and Spellerberg, I.F. (1983), 'Ecology and conservation of the sand lizard (*Lacerta agilis* L.) habitat in southern England, *Journal of Applied Ecology,* **20,** 417–37.

Houston, J. (1997), 'Conservation management practice on British dune systems', *British Wildlife,* **8(5),** 297–307.

Humphries, D. (1997), 'The metropolitan domestic cat: a survey of the population characteristics and hunting behaviour of the domestic cat in Australia', http://www.-petnet.com.au/reark/reark.html (25/7/97).

Inselman, B.R. and Flynn, J.P. (1973), 'Sex-dependent effects of gonadal and gonadotrophic hormones on contrally-elicited attacks in cats', *Brain Research,* **60,** 1–19.

Jackson, H.C. (1978), 'Low May sunshine as a possible factor in the decline of the sand lizard *Lacerta agilis* L. in north west England', *Biological Conservation,* **13,** 1–2.

Jackson, H.C. (1979), 'The decline of the sand lizard *Lacerta agilis* L. population on the sand dunes of the Merseyside coast, England', *Biological Conservation,* **16,** 177–193.

Jackson, H.C. and Yalden, D.W. (1977), 'Study of the habitat requirements of the Merseyside population of the sand lizard', unpublished manuscript, NCC, London.

Joint Nature Conservation Committee (1994), *Sand Dune Vegetation Survey of Great Britain. Part 1 – England,* Joint Nature Conservation Committee, Peterborough.

Jones, C.R., Houston, J.A. and Bateman, D. (1993), 'A history of human influence on the coastal landscape', in D. Atkinson and J. Houston (eds.), *The Sand Dunes of the Sefton Coast,* National Museums and Galleries on Merseyside.

Kenward, R. (1993), *Wildlife Radio Tagging,* Academic Press, London.

Langton, T. (1989), *Snakes and Lizards,* Whittet Books, London.

Laundre, J. (1977), 'The daytime behaviour of domestic cats in a free roaming population', *Animal Behaviour,* **25,** 990–98.

Leyhausen, P. (1979), *Cat Behaviour,* Garland STPM Press, New York and London.

Liberg, O. (1980), 'Spacing patterns in a population of rural free roaming domestic cats', *Oikos,*

35, 336–49.

Liberg, O. (1984), 'Food habits and prey impact by feral and house-based domestic cats in a rural area in southern Sweden', *Journal of Mammology,* **65(3),** 424–32.

Llewellyn Smith, J. (1997), 'The charge is murder: but how guilty is puss?', *The Sunday Telegraph,* 11 May, 25.

Lups, P. (1972), 'Untersuchungen an streunenden Hauskatzen im Kanton Bern', *Naturhistorisches Museum Bern Kleine Mitteilungen,* **4,** 1–8.

MapInfo Corporation (1996), *MapInfo Professional™ Reference 2nd Ed,* MapInfo Corporation, Troy, New York.

McDonald, M. (1997), 'A preliminary investigation of the potential threat of the sand lizard *Lacerta agilis* L. predation by domestic cats', undergraduate dissertation, Department of Environmental and Biological Studies, Liverpool Hope University College.

Nature Conservancy Council (1983), *The Ecology and Conservation of Amphibians and Reptile Species Endangered in Britain,* Nature Conservancy Council, Peterborough.

Nature Conservancy Council (1986), *The Conservation of Endangered Amphibians and Reptiles,* Nature Conservancy Council, Peterborough.

Nicholson, A.M. (1980), 'The ecology of the sand lizard *L. agilis* L. in southern England and comparisons with the common lizard *Lacerta vivipara* Jacquin', PhD thesis, Southampton University.

Panaman, R. (1981), 'Behaviour and ecology of free ranging female farm cats (*Felis catus*)', *Zeitschrift für Tierpsychologie,* **56,** 59–73.

Parmalee, P.W. (1953), 'Food habits of the feral house cat in east-central Texas', *Journal of Wildlife Management,* **17,** 375–76.

Prestt, I., Cooke, A.S. and Corbett, K. (1974), 'British amphibians and reptiles', in D.L Hawksworth (ed.), *The Changing Flora and Fauna of Britain,* Academic Press, London.

RSPCA Australia Incorporated (1996), 'Control of cats', http://www.ezycolour.com.au/rspca/cat.html (25/7/97).

Sandstrom, S.R. (1997), 'Protecting wildlife from domestic cats', http://www.rl.fws.gov/sfbwr/cats.html (25/7/97).

Simms, C. (1966), *The Status of Amphibians and Reptiles in the Dunes of Southwest Lancashire, 1961–64,* Lancashire and Cheshire Fauna Society Report No. 36, 7–10.

Smith, M. (1951), *British Amphibians and Reptiles,* Collins New Naturalist Series, Collins, London.

Stafford, P. (1989), *Lizards of the British Isles,* Shire Natural History Publications Number 46, Aylesbury.

Tabor, R. (1986), *The Wild Life of the Domestic Cat,* Arrow, London.

Tabor, R. (1995), *Understanding Cats,* David and Charles, Devon.

Turner, D.C. and Bateson, P. (eds.) (1988), *The Domestic Cat: The Biology of its Behaviour,* Cambridge University Press, Cambridge.

Turner, D.C. and Meister, D. (1988), 'Hunting behaviour of the domestic cat', cited in Turner and Bateson, 1998.

White, G.C. and Garrott, R.A. (1990), *Analysis of Wildlife Radio Tracking Data,* Academic Press, London.

BRYOPHYTE CONSERVATION IN THE MANAGEMENT OF DUNE SYSTEMS – A CASE STUDY OF THE SEFTON COAST

D.H. WRENCH

Sefton Coast Life Project, England

Introduction

Bryophytes – mosses, liverworts and hornworts – are a group of plants that have adapted particularly well to the dynamic environment of sand dunes. In areas where rabbit grazing is intensive mosses can dominate the ground cover of fixed dunes even to the exclusion of vascular plants. It is in the dune slacks, however, that bryophytes are at their most diverse. Dune slacks provide a range of microhabitats for many specialist dune species. The range of habitats also has relatively rapid temporal variation with each stage in the successional development supporting a slightly different assemblage. Perhaps the most specialized species are those that are first to colonize.

Rare Bryophytes of the Sefton Coast

The Sefton Coast, northwest England, is historically well known as a stronghold for several rare bryophytes. Material collected by F.P. Marrat in 1854 from Sefton was found to be a species new to science, *Bryum marratii* (Savidge et al., 1963). A year before the same recorder also found *Bryum calophyllum*, a species new to Britain. Eight Red Data Book species are recorded in addition to many other nationally scarce species (see Table 1). Nomenclature follows Blockeel and Long 1998.

Petalwort *Petalophyllum ralfsii* is a thallose liverwort and is one of 20 species of bryophytes listed in Annex II of the Habitats Directive. It is one of three bryophytes in the UK for which candidate Special Areas of Conservation have been selected. Petalwort is a rare liverwort of short compacted turf in damp coastal dune slacks. It has a scattered distribution around the UK coast with some of the largest populations found in Wales; some sites are known to support hundreds or even thousands of plants. It is listed at Braunton Burrows, Burry Inlet Dunes, Anglesey Coast Dunes, Kenfig, North Northumberland Dunes and Penhale Dunes candidate Special Areas of Conservation. Recent surveys (Holyoak, 1998; Rothero, 1998; Hurford, in press) have updated current distribution and knowledge of its population dynamics.

Seven of the Red Data Book species found on the Sefton Coast are from the genus *Bryum*. Many species within this large group of mosses are difficult to identify, requiring microscope work and fruiting material. However, a few of the dune species can be identified with ease by non-specialists. For example, of all the Red Data Book species listed in Table 1 half can be identified without need for microscope work or fruiting material. Of these both petalwort and *Bryum neodamense* can be easily identified in the field and, given practice, without a hand lens. Species that require specialist identification skills can be difficult for dune managers when assessing management works. Guidelines are needed to show how management techniques that benefit dune bryophytes can be incorporated into more common sand dune management practices. These need to be accompanied with up-to-date and accurate information whenever possible. To assess the success of any management works for these species, detailed survey work by an experienced bryologist would be required. Basic monitoring, however, could provide a good measure of the habitat condition and any potential threats to populations.

Most of the dune bryophytes listed in Table 1 are typically found growing on sparsely vegetated damp sand. This is a relatively rare habitat in Sefton with a recorded maximum of only approximately 6 per cent of the sand dune system (data extracted from National Vegetation Classification surveys of varying dates between 1988 and 1997, held by Sefton Coast Management Scheme). Creation of new damp and open sand dune habitat depends upon either human intervention or natural processes such as blowouts and the formation of parallel dune slacks through dune accretion. Many of these dune specialists will depend in the long term upon management that either promotes natural geomorphological processes or imitates these processes. In practical terms, the latter option would be prohibitively expensive in the long run and unlikely to mimic the full range of habitats that natural processes, through time and chance, can create.

Management Issues

Current management of dune slack habitat on the Sefton Coast rarely shows any consideration toward the conservation of bryophytes. This may be a consequence of lack of recording effort and lack of awareness of the conservation requirements of bryophytes, but may also reflect the level of importance given to bryophytes by conservation managers compared with other groups such as vascular plants, herptiles, and birds. The conservation of the natterjack toad, for example, is one of the prime reasons for the management of many dune slacks in the seaward dunes of the Sefton Coast. This often involves reprofiling with heavy machinery, the excavation of new scrapes in dune slacks, and the use of herbicides to treat scrub regrowth. While this form of management is likely to continue in the short term, it is hoped that future conservation effort will be able to work more with geomorphological processes. Dune

Table 1. Rare bryophytes recorded on the Sefton Coast

Red Data Book species	Status
Bryum calophyllum	Vulnerable
B. knowltonii	Vulnerable
*B. mamillatum**	Critically endangered
B. marratii	Endangered
*B. neodamense**	Endangered
B. uliginosum	Critically endangered
B. warneum	Vulnerable
*Petalophyllum ralfsii**	Vulnerable
Nationally scarce and other notable species	
Amblyodon dealbatus	Scarce
Brachythecium mildeanum	Scarce
Campyliadelphus elodes (*Campylium elodes* Kindb.)	Scarce
Catoscopium nigritum	Scarce
Cephaloziella rubella	Not scarce
Drepanocladus lycopodioides	Scarce
Drepanocladus sendtneri	Scarce
Hamatocaulis vernicosus (*Drepanocladus vernicosus* Warnst.)*	Data deficient species
Meesia uliginosa	Scarce
Moerkia hibernica	Scarce
Tortula wilsonii (*Pottia wilsonii* Br. Eur.)	Scarce
Reboulia hemisphaerica	Not scarce
Rhodobryum roseum	Not scarce
Riccardia incurvata	Scarce
Riccia cavernosa	Scarce

* Species listed on Schedule 8 of the Wildlife and Countryside Act, 1981

management in Sefton that encourages natural processes includes the introduction of grazing and the zoning of recreation. For example, restriction of beach parking away from the dune front and to specific sections of the beach has reduced damage to embryo dunes and allowed a rapid seaward development of the dune front. Most dunes in Britain are mature and in a phase of sediment recycling (Ritchie, this volume). Natural dune slack creation and development should therefore be encouraged as a desirable process.

Footpath management is also an important consideration, particularly in areas of little or no grazing. Light levels of trampling can keep some dune slack vegetation open in structure that may otherwise become dominated by scrub and coarse grass species. While it is necessary to ensure heavy trampling does not destroy valuable habitat, the construction of boardwalks should only be undertaken for overriding recreational management reasons. Boardwalks cover valuable habitat, remove beneficial trampling pressure and encourage scrub to spread right up to the edge of the boards.

The most critical factor on the Sefton Coast for several of the Red Data Book bryophytes is water quality. This is an issue in the Southport Sand Dunes and Foreshore SSSI. A low-lying dune area near an amenity lake is currently the stronghold for petalwort and two other Red Data Book bryophytes. Sand accretion to the north of the vehicular access point to Ainsdale beach repeatedly blocks an outflow pipe from the lake. This leads to episodes of the flooding of large areas with eutrophic water. Aside from the direct effects this may be having on bryophytes, indirect effects include an increased vigour of competitive vascular plants and growth of an algal mat that coats the ground when the water subsides. This algal mat then either hardens on drying or is puddled into a muddy mess. Species such as petalwort may be able to cope with such treatment in the short term by surviving as tubers beneath the sand but species of *Bryum* are less likely to survive. High levels of nutrients in this water (Sefton MBC, 1987) are also likely to favour fast-growing species, leading to the shading out of bryophytes and other less vigorous species. The effects of the relatively high levels of hydrocarbons present in this water are not known.

Monitoring

In 1995 two surveys by Newton (1995a; 1995b) of Red Data Book bryophytes and stoneworts on the Sefton Coast found only four of the seven previously recorded Red Data Book bryophytes. Difficulties in identification may account for some of these species not being seen. However, it seems likely that *Bryum marratii* is now extinct on the Sefton Coast. The reports also gave management recommendations for these species. Prime among these was control of scrub, particularly sea buckthorn, *Hippophae rhamnoides*. In situations where cut material cannot be removed from the site, it was suggested that appropriate dump sites should be found away from potential rare bryophyte habitat. Management of slacks for the conservation of the natterjack toad should be carefully planned in order to limit disturbance to the bryophyte interest. In fact, as long as existing rare bryophyte colonies are not damaged, management work for natterjack toads is also beneficial to bryophytes in that an early successional stage of slack development is maintained with open vegetation and no scrub.

Further survey work (Wrench, 1998) was carried out to map precisely the distribution of petalwort and to estimate the number of individual thalli. Colonies of petalwort were mapped and the information transferred to a Geographic Information System (GIS) using aerial photographs as a base. This allowed accurate relocation of colonies and the ability to combine records from different surveys into one dataset. To assist in relocation, photographs were taken of each colony (see, for example, Plate 1) with the outer perimeter of each indicated by white markers.

Plate 1. *An example of optimum habitat for a range of dune bryophytes.*

In sand dune systems the rapidly changing conditions and high environmental stress require a recording system that can provide current and accurate data on bryophytes to dune managers and preferably provide detailed location maps and management advice. Such a system should be easily adaptable and updatable with a level of detail that can translate the information gathered in a manner that is accessible to land managers. Although such systems are already in use for some dune systems, such as the Sefton Coast, until they are used on a regular basis by land managers and planners habitat deterioration or direct damage to bryophytes and other species of interest will continue, due in part to a simple lack of easily accessible information.

References
Blockeel, T.L. and Long, D.G. (1998), *A Check-List and Census Catalogue of British and Irish Bryophytes*, British Bryological Society, Cardiff.

Holyoak, D.T. (1998), *Petalwort* Petalophyllum ralfsii: *Report to Plant Life on Work Carried Out during 1997*, Report No. 91, Plantlife, London.

Hurford, C. (in press), *Draft Report of 1993* Petalophyllum ralfsii *Survey at Three Welsh Dune Systems*, Countryside Council for Wales, Bangor.

Newton, M. (1995a), 'Sefton Coast Dune System: Survey of Red Data List Bryophytes and Stoneworts', JNCC No. 239, unpublished.

Newton, M. (1995b), 'Sefton Coast Dune System: Survey of Red Data List Bryophytes and Stoneworts', JNCC, unpublished.

Rothero, G. (1998), 'Baseline Survey of *Petalophyllum ralfsii* and Preliminary Survey of Red Data Book Species of *Bryum* at Achnahaird Bay, Wester Ross', Scottish Natural Heritage, unpublished.

Savidge, J.P., Heywood, V.H., and Gordon, V. (1963), *Travis's Flora of South Lancashire*, The Liverpool Botanical Society, Liverpool.

Sefton MBC (1987), *Assessment of Water Quality – Ainsdale and Birkdale Hills*, internal report, Sefton MBC.

Wrench, D.H. (1998), 'The Sefton Coast Dune System: Survey of *Petalophyllum ralfsii* (Petalwort)', Sefton MBC, unpublished.

THE SPATIAL AND TEMPORAL EFFECTS OF GRAZING ON THE SPECIES DIVERSITY OF SAND DUNES

L.A. BOORMAN and M.S. BOORMAN
LAB Coastal, England

ABSTRACT

The maintenance of potentially high levels of plant species diversity in a sand dune flora is usually dependent on the existence of some control to limit the growth of rank vegetation. This very often takes the form of grazing by some kind of domestic animal. However, the relationship between the nature and intensity of grazing and plant species diversity is a complex one. For any given grazing regime the response of the flora can vary in both space and time. This makes the practical use of grazing as a management tool a complex matter.

The vegetational responses vary quite specifically with the particular sand dune plant community concerned and the rate of response is generally slower with the more mature closed plant communities characteristic of the older dunes. For any one plant community the response tends to be at least at two levels: short-term changes often apparent after a few months and long-term changes often only detectable after a number of years. In some respects the standing crop can be a measure of the levels of grazing needed but certain communities tend to have higher levels of inactive standing crop and therefore a lower index of grazing need. It is also important to take account of the various ways that soil nutrient status can be affected by grazing management regimes and the fact that there can often be deleterious effects especially at high grazing intensities.

The patterns of response of vegetation under the influence of particular grazing animals have generally been well documented. There is some evidence that the benefits are greatest where there is a mixed grazing regime; however, the potential benefits of such regimes have to be offset against the practical difficulties of managing mixed grazing. A further factor that has to be considered is the local climatic situation, particularly temperature and rainfall and the consequent soil water regimes. The onset of the effects of climate change and of sea level rise are further factors that must be taken into account in the future.

The range of complexities in the response of sand dune vegetation to different forms of grazing management under varying conditions is illustrated by examples from a number of sites around the British Isles. A number of suggestions are made to improve the selection of the most appropriate regime for particular circumstances.

Introduction

Species-rich grasslands form an attractive part of the natural vegetation of the British countryside; however, in recent years they have come under severe pressure and are now increasingly rare. Sand dunes form an important refuge for this type of vegetation and sand dune floras often have high species diversity which includes a wide range of attractive and unique plant species. Many of the plant species involved are small in size and have relatively low competitive ability. Because of this the maintenance of these high levels of plant species diversity in a sand dune flora is usually dependent on the existence of some control to limit the growth of rank, coarse vegetation (Hewett, 1985; van der Meulen and van der Maarel, 1993; Boorman, 1993; Boorman et al., 1997). This control can take various forms including the existence of very low soil nutrient status or the use of some form of artificial cutting. Very often it takes the form of grazing by some kind of domestic or wild animal. However, the relationship between the nature and intensity of grazing and plant species diversity is a complex one (Ranwell and Boar, 1986; Boorman, 1989; van Dijk, 1992). For any given grazing regime the response of the flora can vary in both space and time. Both temporal and spatial variation are essentially non-linear and both are vulnerable to changes in external environmental influences. This makes the practical use of grazing as a management tool a complex matter.

The Rate of Response to Changing Pressures

The vegetational responses can vary quite specifically with the particular sand dune plant community concerned. The rate of response is generally slower with the more mature closed plant communities which are characteristic of the older dunes (Boorman and van der Maarel, 1997; Boorman et al., 1997). A high nutrient status of the dune greatly increases the need for a controlling factor limiting the growth of the more vigorous plant species. A low nutrient status will also reduce the rate of response to any changing circumstances. For any one plant community the overall rate of response tends to be in the form of two contrasting stages. First of all there are usually short-term changes over a period which may be as short as one or two months or may persist for a few years. Then long-term changes gradually appear. These are often only detectable after a number of years and may continue for decades (van der Maarel, 1966; 1975). The monitoring and interpretation of long-term changes is difficult both from the practical point of view and because of the effects of long-term variations in background environmental factors.

Studies made of the diversity of the rabbit-grazed dunes at Holkham, Norfolk (Boorman, unpublished data) showed these different rates of response (Fig. 1). Following

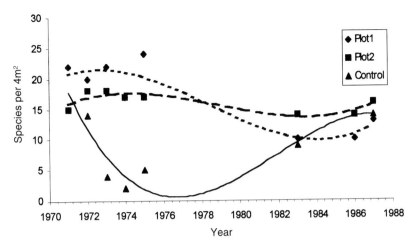

Figure 1. *Effect of excluding rabbits from sand dune grassland; exclosures at Holkham, Norfolk.*

the fencing of selected areas to exclude rabbits there was an initial small increase in species diversity which lasted several years. This occurred at a time when an increasing rabbit population was causing a major decrease locally in plant species diversity. Fifteen years later the much lower numbers of rabbits allowed plant species diversity to recover at least to the levels found in the relatively overgrown plots from which rabbits had been excluded (Boorman, 1989). The various changes that were observed indicated the need for caution in interpreting data collected over limited periods.

Different Grazing Animals

There has been no comprehensive survey of the occurrence of grazing animals on all British sand dune areas but a survey of 48 sites in 1985–87 (Boorman, 1989) confirmed that wild rabbits were the most commonly found grazing animal, occurring in three quarters of the sites (Table 1). Just under a third of the sites were more or less ungrazed but inevitably this had a significant observable effect on plant species diversity. Cattle and sheep grazing each occurred in about one in three of the sites, while mixed cattle and sheep grazing occurred in five sites. Grazing by ponies, goats or hares was only recorded in a very few sites. It is possible that local grazing by deer and wildfowl can occasionally also have some effect.

Different Plant Communities

Sand dunes are a highly dynamic complex of contrasting plant communities. The pioneer yellow dune communities tend to have the lowest species diversity and usually

Table 1. Proportions of British sand dune areas showing signs of grazing by different animals, 1985–87. (Source: Boorman, 1989)

Grazing animal	Percentage of sand dune areas surveyed
Ponies	2.6
Cattle	12.1
Sheep	13.3
Rabbits	72.2
Hares	1.0
None	31.3

the greatest vegetation height. Plant height tends to be a necessary response for survival in areas where there can be rapid accretion of wind-blown sand. Successive stages in dune maturity, particularly the development of dune grassland and damp slack communities, tend to be accompanied by an increase in species diversity together with a decrease in the mean height of the vegetation (Boorman, 1989). The communities associated with wet slacks tend to have a rather lower species diversity and many of these semi-aquatic communities include tall-growing emergent species, so the mean vegetation height is greater. Although dune heath forms unique and fascinating plant communities the diversity is relative low and the common occurrence of species of *Erica* or *Calluna* generally results in quite tall vegetation.

The Effect of Standing Crop

Grasslands are often highly productive plant communities. The productivity of a fertile agricultural grassland can result in an aerial biomass that can be as high as 1.5kg m^{-2} (Jackson and Williams, 1979). Usually, however, this level of production is by one or two species only and most other species are excluded by the intense competition that results from the vigour of the plant growth. Various different studies and observations all indicate that the greatest species diversity occurs when there are relatively low levels of productivity (Klinkamer and de Jong, 1985; Crawley, 1986), levels which result in an aerial biomass of the order of 300g m^{-2}. Conversely, at very low levels of biomass diversity is also low as then all plant species tend to be limited in their growth and survival.

The existence of an *optimum* aerial biomass for maximum species diversity is shown diagrammatically in Figure 2. The precise level of this optimum will vary depending on soil and climatic factors. In terms of practical dune management these factors are not ones that can readily be altered.

In some respects the standing crop can thus be a useful measure of the levels of grazing needed. Grazing management should therefore aim to limit aerial biomass to

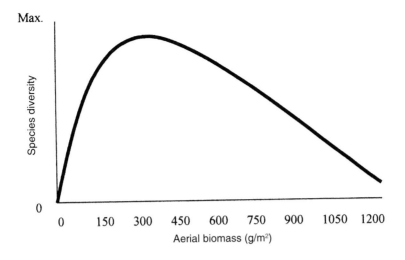

Figure 2. The effect of aerial biomass on the diversity of dune grassland.

around this optimum value. But, as already described, certain sand dune plant communities tend to have higher levels of inactive standing crop than others. This therefore implies a rather lower index of grazing need than at first apparent. It is also important to take account of the various ways that soil nutrient status can be affected by grazing management regimes (Boorman and Fuller, 1982; van Dijk, 1992; Boorman and van der Maarel, 1997) and the fact that there can often be deleterious effects especially at high grazing intensities which can result in the eutrophication of the soil.

The Selection of Grazing Animals

The patterns of response of vegetation under the influence of different grazing animals have generally been well documented (Boorman, 1989; van Dijk, 1992) but less clear are the interactions that can occur between different grazing animals. There is some evidence that the benefits are greatest where there are mixed grazing regimes (Oosterveld, 1985; Boorman, 1993). However, it is often difficult to assess whether the use of more than one type of grazing animal has more or less effect than the use of the optimum grazing intensity. The different grazing preferences of, for example, cattle and sheep certainly suggest that such mixed grazing would be beneficial (van Dijk, 1992); however, the potential benefits of such regimes have to be offset against the practical difficulties of managing mixed grazing. A further factor that has to be considered is the local climatic situation, particularly temperature and rainfall and the consequent soil water regimes. The onset of the effects of climate change and of sea level rise are further factors that must be taken into account in the future.

Grazing Intensities

The selection of the optimum grazing intensity in terms of the rate of stocking is a crucial factor. It is vital to appreciate that it is not a constant value; it should be seen as a variable to be derived from the current state of the vegetation. The intensities needed to maintain a well-managed system are very much lower than those needed to restore diversity to a neglected and overgrown system.

Indications are that the species diversity of a well-managed dune grassland can be maintained by the grazing of cattle at as low a density as 0.06–0.3 beasts per hectare and dune heath by sheep grazing at stocking rates of 0.2–1 individual per hectare (Bakker et al., 1979; Boorman, 1991). In general larger grazed areas (>100ha) can support rather higher rates of stocking than can small areas (<10ha). If, however, the vegetation is rank and neglected then for an initial period very much higher stocking rates are needed, of the order of three or four times the above figures.

In addition to the mean annual grazing intensity decisions have also to be made as to whether grazing should be applied throughout the season or just for selected periods. This has to be a local decision taking into account the availability of stock, the state of the ground at critical times of the year and the availability of sufficient green natural vegetation. It is very important to note that there are no benefits to be had from grazing dunes if there is no fresh vegetation for the stock to eat; the results are likely to be significant damage caused by trampling. It is vital to avoid importing feed to stock which are kept in a natural area as the only result will be to increase the nutrient level of the soil with a consequent loss in species diversity.

Conclusions

For the successful grazing management of sand dunes six points need to be carefully considered. These are:

1. To identify the management needs appropriate to particular individual plant community and plant species;
2. To identify the likely rates of change that are to be expected in the vegetation;
3. To choose the type of animal most appropriate to the local situation;
4. To identify grazing management needed to produce optimum state of the vegetation (biomass) from the present situation;
5. To identify the grazing management needed to maintain this situation when it is reached;
6. To recognize and act on the need for continued monitoring as local and global environmental conditions change.

References

Bakker, T.W.M., Klijn, J.A. and van Zadelhoff, F.J. (1979), *Duinen en duinvalleien: een land-schapsecologische studie van het Nederlandse duingebied*, Pudoc, Wageningen.

Boorman, L.A. (1989), 'The grazing of British sand dune vegetation', *Proceedings of the Royal Society of Edinburgh*, **96B**, 75–88.

Boorman, L.A. (1991), *The Grazing Management of Sand Dunes*, Joint Nature Conservation Committee, unpublished report.

Boorman, L.A. (1993), 'Dry Coastal Ecosystems of Britain: Dunes and Shingle Beaches', in van der Maarel, 1993, 197–228.

Boorman, L.A. and Fuller, R.M. (1982), 'Effects of added nutrients on dune swards grazed by rabbits', *Journal of Ecology*, **70**, 345–55.

Boorman, L.A. and van der Maarel, E. (1997), 'Dune grassland', in van der Maarel, 1997, 323–44.

Boorman, L.A., Londo, G. and van der Maarel, E. (1997), 'Communities of dune slacks', in van der Maarel, 1997, 275–95.

Crawley, M.J. (1986), *Plant Ecology*, Blackwell Scientific Publications, Oxford.

Dijk, H.W.J. van (1992), 'Grazing domestic livestock in Dutch coastal dunes: experiments, experiences and perspectives', in R.W.G. Carter, T.G.F. Curtis and M.J. Sheehy-Skeffington (eds.), *Coastal Dunes*, Balkema, Rotterdam.

Hewett, D.G. (1985), 'Grazing and mowing as management tools on dunes', *Vegetatio*, **62**, 441–48.

Jackson, M.V. and Williams, T.E. (1979), 'Response of grass swards to fertilizer N under cutting or grazing', *Journal of Agricultural Science*, **92**, 549–62.

Klinkamer, P.L.G. and de Jong, T.J. (1985), 'Shoot biomass and species richness in relation to some environmental factors in a coastal dune area in the Netherlands', *Vegetatio*, **63**, 129–32.

Maarel, E. van der (1966), 'Dutch studies on coastal sand dune vegetation especially in the Delta region', *Wentia*, **15**, 47–82.

Maarel, E. van der (1975), 'Observations sur la structure et la dynamique de la végétation des dunes de Voorne – Pays Bas', *Coll. Phytosociologique*, **1**, 167–83.

Maarel, E. van der (ed.) (1993), *Dry Coastal Ecosystems: General Aspects*, Ecosystems of the World 2a, Elsevier, Amsterdam.

Maarel, E. van der (ed.) (1997), *Dry Coastal Ecosystems: General Aspects*, Ecosystems of the World 2c, Elsevier, Amsterdam.

Meulen, F. van der and van der Maarel, E. (1993), 'Dry coastal ecosystems of the central and southwestern Netherlands', in van der Maarel, 1993, 271–315.

Oosterveld, P. (1985), 'Grazing in dune areas: the objectives of nature conservation and the aims of research for nature conservation management', in P. Doody (ed.), *Sand Dunes and their Management*, Focus on Nature Conservation No. 13, Nature Conservancy Council, Peterborough.

Ranwell, D.S. and Boar, R. (1986), *Coastal Dune Management Guide*, Institute of Terrestrial Ecology, Norwich.

Section 3

PEOPLE AND DUNES

HABITAT RESTORATION AND PUBLIC RELATIONS: A RESTORATION PROJECT IN THE AMSTERDAM WATERSUPPLY DUNES

LUC H.W.T. GEELEN
Amsterdam Water Supply, The Netherlands

ABSTRACT

Wet dune slacks are considered especially valuable ecosystems for conservation of natural habitats and of wild fauna and flora. Wet dune slacks are among the most seriously threatened natural habitats in the Netherlands.

In 1994 a restoration project was carried out in the Amsterdam Watersupply Dunes. Three main objectives of the restoration project were:

1. To create a dune ecosystem which can develop as naturally as possible, with special regard to an undisturbed groundwater system and with opportunities for geomorphologic processes;
2. To offer more opportunities for native and threatened species, namely species of moist and nutrient-poor dune slacks and pioneer vegetation;
3. To restore the natural beauty of the dune landscape.

In an area of 24ha some 270,000m³ (430,000 tons) of sand was used to fill in an extraction canal and to restore the original dune landscape.

Landscape restoration on this scale is a new phenomenon in the Dutch coastal dunes. Such a project appears controversial since nature management has always protected the existing nature values and landscape against destruction. For reasons such as these a publicity programme was drawn up at the beginning of the project. The purpose was to clarify the benefits of the project for nature and landscape and to explain the *unnatural* methods used. The provision of high-quality information, based on the results of clear research, seemed to create the platform for public support for this restoration project.

Introduction: Context and Site Description

The Dutch dunes are relatively well protected as nature reserves and drinking water collection areas. The combination of waterworks drawing water from the dunes and the development of polder areas in the hinterland caused a lowering of the groundwater table in large parts of the mainland dunes. Wet dune slacks are among the most seriously threatened natural habitats in the Netherlands (van Dijk and Grootjans, 1993).

Because large parts of the hinterland lie below sea level we were, and perhaps still are, rather over-concerned about the need to stop blowouts and moving dunes. A strict regime of planting marram *Ammophila arenaria*, sea buckthorn *Hippophae rhamnoides* and other species caused the fixation of the dunes with a consequent loss of pioneer vegetation.

These national problems can be recognized in the case study area. The Amsterdam Watersupply Dunes are about 3300ha stretching 8km along the North Sea coast with a width varying from 1.5 to 5km (see Fig. 1). In 1853 the Amsterdam Water Supply started to exploit drinking water in the area. As a result of the increased extraction, the groundwater reservoir in the dunes decreased. To counteract this, artificial recharge with purified water from the Rhine has been practised for the production of drinking water since 1957 in the northern part of the site. This infiltration did not, however, lead to the large-scale regeneration of former dune slacks.

Between 1988 and 1992 the Amsterdam Water Supply carried out an ecohydro-

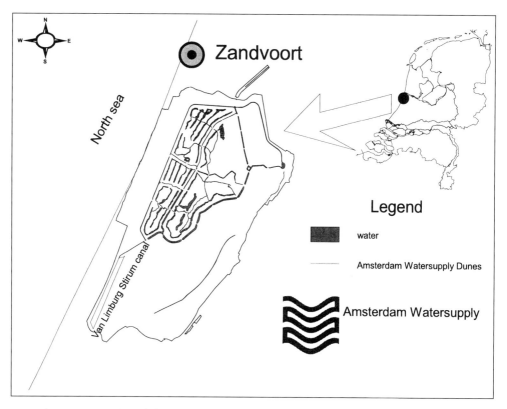

Figure 1. *Location of the Amsterdam Watersupply Dunes along the North Sea coast.*

logical study into the connection between drinking water collection and nature management in this dune area (van Til and Geelen, 1995). In 1994, the water company, the municipality of Amsterdam and the provinces of South and North Holland decided to reduce water collection activities in the eastern part of the area and to stop it completely in the western part of the area. The Van Limburg Stirumkanaal passed through this area (see Fig. 1). In the southern part (2.5km) nature was to be given free rein.

The inland side of our dune area is densely populated. Within a radius of 50km there are 6 million inhabitants (Ehrenburg and Baeyens, 1992).

The Restoration Project

Three main objectives of the restoration project were:

1. To create a dune ecosystem which can develop as naturally as possible, with special regard to an undisturbed groundwater system and with opportunities for geomorphologic processes;
2. To offer more opportunities for native and endangered species, namely species of moist and nutrient-poor dune slacks and pioneer vegetation;
3. To restore the natural beauty of the dune landscape.

The main criteria for the restoration scheme were the predicted groundwater table, the geographic relief in the immediate surroundings of the canal and the available and required amount of sand.

Hydrological modelling of the groundwater table in a scenario without water extraction showed that the water table would rise 1.5–2m in the neighbourhood of the canal. These predictions gave an indication of the surface height to be created for the new dune slacks. A study of old aerial photographs together with some fieldwork showed that excavated sand from the canal was dumped in dune slacks in the direct vicinity of the canal. On the western side of the canal there appeared to be a dumping ground measuring 19ha. The dumped layer had an average thickness of more than 1m. Because of this, only small parts of the original dune slacks on the western side of the canal could regenerate with a rising groundwater table.

We computed a digital elevation model with the existing and planned heights in the geographic information system (GIS) ARCINFO to get a good impression of the existing and proposed relief. By comparing these models with the results of the hydrological modelling, which were also available in the GIS, the hydrological situation in the reformed dune slacks could be computed. By comparing the two digital elevation models an initial rough estimate of the amount of sand to be moved could also be computed. So thanks to GIS techniques we could formulate a well-founded and well-illustrated restoration plan (Geelen, Cousin and Schoon, 1995).

The project was carried out in three stages. The first stage of the work involved removing fish from the canal, adjacent vegetation (mainly the dense sea buckthorn scrub) and a road parallel to the canal. In the second stage the topsoil, with the remaining roots, was excavated and deposited on the canal floor. After this, the original white dune sand was shovelled back and the landscape was restored. Almost 270,000m^3 of sand was replaced. In the modelling and planning phase there is always uncertainty about the accuracy of the modelling predictions. For this reason the sand was not fixed, allowing geomorphologic processes to make the finishing touches.

Publicity Campaign

Before and during this project we paid a lot of attention to a publicity campaign. Landscape restoration on this scale is a new phenomenon in the Dutch coastal dunes. Such a project appears controversial in a nature reserve since nature conservation has always protected existing nature and landscape values against destruction. The visitor sees the existing landscape and vegetation as valuable nature and in most cases does not make a distinction between the value of common and endangered species. No one remembers the landscape of the early twentieth century that ecologists want to restore.

Habitat restoration projects are rather expensive. Everybody knows the first question decision-makers ask: is it worth the money? In addition farmers and villagers in the adjacent polders were afraid that this project would cause groundwater tables to rise on their properties. It was also clear that visitors looking for peace and quiet would react adversely to the large and noisy machines disturbing the tranquillity of the dunes.

A publicity programme was drawn up at the beginning of the project to explain all these issues. In the initial stages of the publicity campaign the first questions to be answered were: Who do we want to reach? Whose support do we need? Essential, of course, was the support in our own organization and of those who take the decisions about money and permits. We had to inform and deliberate with our neighbours,

Plate 1. Work in progress.

Table 1. Target groups and information methods used in the Van Limburg Stirum habitat restoration project. Asterisks indicate that the technique was used with the target group; two asterisks indicate that this was a major method with the target group. See text for explanation of 'visitor: landlord'

Target group \ Technique	Reports	Lectures	Newspapers	Newsletters	Field trips	Information evening	Leaflets	Information boards	Visitors' centre
Colleagues	**	**		*	*				
Politicians	**	**	*	*					
Farmers		*	*	*		**	*		
Villagers			*	*		**			
NGOs	*	*	*	*	*	**			
Visitor: nature lover			*	*			**	*	*
Visitor: ordinary			*	*				**	*
Visitor: landlord			*	*	**			*	*

farmers and villagers. We also consulted NGOs, bird-watchers and nature conservation groups who can appear to be formidable opponents to restoration projects in nature reserves. And, last but not least, we had to inform our visitors, whose tastes differ widely. Different visitor groups have different demands. Different categories can be distinguished: pleasure seekers, nature lovers and naturalists, for example. A special category important for restoration projects may be categorized as 'the landlords'. These people are regular visitors who have been coming to the area for a long time, behave as if they own the place, are sad that things will never be the way they used to be and claim that wildlife managers and ecologists do not know what they are talking about. This category, scientifically insignificant in number, can cause a lot of trouble, for example by legal procedures and putting the police onto the practical work supervisors.

Different people and targets require different methods of providing information. Information evenings took place with company employees, local bulb-growers, pressure groups, local villagers and bird protection groups. Both regional and national newspapers were informed. The national press was used to reach opinion leaders and decision-makers, and the regional press because 80 per cent of visitors to the dunes live within 15km of the area. These visitors were also targeted by direct mail. Information officers and leaflets informed visitors in the visitor centre. Project information boards were set up on the site to inform visitors.

The project was completed without delay. In our opinion high-quality provision of information based on the results of clear research created the base for public support for this restoration project. During the work, and in the first months after completion, many people visited the area. Not enough attention, however, was paid to the aftercare of the restoration project. In 1997 bulb growers in the region started a media campaign. They claimed that reduced bulb yield was caused by rising groundwater tables due to our restoration project. Although we could prove that this was not the case and that frost damage was the real cause, we lost a lot of our public and political support. The lesson we learned was that we should not relax too early. We won an ecological battle, but we can still lose the war! The loss of political confidence affects subsequent restoration projects. We try to stay on speaking terms with opponents of our project and we will see if we can reach some mutual agreement on further restoration projects and the monitoring of possible effects.

Concluding Remarks

Restoration projects need a multidisciplinary approach. Scientists, site managers and information officers should all play their part. Sound project management techniques should be followed to ensure success. The blueprint for the information campaign should answer the why-who-what-how questions. Why do you need an information campaign? Who do you want to reach? What do you want to tell them? How are you going to do that? In the long term a consensus approach will be most effective for habitat restoration projects. Time that is needed for consensus-building in early stages can be reclaimed afterwards.

This is not the final word on habitat restoration and public relations. A conceptual framework on habitat restoration projects which is both scientifically defensible and useful for site managers is needed. The question is how publicity campaigns can close the gap between science-based and politics-based decision-making.

References

Dijk, H.W.J. van and Grootjans, A.P. (1993), 'Wet dune slacks: decline and new opportunities', in E.P.H. Best and J.P. Bakker (eds.), *Hydrobiologia,* **265**, *Netherlands-Wetlands*, Kluwer Academic Publishers, Dordrecht, 163–82.

Geelen, L.H.W., Cousin, E.F.H. and Schoon, C.F. (1995), 'Regeneration of dune slacks in the Amsterdam Waterwork Dunes', in M.G. Healy and J.P. Doody (eds.), *Directions in European Coastal Management*, Samara Publishing, Cardigan, 525–32.

Ehrenburg, A. and Baeyens, G. (1992), 'Landscape ecological mapping as a basis for management of the Amsterdam Waterwork Dunes', in R.W.G. Carter, T.G.F. Curtis, and M.J. Sheehy-Skeffington (eds.), *Coastal Dunes: Geomorphology, Ecology and Management for Conservation*, A.A. Balkema Publishers, Rotterdam, 407–18.

Til, M. van and Geelen, L.H.W.T. (1995), 'Ecohydrological modelling of regeneration of dune slacks in the Amsterdam Waterwork Dunes applying a GIS', in A.H.P.M Salman, H. Berends and M. Bonazountas (eds.), *Coastal Management and Habitat Conservation*, EUCC, Leiden.

THE CONSERVATION OF MACHAIR IN SCOTLAND: WORKING WITH PEOPLE

STEWART ANGUS

Scottish Natural Heritage Advisory Services, Scotland

The author was Area Manager for Scottish Natural Heritage (SNH) in the Western Isles for 1992–97. The views presented in this paper are not necessarily the views of the present Area staff of SNH in the Western Isles or of SNH as an organization.

ABSTRACT

Machair is a habitat complex restricted globally to northern Scotland and north-west Ireland. In terms of natural heritage conservation, it is rich and varied, with much of the interest directly or indirectly dependent on active management for traditional agriculture. Although the machair areas are covered by national and European designations and legislation, conservation of this natural heritage can only be achieved by working closely with local people, a process which cannot be isolated from a wider context of land tenure, culture, political and socio-economic factors. The statutory natural heritage conservation body, Scottish Natural Heritage, requires a particularly high level of awareness of these subjects and an approach which is both understanding and pragmatic in order to achieve its objectives for machair grassland.

This paper examines the conservation of machair in this varied context, making particular reference to the Western Isles of Scotland, where machair is particularly well developed.

The maxim 'think globally, act locally' is highly applicable to this situation.

Introduction

Machair is a coastal grassland plain established on shell sand, confined globally to the northwest of Scotland and the northwest of Ireland (Ritchie, 1976; Angus, 1994). Its distribution in Scotland is shown in Figure 1. The grassland is almost invariably an integral part of a wider machair system incorporating marsh, dune, saltmarsh, loch and even brackish waters, all of which are affected, to varying extents, by the blown sand which is an essential feature of this dynamic formation. The habitat has geomor-

Figure 1. Distribution of Machair in Scotland.

- ● Machair
- ✦ Probable Machair
- ○ Possible Machair

phological, landscape, botanical, ornithological, invertebrate and archaeological interest. This is reflected in the high proportion of the machair of both countries which is formally designated for the conservation of these features. The designations are multiple and overlapping, leading to bureaucratic complexity and potential misunderstanding.

Machair grassland is listed on Annex 1 of the Habitats and Species Directive (with Priority status in Ireland). Scottish machair grassland is unusual among habitats of high conservation importance in that the best sites are not only inhabited and managed for agriculture, but the habitat quality is directly dependent upon traditional land management, with rapid and substantial reduction in wildlife interest where management is reduced. The level of management input tends to be closely tied to factors such as age profile of the community, availability of part-time or full-time employment, and wider agricultural pressures. Machair quality is therefore inextricably linked to socio-economic factors, so that its management and conservation must be examined against this broader background.

That the objectives of the Habitats and Species Directive and the Birds Directive are often compatible with those of crofting greatly facilitates the conservation process, but any discussions have a context that stretches geographically well beyond the confines of machair. They must accommodate attitudes to conservation and conservationists generally, as well as to Scottish Natural Heritage (SNH) in particular (because not everyone recognizes this distinction), and these in turn are determined partly by the media, by local word of mouth, and by personal experience of individual staff in the conservation bodies. In working with local people to conserve the machair, the conservationist has no option but to do so against this wider context, and some of the more pertinent aspects of this background are described below.

There are six candidate Special Areas of Conservation (cSACs) for machair in Scotland, mostly as aggregate sites: South Uist, North Uist, Tiree, Coll, Monach Isles and the Oldshoremore/Sandwood systems. All cSACs are also notified as Sites of Special Scientific Interest (SSSIs) and are part of a wider SSSI network which includes machairs of national but not international importance. Machair SSSIs invariably include much of the wider machair system, and usually also elements of the transitions to inland habitats. It is estimated that some 50 per cent of the Scottish machair resource is covered by cSAC and possibly more than 80 per cent by SSSI.

Unless otherwise indicated, 'machair' is used below to indicate the machair grassland plain rather than the machair system as a whole.

Land Tenure

Most machair land in Scotland is held in crofting tenure, divided into 'township' units corresponding to a village and its land. The croft land itself is usually within a block of individual crofts, each (usually) with a croft house inhabited by the crofter and family. In addition to this 'in-bye' land, each croft has shares in a larger area of the township which is tenanted in common, and administered in compliance with the various agricultural Acts by a Common Grazings Commitee. In some townships, much or even all of this common land has been 'apportioned' to individual crofts so that each part is effectively managed by an individual as an integral part of the croft. In order to be recognized under crofting legislation, an apportionment must be fenced. Land management is usually conducted at the township level, including much non-machair land. In some townships, individuals may inherit a number of crofts and absentee crofters may sub-lease their holding to active crofters. Although it is not unknown for a single individual to manage a large area of land, the different crofts involved cannot be legally amalgamated, nor can a single croft be split. Crofts are rarely owned by the crofter, and more often than not they are rented from large estates who retain only the sporting and mineral rights, though the crofter retains the right to extract sand and gravel for his or her own agricultural use.

The present crofting system dates from the first crofting legislation of 1886, which gave crofters security of tenure for the first time. This legislation was very hard-won, following many decades of intense privation, including famine and enforced clearances, during a time when most local people seemed to lack even civil rights and were largely at the mercy of landowners. Though the island clearances are less well-known than those of Sutherland, the actions of Gordon of Cluny, who owned the islands of Benbecula, South Uist and Barra, have been described as 'unrivalled in brutality', and anyone who refused to board the emigrant ships was 'hunted down with the aid of dogs, bound and despatched willy-nilly' (Hunter, 1976). It may seem very long ago, but these barbaric events have only recently been recognized by academic historians, and the sense of injustice, sustained by a powerful folk memory, has contributed to a bond between the crofter, his or her land and its associated rights, which is intensely strong. Anything which is perceived, rightly or wrongly, as impinging on this relationship may be regarded with great suspicion or even resentment.

Despite the proliferation of individuals with multiple formal and informal tenancies, there are very few full-time crofters. The crofting system was designed to be part-time, and it is significant that the most active crofters tend to be those in other part-time or even full-time employment. Where such employment is scarce, as in the remoter parts of Lewis and Harris, emigration results in a depletion of the township population and an increase in the average age, with a consequent reduction in agricultural activity. In Lewis and Harris this has resulted in a dramatic reduction in the habitat quality of machair, and it is of some concern that in areas such as Uig in southwestern Lewis, where rural employment in fish farming has restored the age profile (though not the population numbers), there has been little evidence of a return to agricultural activity. Though it should theoretically be possible to bring the machairs of Uig back into cultivation and thus bring back the wildlife, it may be that continuity of agricultural activity is more important than is currently realized.

Crofters are often portrayed as beneficial custodians of the countryside and in the case of machair this is usually true, the few exceptions being related mainly to excessive or badly planned sand extraction or the use of land for non-agricultural activity such as caravanning. Crofting tenure includes certain 'rights', some real and others perceived, and some of these, such as extraction of sand for agricultural purposes (which can include construction of the croft house or access road) are 'permitted development' in terms of the planning legislation.

Though the SSSI legislation provides for compulsory land purchase if all other avenues of achieving conservation objectives fail, this would make little, if any, difference on croft land, as all rights relating to crofting are held by the crofter, not the landowner, and a second legal process of decrofting would be required to gain any control over land management. In practice, this would be widely regarded as draconian, inviting comparisons with the hated clearances, to the extent that it could not even be contemplated, and more often than not, negotiation and persuasion, with the opportunity of offering management agreements or other financial incentives, are the only ways in which conservation can be augmented in crofting areas.

Agriculture

The in-bye lands of the crofts proper usually extend linearly seawards, though much of the machair is common grazings. Traditionally, parts of the machair grassland were cultivated in a rotation of three years or thereabouts, with fallow following cultivation. Small patches were used, and these were unfenced. Livestock therefore had to be excluded during the growing season to avoid damage to the crops, and this was associated with transhumance, whereby the cattle were driven to the moorland in May, where they remained until after the harvest, coincidentally allowing them to benefit from some trace elements which are missing from the machair. Transhumance had more or less disappeared from northwest Scotland by the 1950s, but where the machair is cultivated, livestock are still excluded during the growing season; where this cannot be applied, the cropped areas must be fenced.

As in many small communities, the mutual cooperation required to work effectively at the township level is not always what it might be, and it is likely that some apportionments arise from difficulties in agreeing on management of the common grazings. The stocking levels or 'soumings' established by regulations are widely ignored but they have never been reviewed to take account of improvements in land and quality of stock. Though fencing causes disruption of an expansive habitat, detracts from its visual quality, and acts as a barrier to bird movements, cattle management is so vital to the future of machair that there may be circumstances where conservationists would accept the fences in order to retain the cattle, though alternatives are explored where possible. Likewise the increase in value of stock as husbandry is improved has led to more requests for outbuildings on the machair; though these tend to be highly intrusive visually, without them the crofter might abandon cattle in favour of sheep, which would be infinitely more damaging to the machair than a large shed.

The main agricultural output of the 'core' machair islands (Uists and Tiree) is store cattle; calves bred in the islands are exported for finishing elsewhere. Though some winter feed is imported from the mainland and some crofters operate 'away wintering' of their stock, the main conservation value of cattle rearing is in the cultivation of traditional fodder crops: a mixture of black oat and rye. The value of cattle grazing, deposition of dung and (occasionally) trampling should also be noted. In the absence of herbicides the patches of 'corn' tend to have a rich wild flora early in the season (most of which dies back prior to the harvest), while adjacent areas of fallow of differing ages support different assemblages of wild flowers. The resulting patchwork is visually spectacular, and the wide range of wild plants supports a wide range of insects, combining to produce a highly varied feeding and breeding habitat for birds. The densities of breeding waders on the Uist and Tiree machairs are unsurpassed in Europe and this richness is directly attributable to human management of land for agriculture. Machair also supports important populations of birds declining in many areas such as corncrake *Crex crex* and corn bunting *Embiriza calandra*, as well as many wildfowl on the associated loch network (Angus, 1994).

The conservation value of machair grassland is inextricably linked to active, skilled crofting. There are several farms on the machair of the Uists where the cultivation methods used tend to be slightly more intensive (though not to the extent of a typical mainland arable farm). Direct agricultural subsidy to the Western Isles was almost half of total output over the period 1995–1997 (Black and Stubbs, 1998), and though this subsidy is informally linked to a form of social support in the rural communities of the Highlands and Islands, agriculture is nevertheless the primary delivery mechanism, and as such the support must be developed within the framework of the Common Agricultural Policy. In the Uists and Barra and in the Argyll Islands, the environmental value of the land is reflected in their designation as Environmentally Sensitive Areas (ESAs), where enhanced payments are available to those who voluntarily undertake to manage their land using methods believed to be particularly environmentally sensitive. Elsewhere, the Countryside Premium Scheme (CPS) operates similar environmental measures, though financial support has so far been rather more difficult to obtain than has been the case in the ESAs.

Where the agricultural activity moves from cattle to sheep, or even towards methods of cattle rearing which do not involve the local growth of winter fodder, there is a corresponding decline in the natural heritage value of the land. With limited return from cattle, especially following the alarm over Bovine Spongiform Encephalopathy (BSE), the incentive to continue with this agriculture is diminishing, and Uist-based staff of Scottish Natural Heritage are currently working with local crofters to devise a scheme to stem the decline in cattle numbers. It cannot overlap with existing agri-support mechanisms and thus has to focus upon the recognized environmental benefits that cattle produce. This is intended to demonstrate that cattle production must be sustained for its conservation value to the machair and will hopefully influence subsequent policy decisions over the wider machair area (John Love, personal communication, 1998).

In association with (and with grant aid from) Scottish Natural Heritage, the Royal Society for the Protection of Birds devised a Corncrake Initiative, whereby payments are available for crofters to delay cutting of hay or silage until a defined date, with additional payments for mowing the crop in a 'corncrake-friendly' pattern which allows birds to escape to adjacent ground. The operation of this voluntary scheme requires very close liaison with crofters, to the extent that the RSPB appoints seasonal officers devoted entirely to this work, and take-up has been gratifyingly high, as has been the success of the scheme. Similar measures are also available through the ESAs and CPS. SNH has recently introduced a pilot approach to Special Protection Areas for corncrake in the Western Isles, where SSSI notification is not involved, yet financial incentives are available for managing the land in such a way as to encourage corncrakes (SNH press release, 6 October 1998).

The winter gales off the Hebrides deposit vast amounts of 'tangle' *Laminaria hyperborea* (a seaweed) on the beaches of the western seaboard. This has long been spread on the machair and ploughed in, where it provides not only much-needed nutrients but also a binding agent, essential in soils which blow all too readily when dry. This work is labour-intensive, and there had been a trend towards commercial fertilizer, but this has been reversed by a range of incentives delivered through mechanisms such as the ESAs.

Tourism

A strategic study of tourism in the Western Isles was carried out by TMS Consultants in 1995 on behalf of the Western Isles Tourist Board, with financial assistance from SNH, the Western Isles Council and Western Isles Enterprise, the Local Enterprise Company. The TMS study revealed that tourism had become one of the main industries of the Western Isles. In 1994 it contributed £12 million annually to the local economy, roughly on a par with Harris Tweed (TMS, 1995). It was then believed to be growing, but experienced the same downturn in 1998 as other areas of the UK.

According to the TMS survey, of visitors questioned in the ferry ports of Stornoway, Lochmaddy and Tarbert, 83 per cent took part in 'countryside/beach walks', 65 per cent in 'bird-watching', 67 per cent in 'watching seashore wildlife', 55 per cent in 'observing wild flowers' and (for comparison) 60 per cent in hillwalking. 'Scenery and landscape' was the most important single factor in attracting tourists to the Western Isles (38.9 per cent – second most important category was 'remoteness' with only 12.4 per cent). Though machair itself cannot be separated from these figures, it is such an integral part of the scenic and recreational setting of the Western Isles that it must be of immense importance to the local economy. More surprisingly, a high percentage (60 per cent) of visitors questioned had some awareness of the meaning of the term 'machair'. This led the Western Isles Tourist Board to highlight this 'unique selling point' of the Western Isles by naming its environmental brochure (one of a series of four) *The Machair Islands.*

As environmental awareness tends to be high among island tourists, such problems as exist are more related to tourism than to tourists. There are minor exceptions to this; for example, where drivers can gain access to the wider machair surface, significant numbers tend to park at the seaward machair edge so as to enjoy the view, and such off-road driving contributes to a wider machair erosion problem (Angus and Elliott, 1992). The most extreme example of this type of damage is possibly on Tiree, where some windsurfers drive to the machair edge, then leap down the eroding machair front to gain access to the beach. In the Western Isles, there has been a cooperative effort backed by EU Objective One funding linking agencies and communities aimed at alleviating minor problems of beach access before they become major, as well as a general attempt to improve *planned* vehicle and pedestrian access to beaches. Otherwise the main problem is wild or poorly organized caravanning, as at Horgabost in Harris. This beach has unusually calm water for the Atlantic coast, and has a car park and access road funded by the EU Access to Beaches Programme to improve access and amenity. Unfortunately the improved access has stimulated a proposal for an organized caravan site covering most of a machair which occupies a highly strategic and conspicuous position in the scenic context of South Harris, which is part of a National Scenic Area. A suggestion from SNH that a nearby disused and unsightly sandpit could be reinstated as a planned caravan site, thereby meeting two objectives, has met with a lukewarm reception locally. There is a real possibility of damaging the wider tourist industry of Harris by providing caravan facilities in such a popular and conspicuous position.

Conservation Designations

In Scotland a single area of land may be covered by National Scenic Area (NSA), Site of Special Scientific Interest, Special Protection Area, Special Area of Conservation, Ramsar site, and even National Nature Reserve. Though it may be argued that in practical terms only two systems of natural heritage designation (NSA and SSSI) operate the entire list, this is not necessarily apparent to a crofter with a fistful of confusing papers.

Island humorists have long held that a croft may be defined as 'an area of land surrounded by regulations'. The multiple designation approach of natural heritage conservation legislation has unquestionably added to the existing bureaucratic burden, but is an inevitable adjunct to the national and European funding mechanisms which have significantly benefited the machair areas, and which increasingly have compulsory environmental conditions.

During the first five years of the existence of Scottish Natural Heritage (1992–97), the media gave considerable attention to the enhancement of the existing SSSI network in the Western Isles to include European designations and a few cases of new SSSIs, also associated with SAC or SPA notification. Journalists have always highlighted the more colourful complaints about the perceived impact of SSSIs, and this is now expected. However, the (unavoidable) timing of this programme of SSSI notification coincided with the lodging of a planning application for a coastal superquarry at Lingerabay, in southeast Harris. Though Lingerabay is some 7–8km from the nearest machair, the superquarry issue came to dominate the entire context of natural heritage conservation in the Western Isles over several years. SNH was the only statutory objector to the application and, despite a 2:1 vote against the application in the Harris plebiscite supervised by the Electoral Reform Society, the local authority was minded to approve the application, and the Secretary of State for Scotland decided to hold a Public Local Inquiry (PLI) into the application. Though conservation bodies other than SNH became involved in the PLI, SNH was the main target for some of the more extreme criticism in a situation which became highly polarized.

During the media coverage of the activities of SNH during this period, most of the SSSI-related attention focused not on the ability of SNH to object to planning applications (a right which any member of the community may exercise) so much as on the list of 'Potentially Damaging Operations' (PDOs) which form part of an SSSI notification and which apply mainly to land use outside normal planning control. Though in practice existing operations are almost invariably granted a 'consent' to permit them to continue without further bureaucracy it must be conceded that the list might *appear* intrusive and restrictive. Though SNH devotes massive amounts of staff time to explaining this procedure to notified parties and attempting to allay (understandable) fears, inevitably some unease persists, and is brought to the attention of the politicians and the press. The media (both broadcast and print) have never expressed much interest in reporting the extensive co-operation between conservationists and community, but have always been ready to highlight apparent difficulties. This attitude on the part of the media, accompanied by a highly charged political situation, resulted in an over-reaction which at times became hysterical. The embers of unease and discontent, which could have been dampened by reporting the real facts, were instead furiously blown upon to produce something of a blaze of unease within the community about the operation of designations, most particularly the PDOs.

During this five-year period, with many hundreds of crofts covered by up to 27 PDOs, the number of applications for consents was very low indeed, and the number of consents declined by SNH in the Western Isles was... one! In short, there has been a negligible negative effect of the PDOs on crofting management in the Western Isles. There has, however, been a negative effect of the publicity on the work of SNH with the community, where the skills of officers have had to be diverted from positive community work to dealing with the heightened unease about designations. The losers in this scenario, as is so frequently the case, are the people of the Western Isles.

With the designation of substantial areas of machair under the European Birds Directive (SPAs) and the Habitats and Species Directive (SACs), European-wide conservation legislation and policies may be applied at the local level. It may be difficult to communicate the unique dependence on active management and the close relationship between general economic activity and habitat quality to policy-makers in

Strasbourg or Brussels, but such communication is required. Pragmatism is more evident in the terms of the two relevant Directives than is generally realized (Anon., 1998), but it would be possible for someone without knowledge of the area and its people to apply the regulations more rigidly than is required to 'protect' machair by halting some initiative which is actually essential to its survival.

Attitudes to Conservation

The economic importance of the natural environment to tourism is now beyond dispute, thanks to the TMS study. Press reports about designations might suggest that local people have a more negative attitude to the environment, but there is now a substantial body of evidence testifying to a deep appreciation of the environment on the part of local people.

Against the background of the superquarry proposal, the Harris community initiated an appraisal of community attitudes and aspirations, part-funded by SNH (Rennie, 1995). Though conservationists had long been aware of a great appreciation of the Hebridean natural heritage on the part of local people, even they were taken by surprise by the strength of these feelings relative to feelings about the community, religion and employment.

Subsequent community appraisals, some with financial support from SNH, were conducted in Uig (southwest Lewis), Barvas (western Lewis), Ness (north end of Lewis), and at the south end of South Uist. The Ness appraisal concluded that the most important reasons for living in Ness were (in order of preference) family connections, marriage, environment, employment, housing, community. In Uig (where there are significant areas of machair), the best things about living there were (also in order of preference): lack of crime, peace and quiet, environment and scenery, good for bringing up children, people, quality of life, sense of community, sense of belonging, Christian community. Similar responses were noted in the other appraisals.

The questions were formulated and in most cases the responses analysed by the communities themselves. Although this means that the answers are not directly comparable the fact that the conclusions are unquestionably community-generated far outweighs this disadvantage. Overall, there is a strong, unequivocal regard for the natural environment, in most cases backed up by a willingness to defend that environment. This defence has been demonstrated many times at the local level, as at Northton (southwest Harris), where the community has excluded vehicles from the machair to minimize erosion, and at Brue (western Lewis), where the grazings committee blocked access to lorries operated by a local contractor who wanted to remove boulders to sell to garden centres, on the grounds that the vehicles would cause machair erosion. In both cases, there has probably been a financial disadvantage to the communities involved, but the appraisals and the Harris plebiscite suggest that communities in the Western Isles place environmental considerations ahead of financial gain. The Barvas study also revealed that the majority of local residents wanted to stop commercial sand extraction, to ban dumping on the machair, and restrict vehicle access to the machair. More surprisingly (for a substantially crofting community), there was a feeling that grazing should be further restricted.

Discussion

Though the agriculture of the machair areas appears to involve methods and machinery which have long been abandoned elsewhere, and there might be a tendency to regard this as an anachronism in a country where food production involves an economy of scale and technology, these methods are particularly suited to the circumstances. The light soils are stabilized and fertilized using a readily available organic medium, and the methods support a rich flora and fauna, as well as a dense rural human population. Though the benefits are evident, the pressures on these islands are many, and their economy is fragile. Monitoring of the machair is essential to ensure that the subtle changes which occur in any society do not result in a loss of machair value, and that any negative change can be addressed. Thus conservationists are obliged to maintain an unusually high level of knowledge of the social and economic aspects of the area, as well as particularly good relations with the community, in addition to the usual skills required of such posts. This is easier in the Western and Northern Isles where conservation workers live in the community but is more difficult in the Argyll islands and some of the northern Inner Hebrides, where conservation officers are based in mainland towns. Indeed, Scottish Natural Heritage is the only government or local government agency which has its Uists base outside the administrative centre of Balivanich in Benbecula; the SNH Uists and Barra office is in Stilligarry, in the heart of the machair of South Uist.

The machair areas are associated with high unemployment. There is a range of initiatives aimed at alleviating this situation, usually operated by committees where SNH is not only represented but active in promoting economic development, to the extent that even committees with little or no environmental remit often seek out SNH staff for their skills and commitment in this field. SNH staff are also in demand locally as academic 'experts', and form a valuable asset to the community in an area where there are presently no universities, though there are plans to link the existing further education college network as a University of the Highlands and Islands.

The exceptionally high conservation value of Scottish machair depends to a large degree on well-informed and carefully targeted advice to land users. Most of all, however, it depends on the land users themselves and, despite the many designations, conservationists can only meaningfully intervene with the consent of local people. It is thus essential that the staff of the conservation bodies devote as much – possibly more – time to listening to local land users and their concerns as they devote to studying the habitat itself. In recent years, the conservation bodies have become increasingly aware of this need, and have appointed staff with a positive and constructive approach to working with communities.

Even the best staff, well armed with information, can only achieve what time, finance and policies permit. Working effectively with communities is immensely time-consuming, in terms of both the length and the amount of time which initiatives may take to come to fruition, but this time is rewarded by the correspondingly greater effectiveness of such co-operation. Inaccurate and irresponsible reporting of the work of the conservation bodies may impact significantly on the already small amount of time available for such work. One example of positive community work is the Western Isles Mink Control Group, devised and constituted (and partly funded) by SNH in response to local aspirations, with its operation significantly delegated to the community. As well as reducing numbers of feral mink in Lewis and Harris and their impact on ground-nesting birds, poultry and wild and farmed fish, the possibility of mink establishing a population on the Uists, where there are internationally important numbers of machair-nesting birds, is significantly reduced (Angus, 1993). SNH staff have also devised a rabbit control programme for parts of the South Uist machair, which will have major agricultural benefits for the community if it is successful.

Policies are usually determined at a national level. The machair is geographically peripheral, and the local staff of bodies such as SNH, the RSPB and the Scottish Office Agriculture, Environment and Fisheries Department, as well as development-oriented staff of the local authorities and Local Enterprise Companies, constantly attempt to inform and persuade the policy makers of the special needs of the crofting areas. Environment features increasingly in the discussions of even the development-related organizations, and generally there seems to be a good awareness of the problems of the machair communities. So long as everyone, not just conservationists, listens to the communities involved, then there is more likelihood of the right policies and adequate finance being applied to their problems.

Crofters and others in the community possess a formidable body of knowledge about the history and ecology of machair and some of the more important species. Occasionally expensive scientific studies are mounted which reach conclusions long known to the local population. In fairness, it should be added that sometimes the scientific analyses overturn local perceptions, but there is an underlying need to devote more time to the collection of local knowledge. While it cannot replace the rigour of scientific study, it can offer useful pointers to the direction and methodology of research, possibly shortening the research process. More importantly, local traditions offer a body of information which no scientific analysis can replicate, yet which can offer valuable insights when applied as a context to conventional scientific research.

The combination of the Harris plebiscite and the community appraisals confirm a very strong local affinity with the natural environment (even if this is not matched by an enthusiasm for natural environment designations), and machair is a component of all areas so far covered by such appraisals. If the politicians and the press have so far failed to appreciate the strength of local environmental feeling, it could be argued that the conservation organizations have also been slow to react, in that they have done little to facilitate or enable the organization of this highly motivated affinity for the natural heritage. It is regrettable that the common factor here is the designation programme. If SNH was not diverting so much of its attention (as it must, by law) to implementing this, then there would be much more effective communication and convergent planning which would benefit both community and environment. That the designations are distracting the press and the politicians is doubly damaging, in that they make significant inroads into the remaining time available for positive conservation. It would be facile, however, to suggest that if the designation programme were to halt or even slow down, all would be well. The increasing number of co-operative groups designed to effect change makes increasing demands not only on the staff of the agencies involved (who are paid to attend the meetings) but on the time of local representatives (who are not paid). As is so often the case in rural communities, the burden tends to fall on the same individuals, while the burgeoning trend towards community consultation (which has still not reached the level it has elsewhere through Agenda 21) is resulting in 'consultation fatigue' in some areas. It is not suggested that this be used as an excuse to avoid consultation, but that the consulting bodies be more co-ordinated in this. With the best will in the world, there will always be communities where internal personality conflicts and pig-headedness cloud the issue, or where some peripheral entrepreneur sniffs an opportunity. Although these difficulties can sometimes be defused by good chairing, there will always be a few cases where the wreckers prevail. These cases should never be used as an excuse for not working with communities, especially for machair, where there is no viable alternative.

It is worth reflecting on the unavoidable truth that while the environmental awareness of local people is unquestionably high, the national policies which affect the machair areas are at least partially (but only partially) driven by the environmental aspirations of urban populations, and these aspirations may be based on a lack of knowledge or simply differences of perception. Killing mink and rabbits may not always be to the liking of those in the cities. Taken to extremes, a city-centred philosophy of conservation of the machair environment rather than one based on the aspirations of the inhabitants could give rise to a 'living museum' in the island countryside. No inducements, however large, aimed at producing such a quaint caricature of the Gaelic peasant could ever succeed. Crofting in the machair areas can be very hard work, and the high involvement of those who already have full-time jobs should signal the importance of something beyond money. There is a spiritual relationship with the land which cannot be adequately described, let alone quantified, but by spending time in these environments and listening to those involved, sensitive, relevant, and therefore *effective* conservation policies may be developed for the maintenance and enhancement of these important areas. In essence, one of the most important tools of conservation is conversation.

Conclusions

- Natural heritage conservation of Scottish machair grassland cannot be separated from the wider machair system, the township as a management unit, and the socio-economic context of the wider areas in which machair occurs.

- Local people have a very high awareness of machair landscape and wildlife and a high commitment to its conservation.
- Directly working with local people is the most effective means of achieving positive natural heritage conservation benefits in the machair areas, but sufficient time and financial resources must be allocated for this work. The perceptions, aspirations and experience of local people must be accorded due respect in all discussions relating to machair management.
- Meaningful conservation of machair requires careful, genuinely integrated planning across a range of remits, and the full involvement of a multitude of public agencies as well as local people, in the true spirit of Agenda 21.
- SPA and SAC designations are inevitable, and offer significant opportunities for the machair areas. The principle of these designations is already internationally accepted; debate should concentrate on their operation.
- Media and political highlighting of perceived (rather than actual) problems with designations has eclipsed not only the benefits of the designations, but the wider benefits of natural heritage conservation. Though local people distinguish between natural heritage designations and the natural heritage itself, this is not necessarily appreciated more widely.
- European funding is vital to the machair areas and is always linked to environment. The machair areas will move forward more quickly and more easily if this is accepted and if a policy of mutual co-operation is pursued: conservation and conservationists have much to contribute if they are not too distracted by misguided allegations.

Acknowledgments

I would like to thank my colleagues in SNH for their helpful comments on previous drafts of this paper, especially Sandy Maclennan (Inverness) and David Maclennan (Stornoway). John Love (South Uist), Bill Taylor (Inverness), Greg Mudge (Fort William), Mary Elliott (Fort William), Lesley Cranna (Skye), Alison Brown (Oban), and John Baxter (Edinburgh) also commented on the paper. I am also grateful to Christine and Zena Macritchie of the SNH office in Stornoway who helped in locating information and to Carol Fraser of the HIE Business Source in Inverness who was most helpful in tracing unpublished reports.

References

Angus, S. (1993), 'A mink control programme for Lewis and Harris', *Hebridean Naturalist*, **11**, 78–84.

Angus, S. (1994), 'The conservation importance of the machair systems of the Scottish islands, with particular reference to the Outer Hebrides', in M.B. Usher and J. Baxter (eds.), *The Islands of Scotland: A Living Marine Heritage*, HMSO, Edinburgh, 95–120.

Angus, S. and Elliott, M.M. (1992), 'Problems of erosion in Scottish machair with particular reference to the Outer Hebrides', in R.W.G. Carter, T.G.F. Curtis, and M.J. Sheehy-Skeffington (eds.), *Coastal Dunes: Geomorphology, Ecology and Management for Conservation*, A.A. Balkema, Rotterdam, 93–112.

Anon. (1998), 'Natura 2000: dispelling some of the myths', *Natura 2000 (European Commission DG XI's Nature Newsletter)*, Issue 5, February 1998.

Barvas and Brue Community Appraisal (1997), unpublished report, April.

Black, S. and Stubbs, R. (1998), 'Highlands and Islands agriculture – quality is the way ahead', *Executive Magazine*, No. 155, July, 18–19.

Hunter, J. (1976), *The Making of the Crofting Community*, John Donald, Edinburgh.

Ness Community Appraisal (1997), unpublished report.

Rennie, F.W. (1995), 'The Harris Integrated Development Programme', unpublished report to the Harris IDP Steering Group.

Ritchie, W. (1976), 'The meaning and definition of machair', *Transactions of the Botanical Society of Edinburgh*, **42**, 431–40.

Southend Community Appraisal (1997), unpublished report.

TMS Consultants (1995), 'Western Isles Tourism Strategic Development Plan', unpublished report to Western Isles Tourist Board, Scottish Natural Heritage, Western Isles Council and Western Isles Enterprise, Stornoway.

Uig Community Appraisal (1996), unpublished draft report.

GOLF COURSES AS CATALYSTS FOR CONSERVATION OF COASTAL HABITATS

DAVID STUBBS

European Golf Association Ecology Unit, England

ABSTRACT

Effective nature conservation in Europe involves much more than protected areas and regulations. Current emphasis in conservation policy recognizes the importance of establishing partnerships and voluntary agreements with land users in different sectors. These points are central to the European Union's Fifth Environmental Action Plan, including application of the Habitats Directive, and are also a strong component of the Pan-European Biological and Landscape Diversity Strategy. Golf courses offer one example of this approach in action.

In Europe there are over 5200 existing golf courses, covering in excess of 250,000 hectares. Although concentrated in the British Isles, they do occur throughout the Continent, from northern Scandinavia to the Mediterranean. They occupy a wide variety of habitats, including coastal dunes, maritime grasslands, heathland, downland and parkland. Coastal situations are particularly favoured, and it was in these areas that the game originally developed. A typical 18-hole golf course may occupy 50–60 hectares of open space, much of which receives little direct management or disturbance. Such sites can contain important fragments of semi-natural vegetation, often with a high biological diversity. On some coastal sites, golf courses have provided a last line of defence against encroaching urbanization, tourist development or intensive agriculture. As managed environments, golf courses are ideal for active conservation measures, managing or restoring habitats, and creating new features.

The European golfing community's contribution in this sphere is through the work of the European Golf Association Ecology Unit and its new campaign 'Committed to Green'. This builds environmental awareness among golfers and golf course managers, and encourages them to participate in developing Environmental Management Programmes for their courses. A key part of this initiative involves liaison with local environmental specialists, to help clubs identify and address conservation priorities.

While remaining aware of potential impacts through disturbance of fragile habitats, abstraction of water and use of fertilizers and pesticides, conservationists should recognize the excellent potential of golf to play a significant role in promoting nature conservation to a substantial and receptive audience outside traditional conservation targets. This is especially important in areas of mixed land use with high nature-conservation value, such as coastal habitats.

Introduction

Golf represents a significant land use in Europe. There are over 5200 golf courses, many on coastal sites, occupying a substantial area of land across a wide range of environments. An average 18-hole golf course may cover between 50 and 60 hectares, giving a continent-wide aggregate of some 250,000 hectares, almost the size of Luxembourg. Within this area less than one third is likely to be intensively managed sports turf, leaving significant parcels of virtually unused land potentially available for conservation management.

This is particularly important in the context of taking conservation into the wider environment and engaging other sectors and land uses in the process. The conservation of biodiversity across the region cannot be achieved solely through protected areas and legislation. Such recognition is implicit in the EU Fifth Action Programme for the Environment, Towards Sustainability (European Commission, 1993), in which emphasis has shifted from a regulatory approach to one embracing partnerships and shared responsibility, in particular across different sectors.

Likewise, the Pan-European Biological and Landscape Diversity Strategy (Council of Europe et al., 1996) adopts a broad-based approach for the conservation of biodiversity, including integrating objectives into different sectors, and emphasizing the importance of raising awareness. Among the key sectors, tourism and recreation are highlighted.

In parallel with these environment-driven initiatives, the golfing community has itself become more aware of environmental issues in recent years. Among all the major sports, golf is the one which is most closely bound up with the natural environment as an inherent feature of the game (Royal and Ancient Golf Club of St Andrews and European Golf Association, 1997). Since January 1994 the European Golf Association (EGA) has maintained a full-time Ecology Unit to serve as a clearing house and provide technical advice on environmental issues related to golf. This novel approach is opening up exciting opportunities to forge effective partnerships between conservationists and golf club managers (see European Golf Association Ecology Unit, 1995a).

Golf Courses in the Coastal Environment

There are many hundreds of golf courses in coastal areas. Most of these are long established, 'links' courses, often dating back over 100 years. The main concentrations of coastal courses are in the British Isles. There are smaller numbers in Scandinavia, The

Netherlands, Belgium, France and Iberia.

In southern Europe the number of coastal golf courses has exploded in recent decades with the advent of golf tourism. These new resorts attract the majority of their golfing clientele during the winter, enabling year-round tourism to flourish in these areas. Such developments have had a major impact on the coastal environment but it would be wrong to single out golf as the only factor. It is a complex equation between socio-economic development and environmental protection, in which golf is but one part. Indeed, in areas of intensive development, golf courses may provide the only sizeable patches of semi-natural green space. Small compensation, perhaps, for the modification of original coastal habitats, but one must evaluate the situation in its wider context.

Established Golf Courses

Links golf expresses the traditional ideal of the game. The image is of natural dune landscapes, interwoven with fairways and greens, using the naturally occurring fine turf fescues (*Festuca* spp.) and bent grasses (*Agrostis* spp.). Originally these courses were virtually unmodified, making use of the natural topography and vegetation. Most of them still retain this essential character today although, after several generations of use as golf courses, they have evolved and transformed to some extent from their initial state.

Nevertheless, a large number of links courses, both in the British Isles and in continental Europe, are recognized for their nature conservation value. They support valuable patches of coastal vegetation, notably dune slacks and fixed dune communities. Rare herpetofauna, birds and invertebrates are associated with links courses.

An important factor is that the presence of golf has safeguarded many sites from more intrusive development, for example agriculture, forestry, caravan parks or urban development. Until the relatively recent advent of nature conservation laws and statutory designations for protected sites, the coastal environment was at the mercy of unrestricted development. The reason that so many links golf courses are of high nature conservation value is twofold:

1. in many cases they represent all that is left of the original natural dune habitats; and
2. their management over the years has been conducive to retaining the site's ecological value.

In other cases, golf courses serve as a vital ecological buffer between urban or agricultural encroachment and areas of coastal habitat. This may give a clue to future development potential.

New Golf Course Developments

Although golf can claim credit for conserving many important coastal sites, this would not be the case if today's nature conservation and planning laws had been in place 100 years ago. Nature conservation is now a land use in its own right and our best natural sites need, and can obtain, protection from any form of development.

In other situations some form of economic use of a site is the best way of maintaining its favourable conservation status. In this respect golf courses have significant potential. The choice of site is critical, however. Modern construction methods are not compatible with pristine, fragile sand dune habitats. One also has to assess whether the water requirements for golf course irrigation would impact on the hydrology of the site in question and its surrounds.

The EGA Ecology Unit (1995b) has produced environmental guidelines for new golf course development. These provide a general approach for all projects, not specifically those in coastal areas, but the same principles apply. The emphasis is on thorough appraisal (environmental, economic and technical) prior to final site selection, or the decision whether or not to proceed with the project. Where development is feasible and environmentally compatible, the guidelines go on to specify appropriate environmental assessment criteria and a step-by-step integration of environmental aspects into the development process.

A good golf course can be designed and built to accommodate existing environmental constraints – the golf course must fit into the landscape, not make the landscape fit the golf course. Once it is established, a golf course has to be managed on a continuous basis. That in turn provides opportunities for effective long-term environmental management of the site. Modern greenkeepers are not merely sports turf managers. They are enthusiastic and professional land managers in the full sense.

Getting the Message Across

Educational opportunities are also apparent. Golf courses are used by many people throughout the year. For many players this is their contact with nature. In some situations, where space and safety considerations permit, there are public pathways through golf courses. When it comes to major championships, the audience is enormous – tens of thousands of people on site, and millions of radio listeners and TV viewers.

The highest-profile golf event associated with the coastal environment is the Open Championship, organized each year in mid-July by the Royal and Ancient Golf Club of St Andrews (R and A) on a links course in either England or Scotland. All of the Open sites are on dune environments, many of them are Sites of Special Scientific Interest, and in some cases Special Areas for Conservation under the Habitats Directive.

One such case is the Royal Birkdale, northwest England, where the 1998 Open Championship was held. The R and A and the host club worked in close co-operation with the two statutory bodies, English Nature and the Environment Agency, as well as with the Sefton Coast Life Project (see Rooney, this volume), to ensure that the site was properly safeguarded. Physical measures were introduced on-site to keep temporary structures and spectators away from the most sensitive habitats.

The 1998 Open Championship was also an occasion for promoting the conservation message to a wide audience. The partners listed above collaborated on a booklet entitled, 'A Hole by Hole Guide to the Wildlife and Conservation of Royal Birkdale'. Copies were given to all journalists, TV and radio commentators attending the event. The environmental information was used as supplementary material throughout the four days of live televised broadcast. In this way the story of dune habitat restoration and the connection between conservation and links golf was broadcast worldwide. Green issues were seen to be part and parcel of making Royal Birkdale a special championship site.

Golf offers many opportunities to promote conservation. There are many diverse initiatives taking place across Europe on this theme. On the other hand, there are still situations arising where golf and the environment are seen as a problem, not as mutually compatible interests. What we must avoid is the traditional, adversarial pressure-group approach. Instead, what is needed is a coherent programme to draw all these strands together to provide a credible measure of good environmental practice in golf. By taking a positive approach one can expect to achieve much greater co-operation from the golfing community.

The 'Committed to Green' Programme

'Committed to Green' is a new initiative by the European Golf Association Ecology Unit, arising from a pilot project co-financed by the European Commission in 1995 (European Golf Association Ecology Unit, 1996). The campaign was launched on 28 September 1997 by European Commission President, Jacques Santer, during the Ryder Cup Matches held at Valderrama, southern Spain. To be 'Committed to Green' is to demonstrate environmental awareness and responsibility. The campaign encourages everyone involved in golf to participate in improving the environmental quality of golf courses.

'Committed to Green' aims to show how well-managed golf courses can be of benefit to the environment and the community. At the same time this will help golf clubs by offering cost-effective means of enhancing environmental quality without compromising turfgrass quality and playing conditions.

There are initially three targets for the 'Committed to Green' campaign:

1. Existing golf clubs: to encourage improved environmental performance;
2. New golf course projects: to ensure they are developed in accordance with appropriate environmental guidelines;
3. Major golf events – to use the strong media interest to promote environmental awareness to support 'Committed to Green' in the wider golfing community.

Real improvements in environmental performance will only be achieved through the active involvement of Europe's 5200 golf courses. The first priorities are to build awareness and to encourage golf clubs to participate. 'Committed to Green' is open to all types of golf facility on a voluntary basis. Emphasis is on continual improvement. Clubs which implement a full Environmental Management Programme can qualify for 'Committed to Green' recognition. To guarantee objectivity and credibility, the setting of environmental criteria and awarding of recognition will be subject to independent verification.

The process follows five steps:

1. Policy. To participate, a golf club will first need to make a policy commitment endorsed by its membership. This should include a statement of intent to improve environmental performance, to establish a 'Committed to Green' team to manage the project, and to carry out an environmental review.
2. Environmental Review. This looks at the current environmental situation and forms the basis for introducing appropriate conservation measures.
3. Environmental Management Programme. A comprehensive, integrated management plan, combining environmental and golfing objectives.
4. Audit. After a maximum of three years, progress will be evaluated to assess whether initial targets have been achieved.
5. Recognition. A 'Committed to Green Award for Environmental Excellence' will recognize and support golf clubs that have made significant achievements across the eight categories listed below.

   Nature conservation;
   Landscape and cultural heritage;
   Water resource management;
   Turfgrass management;
   Waste management;
   Energy efficiency and purchasing policies;
   Education and the working environment;
   Communications and public awareness.

The EGA Ecology Unit is currently working with the International Institute for Industrial Environmental Economics at Lund University, Sweden, to benchmark Environmental Performance Indicators for the 'Committed to Green' programme. These will form an objective basis for measuring progress, both internally and between clubs.

National Ecology Officers attached to their respective National Golf Federations will provide the direct interface with golf clubs, thereby ensuring a relevant, local approach to implementing good environmental practice. Clubs which eventually achieve recognition at the European level will be expected to continue to follow the programme in order to maintain their 'Committed to Green' status.

Conclusions

The full environmental load of a golf course operates across a wide range of subjects. The intention behind 'Committed to Green' is to draw all these different strands together into a coherent, integrated approach to environmental management. This operates at the golf course level and at the whole golf club level, and also provides for interaction with local community interests. For the first time we have a holistic approach to environmental issues related to golf courses. Instead of conservation being perceived as a marginal interest, it can now be seen as an integral part of the operational management of a golf facility. This draws in external relationships plus local socio-economic and environmental factors.

This is especially important in coastal areas, where golf courses have long been established, and may be a primary economic driver for the local tourism industry. The various activities on coastal golf courses cannot be treated in isolation. To achieve sustainability in coastal zone management, one must engage all land users and local interests.

The golf community is taking environmental issues seriously. There is still much to be done, and we must remain aware of potential impacts through damage to fragile habitats, abstraction of water and over-use of fertilizers and pesticides. However, it is important for conservation interests to recognize that golf clubs can be effective partners in environmental protection.

Golf can be a catalyst for promoting nature conservation and general environmental awareness to a substantial and receptive audience outside traditional conservation targets. 'Committed to Green' offers the vehicle to make this happen.

References

Council of Europe, United Nations Environment Programme, European Centre for Nature Conservation (1996), *The Pan-European Biological and Landscape Diversity Strategy.*

European Commission (1993), *The Fifth Action Programme for the Environment, Towards Sustainability.*

European Golf Association Ecology Unit (1995a), *An Environmental Strategy for Golf in Europe*, Pisces Publications, Newbury.

European Golf Association Ecology Unit (1995b), *Environmental Guidelines for Golf Course Development in Europe* (revised edition), EGA Ecology Unit, Brussels.

European Golf Association Ecology Unit (1996), *An Environmental Management Programme for Golf Courses*, Pisces Publications, Newbury.

European Golf Association Ecology Unit (1997), *The Committed to Green Handbook for Golf Courses*, Pisces Publications, Newbury.

Royal and Ancient Golf Club of St Andrews and European Golf Association (1997), *A Course for All Seasons: A Guide to Good Management*, Royal and Ancient Golf Club, St Andrews.

MILITARY LAND USE, SAND DUNES AND NATURE CONSERVATION IN THE UK

COLONEL JAMES BAKER
Ministry of Defence, UK

Introduction

The Ministry of Defence (MOD) estate has developed over the last 150 years in response to contemporary military requirements. Principally this has involved progressive increases in the range and killing power of weapons and advances in the manoeuvrability of vehicles. The current estate consists of an area of nearly 240,000ha involving some 3200 separate sites. They range in size from the 38,000ha expanse of Salisbury Plain training area in southern England to individual small buildings and communication masts. The estate includes over 200 Sites of Special Scientific Interest and thousands of ancient monuments, many of them scheduled. These and other designations present the MOD with the major challenge of the management and use of a considerable natural and cultural resource.

Following the Report of the Defence Lands Committee 1971–73 (The Nugent Report), a full-time Conservation Officer was appointed. In response to the ever-increasing emphasis on conservation matters, the number of staff in the MOD Conservation Office has increased. The office is responsible for looking after what is one of the finest estates for wildlife in any single ownership in the United Kingdom. The office also holds a watching brief over areas of private land wherever troops are training within or outside the UK.

The widespread deployment of the three Services (Army, Royal Air Force and Royal Navy) across the country means that the estate now includes examples of the main indigenous habitats, including sand dunes, heathland and saltmarshes, and with them, an immensely wide diversity of species, many of which are rare or endangered. Due to the nature of military training many of these areas have escaped intensive farming, agrochemical sprays and major development.

The Conservation Office

The Conservation Office has several main areas of responsibility:

- co-ordination of the conservation of natural history and archaeology, mainly through the activities of local military conservation groups;
- co-ordination of integrated conservation management plans;
- education of military and civilian staff and increasing the awareness of the general public of the MOD's conservation effort;
- liaison with other government departments, statutory bodies (e.g. English Nature) and non-governmental organizations (e.g. RSPB);
- advising MOD ministers on conservation policy, in conjunction with other MOD branches;
- supporting the Services Branch, British Deer Society.

The MOD Conservation Office maintains close links with the regional network of Defence Land Agents, who provide the main local advisory and estate management service to the military users of the estate. The Conservation Office also organizes a committee that brings together the three Services, the Territorial Army (responsible, for example, for the Altcar Rifle Range estate on the Sefton Coast) and the Defence Evaluation and Research Agency (responsible, for example, for the Eskmeals Range in Cumbria). Another major committee brings together the MOD and the national statutory bodies to exchange ideas and views.

Land Management Priorities

The MOD retains land for military training in the interests of national defence. Although this must remain the primary purpose for land use, the MOD places priority on planned and sympathetic management of the estate. Conservation of natural history, archaeology and the environment is afforded an equal priority with agriculture and forestry. Public access is also important and is allowed wherever it is compatible with operational, safety, security and conservation needs, and those of tenant farmers.

Nature Conservation Designations

MOD conservation policy takes account of all relevant UK and European legislation, including the UK Wildlife and Countryside Act 1981, as amended, and the UK Conservation (Natural Habitats etc.) Regulations 1994, which incorporates the EU Birds Directive and Habitats Directive.

The MOD owns or leases many sites that have been designated as areas of high conservation value under these and numerous other forms of legislation, including Sites of Special Scientific Interest, Areas of Outstanding Natural Beauty, National Nature Reserves, Ramsar (Wetlands of International Importance) sites, Special Protection Areas and candidate Special Areas of Conservation (cSACs), and it seeks to carry out its operations within the prescriptions of these designations.

Conservation Groups

Establishing an appropriate balance is achieved through local conservation groups. These groups cover the main sites where there is a significant conservation value and are an essential part of the MOD conservation effort. The groups contain both military and civilian personnel. The conservation group for the Altcar Rifle Range estate on the Sefton Coast is a good example of such a group. Chaired by a retired Officer, it brings together the estate managers and statutory agencies, naturalists and local authority conservation officers to work towards a conservation strategy for the area. The group helps with surveys, recording and monitoring, all information being added to a site dossier, a compilation of relevant information for the conservation of the site. Throughout the defence estate the advice of conservation groups is crucial to the active management and planned use of the land.

Relationships with Statutory Bodies

Declarations of Intent have been agreed with English Nature, Scottish Natural Heritage and the Countryside Council for Wales. These Declarations recognize formally the MOD's difference from other landowners in that the primary role for its land must be the preparation of defence forces. They also reaffirm the MOD's commitment to preserve or, wherever possible, enhance the natural history value of its sites of significant interest. The MOD has also signed a Declaration of Commitment to National Parks.

In addition to the close links with the statutory bodies, the MOD maintains contact and liaison with many other national and international wildlife and countryside organizations and is willing to take part in national habitat and species initiatives. The Conservation Office maintains contact with its many conservation groups and interested bodies through the publication of *Sanctuary*, its annual conservation magazine.

Conservation in Action

Some examples of practical work from dune sites around the UK are highlighted below to give some idea of the issues facing the conservation management of the defence estate and to demonstrate that these are little different from the actions taken by many conservation bodies in the management of nature reserves.

Altcar Rifle Range, Sefton Coast, England

Altcar Rifle Range estate covers 250ha of beaches, dunes, fields and small woods. The site, owned and managed by the Territorial Army, is part of the Sefton Coast cSAC and contains priority dune habitats. In 1977 a conservation group met and a management plan was prepared. With the support of the Sefton Coast Life Project (see Rooney, this volume) all the actions proposed in the plan have now been completed and work has started on a revision of the plan. Conservation management work has included the removal of sea buckthorn *Hippophae rhamnoides* scrub, the mowing of dune slacks to encourage wildflower populations and the creation of new slack habitat by turf stripping and re-contouring. The site is a key area for the rare natterjack toad *Bufo calamita* and habitat management has benefited this species.

Royal Air Force, Air Weapons Range, Tain, Scotland

The site is a busy NATO range in northeast Scotland. It lies on the peninsula of Morrich More, one of the largest dune systems in the UK. The area contains a fore-dune and beach complex, brackish pools, shifting dunes, dune grassland and, on the older dunes, dune heathland. Morrich More is designated as a cSAC with the presence of eight habitats rare or threatened in Europe. One of these, dunes with juniper *Juniperus communis* thickets, found on the range, is regarded as being the most outstanding example of its kind in the UK.

The juniper extends over about 20ha, and gorse *Ulex europaeus* is also present. Research is being carried out by Scottish Natural Heritage to assess the causes of a noticeable decline in the health of the habitat. The response of the juniper and the rest of the vegetation to excluding sheep and rabbits is being assessed.

Today many other coastal dune systems in the UK have been transformed, either by tree planting or conversions to caravan parks or golf courses. Although the intermittent bombing with inert ordnance causes some insignificant damage and disturbance to a small area around the targets, the existence of the range has afforded special protection by reducing human disturbance. It has also prevented any development of this unique and important wildlife refuge.

Magilligan Training Centre, Northern Ireland

The Magilligan dunes are one of two large, relatively undisturbed dune systems in Northern Ireland. Most of the area is designated as an Area of Special Scientific Interest (ASSI) and now also a candidate Special Area of Conservation.

The beach and dune formation process are of great scientific interest. Magilligan Point is one of the largest depositional shoreline features in the British Isles and is significant at the international level. There are five main categories of land use in the Magilligan area: conservation of the scientific interest, education and research, military training, recreation and farming.

Conservation management includes grazing with domestic stock, rabbit grazing, scrub control, information and education. With the dramatic reduction in rabbit grazing caused by myxomatosis, large tracts of the dunes are now in the process of becoming covered by dense thickets of blackthorn *Prunus spinosa*, burnet rose *Rosa pimpinellifolia* and bramble *Rubus fruticosus* agg. These are natural components of the dune system but due to the absence of significant grazing have spread more vigorously. A combination of sheep and cattle grazing has proved successful in restoring habitat diversity.

Another major dune site in Northern Ireland managed by the MOD is the Ballykinler dunes lying across Dundrum Bay from the Murlough National Nature Reserve. The site has extensive areas of dune heath, a rare habitat in the UK. The MOD recognize that both dune sites are part of the same coastal system and therefore should be managed sympathetically. By looking at the experience at Magilligan the conservation group have discussed the need for grazing and scrub control on this important site.

West Freugh, southwest Scotland

The MOD airfield site at West Freugh lies adjacent to Torrs Warren, the largest acidic dune system on the west coast of Scotland. Parts of the site have been planted with conifers. The present conservation management of the site is centred around controlling invasive plant species. This has involved the removal of Corsican pine *Pinus nigra* derived from the nearby plantation and controlling the spread of bracken *Pteridium aquilinum* into the heathland by spraying with herbicide.

Strandline invertebrates

Like many MOD coastal sites, the beach at West Freugh forms part of the Danger Area and access is restricted. This allows the strandline and foredune communities to remain relatively undisturbed and support a specialized invertebrate fauna including species such as the woodlouse *Armadillidium album*. This species is also found in good numbers along the strandline at Altcar Rifle Range on the Sefton Coast, together with the nationally rare sandhill rustic moth *Luperina nickerlii gueneei*. On the Castlemartin ranges in Pembrokeshire, south Wales, the conservation group have found a good population of the strandline beetle *Nebria complanata* on the MOD beach at Frainslake Sands whereas almost no individuals are present on the adjacent public beaches at Freshwater West. The accessible beaches are used for recreation, and activities such as barbecues can be very damaging by removing all driftwood. Many local authorities, often responding to requests from local people, have regular and efficient clean-ups but these too can leave a sterile and lifeless strandline. It is also the lack of interference, therefore, which makes many MOD sites a sanctuary for nature.

As these examples demonstrate, the MOD is responsible for the conservation of some of the finest coastal landforms and habitats in the UK. Many of the properties have been in military use for over a century and this land use has undoubtedly protected many sites from development or loss of natural features. Now, in partnership with the

Table 1. Examples of MOD sites with coastal dune interest

MOD site	cSAC	County
England		
Penhale	Penhale Dunes	Cornwall
Barrow-in-Furness	Morecambe Bay	Cumbria
Eskmeals	Drigg Coast	Cumbria
Braunton Burrows	Braunton Burrows	Devon
Wainfleet	The Wash and North Norfolk Coast	Lincolnshire
Altcar	Sefton Coast	Merseyside
Boulmer	Berwickshire and North Northumberland Coast	Northumberland
Scotland		
Barry Buddon	Barry Links	Angus
Kinloss	Culbin Sands/Findhorn Bay	Moray/Nairn
Tain	Morrich More and Dornoch Firth	Highland
Benbecula, Hebrides	South Uist Machair	Western Isles
West Freugh	Torrs Warren–Luce Sands	Dumfries and Galloway
Cafd and Mould Site, Tiree	Tiree Machairs	Strathclyde
Wales		
Castlemartin	Limestone Sea-Cliffs of South West Wales	Dyfed
Pendine	Burry Inlet Dunes	Dyfed

conservation agencies, the MOD is taking its national and international responsibilities seriously and, through conservation management plans, will carry out the necessary practical habitat action to ensure the conservation of rare habitats and species.

The Future

With over 70 per cent of the Army now stationed in the UK, compared to the 1990 figure of 49 per cent, pressure on the existing training areas has shown a marked increase, making the sympathetic management of land crucial. At the same time, mounting legal and moral pressure to conserve the unique natural history resources of the defence estate has placed a greater emphasis on long-term management. It is now MOD policy to create enduring integrated management plans, which will incorporate the individual needs of military training with the conservation interests of the area. This policy will be implemented over the coming years.

The MOD Conservation Office acts as the central co-ordinator, with top priority being given to areas designated under the Habitats Directive. Plans for all SSSIs and other areas of high natural history value will follow.

The large number of MOD sites selected for consideration as candidate Special Areas of Conservation reflects the high regard for the conservation value of the Defence Estate (see Table 1 for dune sites). Conservation management plans have already been produced for many areas and the aim is to produce similar plans for the remainder of our designated sites. The production of these plans is led by the local Defence Land Agent, who has the best knowledge of the area, with advice sought from the statutory bodies and conservation group members. The completed plans will carry the endorsement of the military authorities. The MOD will continue to support the UK government's initiatives on Biodiversity and Sustainable Development and will, where possible, integrate the recommendations into future management plans.

Acknowledgments

The information used in the preparation of this article is derived mainly from *Conservation on the Defence Estate: The Role of the Ministry of Defence Conservation Office* (Ministry of Defence, 1996).

The input of the following authors is also acknowledged: Jack Donovan ('The Strandline Beetle', *Sanctuary*, 27), Major J. Nethercott ('Magilligan', *Sanctuary*, 27), Tom Dewick ('Morrich More', *Sanctuary*, 27), Major T.J. Farrington and David Simpson ('Altcar Rifle Range', *Sanctuary*, 27), Ian Langford ('West Freugh', *Sanctuary*, 25) and Paul Corbett ('Ballykinler Dunes', *Sanctuary*, 26).

PUBLIC PERCEPTION OF NATURE MANAGEMENT ON A SAND DUNE SYSTEM

S.E. EDMONDSON and C. VELMANS

Environmental and Biological Studies, Liverpool Hope University College, England

ABSTRACT

The interventionist management strategies to promote dune dynamics on the Sefton Coast have elicited considerable controversy among local people. The apparently destructive practices such as clearing scrub and, particularly, clear-felling large areas of pine woodland have been widely criticized in the local press and the issue has consumed a significant amount of management time in conflict resolution. This paper reviews reports of the issue in the local press, and uses questionnaires to assess the perception and understanding of sand dune ecology and management. Results indicate that although people have a reasonable knowledge of the presence and importance of rare dune species, understanding of their ecology and management is low. Negative headlines in the local press have the potential to reinforce these misunderstandings. Nature managers face a significant challenge in consulting the public, and explaining and justifying their management decisions.

Introduction

Aim

This paper aims to investigate public understanding of interventionist conservation management on the Sefton Coast, an internationally important coastal sand dune system in Merseyside on the northwest coast of England.

The Sefton Coast

The Sefton Coast is a large, internationally important dune system in Merseyside on the northwest coast of England. Its nature conservation significance is recognized by a number of local, national and international conservation designations, the latter consisting of one Ramsar site, one Special Protection Area and one candidate Special Area of Conservation (Edmondson, 1997). There are two National Nature Reserves and two Local Nature Reserves entirely on the sand dune system; thus much of the area is managed with nature conservation as the primary land use. Furthermore, all the major landowners on the sand dune system are brought together under the Sefton Coast Management Scheme which aims to co-ordinate management into a whole system approach with nature conservation as a high priority (Houston and Jones, 1987).

Management of coastal dune systems

Management of coastal dunes in the last decade has increasingly recognized the need to maintain a dynamic landscape, a movement catalysed by the international conference in Leiden in 1987 (van der Meulen et al., 1989). Landscape and habitat heterogeneity is increased by disturbance (Forman, 1995) which is a characteristic of dynamic dunes. The progress of succession towards a more stable climax community without disturbance causes a decrease in habitat diversity and allows competitive exclusion of the specialist dune plants and animals by more common and widely distributed species (Edmondson et al., 1993). Sand lizards *Lacerta agilis,* natterjack toads *Bufo calamita* and petalwort *Petallophyllum ralfsii* (a liverwort) which are protected under the Habitats Directive have important strongholds on the Sefton Coast dune system. All three have requirements for habitat features of the early stages of dune or dune slack succession (for sand lizards see Cooke, 1991; for natterjack toads see Beebee and Denton, 1996; for petalwort see Jones et al., 1995) which are created by active aeolian processes. Nature conservation management therefore necessitates the maintenance of the geomorphological and biological processes that maintain a dynamic dune system. These processes constitute the natural disturbance regime which creates a landscape mosaic of continuously varying patches, within which a diversity of habitats and species can co-exist. Thus it is important to recognize the concept of a predominantly non-equilibrium nature (van Zoest, 1992), contrasting with previous conservation effort to preserve the static landscape.

The specialist plants of a mobile dune system are successful because of their highly developed stress tolerance strategies. These enable them to capitalize on the resources of a stressful environment with little competition from other species (Grime, 1979). Highly adapted stress-tolerant plants cannot, however, survive high levels of disturbance caused by human impact. Thus where high visitor numbers coincide with mobile dunes, trampling can cause serious damage. Loss of the vegetation cover (dominated by *Ammophila arenaria*), large-scale wind erosion causing deflation of the frontal dunes and the initiation of large transgressive sand sheets may ensue (Pye, 1990). These scenarios, which result from disturbances far in excess of the natural dynamic, cause habitat loss, threats to coastal defence and loss of landscape and amenity value. European coastal populations have a long history of fighting against drifting sands, thus dune restoration and stabilization techniques are well developed (see Agate, 1986). Given adequate financial resources, repair of damaged dune systems can be achieved. The Sefton Coast Management Scheme was successful in implementing such dune restoration schemes in the 1980s, where major visitor access routes had caused massive deflation hollows in the frontal dunes (Plater et al., 1993).

People, both visitors and residents, are an important factor in the landscape ecology and the management of most coastal dunes. High visitor numbers are attracted to the natural landscape, the wide sandy beaches and the sea. For the same reasons, dune coasts constitute a very attractive environment for housing. The Sefton Coast is no exception. Very high visitor numbers are recorded on the coast (Wheeler et al., 1993) and housing development has destroyed a large area of former dune habitat (Jackson, 1979). Whether they are visitors or residents, recreational users of the coast will encounter nature management operations, either simply by their observations of what is happening or because their activity is directly affected by those management operations. Walkers may, for example, be excluded from newly restored dune areas or be asked to follow only specified footpaths through the dunes. The co-operation of these site users is essential, and raising public awareness forms an important part of management strategy. If a sand dune system has clearly been damaged, most people will understand the need for management intervention to restore the situation. However, if the management aim is to reintroduce disturbance patterns as part of a dynamic dune approach, the techniques employed may appear to site users to be very damaging and contrary to perceived nature conservation needs. Woodland felling, scrub clearance, turf stripping and reprofiling by bulldozers have all been used in an attempt to reverse the trends towards stable, senescent dunes and to provide habitats for specialist dune species. These techniques are overtly destructive and likely to be less easily understood by the public. Adverse reaction in the local press to some dune management operations provides the evidence for this.

Large areas of coniferous woodland were planted on the Sefton Coast between 1890 and 1920 (Wheeler et al., 1993) in an attempt to stabilize the dunes and protect property from shifting sands. The woodlands now constitute a serious threat to the more recently recognized need for dynamic dunes by their stabilizing, sheltering and drying effects. A decision was taken by English Nature (the UK government nature conservation body in England) to implement the Open Dune Restoration Project (see Simpson and Gee, this volume) in a phased programme starting in 1994, involving the clearfelling of a large block of pine woodland on Ainsdale Sand Dunes National Nature Reserve. The area cleared of pines was a wide belt on the frontal, more seaward dunes. Previously open dunes and dune slacks sheltered by the seaward pine plantations had developed large areas of birch (*Betula* spp.) and other scrub species and these woody species were also cleared as part of the project (see Simpson and Gee, this volume). A large, continuous belt of coniferous woodland on the more landward dunes has been retained as carefully managed woodland. Red squirrels, *Sciuris vulgaris,* a rare and legally protected species in the UK, occur in the pinewoods (see Shuttleworth, this volume). Management for species typical of open, mobile dunes such as sand lizards and natterjack toads must therefore be balanced against ensuring secure woodland habitat for the squirrels.

Despite a programme of consultation with councillors and some public education, the clearfelling programme has become an important issue to local people and has elicited strong criticism in the local press. As a result the issue has consumed a significant amount of management resources and time in conflict resolution and conciliation.

In the light of these conflicts, a preliminary survey of public knowledge and understanding of dune ecology and management was undertaken with the aim of analysing the source of the conflict and informing future management. The results of this investigation are presented here and are linked to an analysis of the reporting of the Open Dune Restoration Project in the local press.

Methodology

Questionnaire survey

The natterjack toad was selected as the focus for questions on rare species. This species is well known locally; it is the symbol of the Sefton Coast Management Scheme and gives its name to a local public house. Walkers on a number of sites on the coast will have encountered interpretation boards that aim to educate the public about the natterjack toads and they feature significantly in locally well-publicized guided walk programmes, site leaflets, and newsletters.

The questionnaire was designed to investigate firstly knowledge and secondly understanding of the dune landscape and its management. Open questions investigated awareness of the agencies responsible for wildlife conservation on the coast, the rare species present, threats to those rare species and views on which species should be conserved. Some closed questions investigated understanding of dune conservation, focusing on the issue of blowing sand and dune stability; for example, respondents were asked whether or not they agreed that bare sand was important for natterjack toads.

The questionnaire was used to survey two user groups, visitors and residents. Visitors were surveyed by interview during the summer months of 1996. A total of 76 interviews took place at the three major road access points to the coast (see Table 1). Each of these was the site of dune restoration work in the late 1970s and early 1980s to repair badly eroded frontal dunes.

Residents were surveyed during the autumn of 1996; 220 questionnaires were posted to households on randomly selected roads in Formby and Ainsdale, two residential areas built on the sand dune system. A total of 115 questionnaires were completed.

Monitoring of the local press

The two major local newspapers of the area were monitored between April 1996 and January 1997. All letters and articles were recorded and a subjective assessment made as to whether they were supportive of, neutral towards, or opposed to the frontal pine woodland clearance. The content was analysed qualitatively to investigate common themes and views.

Table 1. Sites used for the visitor questionnaire survey

Site	Managing agency	Notes	Number of interviews
Victoria Road, Formby	The National Trust	Large car park with access to the beach and dunes. Pine woodlands with red squirrels, *Sciuris vulgaris*	30
Lifeboat Road, Formby	Sefton Metropolitan Borough Council Ranger Service	Large car park with access to the beach and dunes	24
Shore Road, Ainsdale	Sefton Metropolitan Borough Council Ranger Service	Small car park. Large area of beach accessible for car parking. Access to beach and dunes. Some tourist facilities – pub, camp-site, children's playground	22
		Total	76

Results

Questionnaires

Results indicate that residents have a significantly greater knowledge of the sand dunes than visitors. They are more aware of the conservation organizations operating on the coast, although a large proportion of visitors knew of the National Trust and the associated presence of red squirrels on the coast (see Table 2). The red squirrels on the National Trust site attract a large number of visitors; the Sefton Coast is the only site in England where red squirrels may reliably be seen (Atkinson et. al., 1993).

When asked to list rare species of the Sefton Coast, residents were significantly more aware of natterjack toads (chi-squared = 40.87, df = 1, p = <0.001) than visitors. More residents knew about sand lizards than visitors but this species was less well known by both groups than natterjack toads. Although there is a high level of awareness of the natterjack toad, no respondents mentioned the lack of the bare sand environment as a factor threatening the survival of the species. There was no significant difference between the visitors and residents with respect to their awareness of the red squirrel (chi-squared = 0.440, df = 1, p = 0.7), over 60 per cent of both groups listing it as a rare species (see Table 3).

When asked which factors they considered to be the main threat to rare species on the coast, people, dune erosion and pollution were the three most common residents' responses, mentioned by 36 per cent, 24 per cent and 23 per cent respectively. The three most common visitor responses were pollution, people and litter, mentioned by

Table 2. Visitor and resident awareness of conservation organizations operating on the Sefton Coast. Respondents were asked to list the organizations that they knew

Conservation organizations	Resident responses (%) n = 115	Visitor responses (%) n = 76
National Trust	52	45
English Nature	12	0
Sefton Metropolitan Borough Council	19	4
Sefton Rangers	25	4
Sefton Coast Life project	2	0

Table 3. Visitor and resident awareness of rare species present on the Sefton Coast. Respondents were asked to list any rare species which they knew

Rare species	Resident responses (%) n = 115	Visitor responses (%) n = 76
Natterjack toad	90	40
Red squirrel	63	68
Sand lizard	38	3

Table 4. The factors most commonly listed by visitors and residents as the main threats to rare species on the Sefton Coast

Factors listed	Resident responses (%) n = 115	Visitor responses (%) n = 76
Pollution	23	43
People	36	42
Litter	0	29
Dune erosion	24	0

43 per cent, 42 per cent and 29 per cent respectively (see Table 4). No respondents mentioned the lack of open, dynamic dune habitat or the development of woodland and scrub, which are well-established causes of the decline of natterjack toads and sand lizards (Beebee and Denton, 1996; Cooke, 1991).

When asked what species were most important to conserve, many visitors tended to be fairly general, giving answers such as 'trees' or 'all species'. A similar proportion of residents and visitors listed the red squirrel (30 per cent and 25 per cent respectively) but far more locals listed the natterjack toad (35 per cent of residents as compared to 3 per cent of visitors).

Testing the contingency of listing natterjack toads as the most important species to conserve, and agreeing that bare sand is important for rare plants and animals, showed no significant relationship (chi-squared = 0.072, df = 1, p = >0.05); thus knowledge that the natterjack toad is an important species is not matched by understanding of its habitat requirements.

Local press

Results of the quantitative survey are shown in Table 5. In total, 21 letters and six articles were published which were generally opposed to the pine woodland clearance. This is balanced by three letters and 12 articles that were either supportive or neutral. The proportion of articles that were assessed as being opposed to the project was inflated by the very negative nature of the headlines. Thus articles which might otherwise be assessed as presenting a balanced view had headlines such as the following:

Table 5. Quantitative survey of the local newspaper coverage of the Open Dune Restoration Project. Dates of published items are listed; a = article, l = reader's letter

Newspaper	Pro-felling	Neutral	Anti-felling	
Formby Times	11/4/96 a	16/5/96 a	18/4/96 l	20/6/96 a
	11/4/96 a	16/5/96 a	18/4/96 l	27/6/96 l
	16/5/96 a	23/5/96 a	18/4/96 l	27/6/96 l
	25/5/96 l	4/7/96 a	16/5/96 l	18/7/96 l
	5/12/96 l	18/7/96 a	23/5/96 l	11/7/96 l
	9/1/97 l		23/5/96 a	11/7/96 l
			23/5/96 l	18/7/96 l
			25/5/96 l	29/8/96 l
			6/6/96 l	28/11/96 l
			13/6/96 a	5/12/96 l
			20/6/96 l	12/12/96 l
				12/12/96 l
The Champion		3/7/96 a	15/1/96 a	
		24/7/96 a	15/5/96 a	
		27/11/96 a	22/5/96 a	
		26/6/96 a	23/10/96 l	
Summary				
Letters	3	0	21	
Articles	3	9	6	

'Call for English Nature's work to be closely watched'

'Town at Risk' (a reference to the perceived threat to coastal defence of the woodland clearance),

'Geophysical meddlers going too far'.

There is clear evidence in the letters that residents have a strong sense of place and are passionately concerned about their local, sand dune environment. They refer to 'generations of local experience', and 'a deep love of our homestead and a concern for all that has happened over the years'. They also refer to past experience of blowing sand in their landscape: 'Sandstorms quickly bury any attempt to create boardwalks. In the past they have buried buildings and damaged the railway.'

There are many examples in the letters of residents' opinions being based on inaccurate facts and misunderstanding, for example the following, which must be written by people who do not know of the severe population decline of the rare species they mention:

'In all those years, Natterjacks and sand lizards have survived in a natural habitat which is perfectly conducive to them'

'The natterjack toads and sand lizards are here because they found the environment conducive to their sustainability as it already WAS – so why change things so drastically'.

Also, there is clear, available evidence to show that the following statements made in published letters are untrue:

'The comment regarding birch overrunning the place is rubbish'

'The argument that the newly created dunes will make a good habitat for the sand lizard is rubbish'.

Much of the press coverage refers to lack of public information about the project prior to implementation:

'this deliberate withholding of the full facts illustrates just how negligible and negative is the role assigned to us, the public, in a so-called democracy'

'A cloak and dagger operation done under the guise of democracy'

'That we were all kept in the dark over this until such mayhem had taken place has sinister undertones'.

A more supportive article points to the consultation meetings held with local councillors in 1989, 1990, 1991, 1994, 1995 and 1996, and suggests that those councillors who were now supporting the protestors failed in their duty to inform their electorate. However, there is no doubt that at least some of the public felt uninformed about the project. This clearly leads to a deep mistrust of the 'experts' who were responsible for planning and implementing the project:

'these people [the experts] have narrow objectives and only a superficial feel for the long-term fragility of the Sefton Coast'

'Soft-soaped by experts'

'so-called experts of the *unaccountable* English Nature'.

These residents clearly believe they have a better understanding of their own environment, and have constructed a clear dualism between themselves (the vociferous element of the local protestors to the project) and the experts who they perceive to be outsiders:

> 'I am one of the very many Formby people sick to death of opinionated outsiders telling us what they think is best for our community and proceeding to carry out their ill-conceived schemes largely without our knowledge and without our approval'
>
> 'Let them go and experiment with their own backyard where it is their amenities that are destroyed and their homes that get flooded or inundated with sand'.

Another line of attack used by protestors is to accuse the managers of misuse of public money:

> 'And so at last, the truth of it is out. The REAL reason for the destruction of 20,000 Corsican pine trees is because English Nature are to receive a massive European handout of up to £850,000'
>
> 'English Nature, funded by huge sums of taxpayers' money is destroying pinewoods'.

One letter refers to the apparent contradiction in spending on a different regional project (the Mersey Forest is a policy to enhance woodland cover in a large inland area of Merseyside and Cheshire): 'millions of pounds are being spent creating the Mersey Forest'.

The issue of the pine woodland clearance continued to be a hotly debated local issue throughout the monitoring period, which towards the end of the year elicited the following comment from a supporter of the project:

> 'unwarranted vituperation and ignorant blustering of a handful of protestors against responsible employees of our Government'.

Discussion

The high level of residents' awareness of nature on the dunes that was found by the quantitative questionnaire is to some extent confirmed by residents' letters published in the press. The awareness demonstrated by the latter, however, gives a biased impression as publication of these letters depends on:

a. people being sufficiently motivated to write the letters;
b. editors considering the issue sufficiently newsworthy and interesting for publication.

The number of articles and letters which were published throughout the monitoring period indicates that editors did consider this to be an important and interesting issue to local people. This constant publicity in the press will have raised public awareness and may have engendered the interest of those who previously had not engaged with the issue. For this reason, the nature of the articles and letters published in the press is very significant to the nature managers who require public support for their work. Thus the large proportion of press items which presented a negative view of the pine woodland felling represent a significant threat to the policies which have been developed by the nature managers.

The red squirrel is well known and apparently held in affection by both locals and visitors. The popularity of such a clearly attractive animal is unsurprising. At Victoria Road on the Sefton Coast it is possible to hand-feed these wild animals which have been conditioned over many years to expect visitors to offer them peanuts; indeed squirrel food is sold at the entrance to the squirrel reserve. Pleasure and satisfaction derived from contact with nature and watching animals are increasingly being reported in the literature (for example Harrison et al., 1987) and the squirrels clearly provide this experience. The squirrels are dependent on the pine woodlands so the need for conservation of this habitat is clear and relatively easily understood by the public. The woodlands represent a significant, fixed and stable element in the landscape that should remain as such. Sustainable management of the pine woodland does require intervention. There is some evidence in the press articles that even this woodland management may be difficult for the public to understand, but it has not become a controversial issue.

The presence of the rare natterjack toad is well known by residents and by a lesser proportion of visitors. Sand lizards are also mentioned frequently in the press letters and articles. This high level of public awareness is, however, clearly not matched by an understanding of the habitat requirements of these two rare animals. This lack of understanding may result partly from people's knowledge that the rare red squirrels are dependent on the pine woodlands. The natterjack toads and the sand lizards are dependent on habitat mosaics which include bare dry sand and open vegetation, and for natterjack toads also young, wet dune slacks, features which are characteristic of a dynamic, changing landscape. In short, what is good for the squirrels is disastrous for the natterjack toads and sand lizards. This contradictory message may account for the failure of any residents questioned to identify the threats to the natterjacks and sand lizards caused by natural succession to scrub and woodland.

Dune erosion was one of the three threats to the natterjack toad most commonly identified by residents. Historically, residents of sand dune coasts have battled against shifting sands to protect property. Implementation of dune restoration schemes to repair the effects of trampling on sensitive, mobile dunes has also been well publicized, and such schemes have been in operation on the Sefton Coast since the late 1970s. Again, therefore, a contradictory message is sent to the public. In some places at some times there is a need to fix and restore mobile sand, in other places at other times it is important to encourage the sand dunes to be unstable and dynamic.

Deep emotional feelings about a local environment have been reported by Erlebnis Wattengebeit (1997) who studied visitor and resident perception of the Wadden Sea islands in The Netherlands. They reported deeper emotions associated with the wildness and isolation of the islands. Possibly the sudden human intervention of the

Open Dune Restoration Project on a previously very quiet and (for the Sefton Coast) isolated area of dunes has affected these deeper emotions felt by the residents.

The Open Dune Restoration Project, in seeking to remove the stabilizing influence of the frontal pine woodland, also challenges the widely publicized and important conservation message that trees are good for the environment. The severe environmental impacts of deforestation in many countries of the world are widely publicized in education programmes, on television and in the press. The UK published a Sustainable Forestry programme in response to its commitment to the Rio Earth Summit outcomes in 1992 (UK Government, 1993). Furthermore, tree planting is a very familiar practical conservation activity in the UK, one topical and local example being the Mersey Forest. Opposition to the Open Dune Restoration Project therefore can be seen as a logical consequence of people's understanding of the importance of trees to the environment.

A further explanation of the opposition to the project is the natural tendency of people to oppose change. In the case of the Open Dune Restoration Project the change from woodland to open dune is rapid, and causes a significant landscape impact. This change is in total contrast to the gradual increase of scrub and woodland on the Sefton Coast dunes which has occurred since the 1950s. Although over a much longer timescale, the landscape change is no less significant, but it has not been commented upon by residents.

The importance of consulting and informing the public about conservation management is clear. Despite some consultation, residents felt that they were not informed about the Open Dune Restoration Project. In a similar situation on a Dutch sand dune area Zwart (this volume) has found that management had not consulted the public sufficiently before implementing interventionist policies, and concluded that future operations would be both facilitated and improved by better public involvement.

English Nature and the European co-funded Sefton Coast Life Project (see Rooney, this volume) both report considerable management time being spent in conflict resolution resulting from the controversy over the Open Dune Restoration Project (personal communication, 1997; 1998). Informing and educating the public is therefore an important management priority. It is possible however that, no matter how effective the public consultation, there will always be a vociferous minority opposed to such overtly interventionist practices. This tendency is indicated by the results of Zwart (this volume), whose data show a small but persistent proportion of questionnaire respondents opposed to nearly all management actions.

Conclusion

The Sefton Coast has been highly fragmented by residential development and its associated infrastructure. Together with the dunes and their associated biota, the residents now form a significant factor in the landscape and this factor should be recognized by the nature managers.

Although the public has a high level of awareness of rare dune animals, under-standing of their ecology and management is poor. The importance of the bare sand environment and the negative impacts of trees on sand dunes are particularly poorly understood. It is suggested that the well-publicized conservation messages of the importance of trees, and more locally of the need to restore badly eroded dunes, are major contributors to this lack of understanding. Nature managers therefore face a significant challenge in explaining and justifying their policies to the public. Public information and consultation is important not only because people have a right to be informed and involved in decision-making, but also because the resolution of conflicts created by lack of information can take up a considerable amount of management resources.

Acknowledgments

The authors would like to thank the staff of the Sefton Coast Life Project, and partic-ularly Rachel Flannery, for their assistance with this research.

References

Agate, E. (1986), *Sand Dunes*, British Trust for Conservation Volunteers, Wallingford.

Atkinson, D. and Houston, J. (eds.) (1993), *The Sand Dunes of the Sefton Coast*, National Museums and Galleries on Merseyside, Liverpool.

Atkinson, D., Bird, M., Eccles, T.M., Edmondson, M.R., Edmondson, S.E., Felton, C., Garbett, M.A., Gateley, P.S., Gunn, A., Hall, R.E., Hull, M., Judd, S., Nissenbaum, D.A., Rooney, P.J., Simpson, D.E., Smith, P.H. and Wood, K.W. (1993), 'Animal and plant groups', in Atkinson and Houston, 1993, 93–125.

Beebee, T. and Denton, J. (1996), *The Natterjack Toad Conservation Handbook*, English Nature and Countryside Council for Wales, Peterborough.

Cooke, A.S. (1991), *The Habitat of the Sand Lizard*, Research and Survey in Nature Conservation No. 41, Nature Conservancy Council, Peterborough.

Edmondson, S.E. (1997), 'Perspectives on nature conservation on Merseyside', in D. Light and D. Dumbrăveanu-Andone (eds.), *Anglo-Romanian Geographies: Proceedings of the Second Liverpool–Bucharest Geography Colloquium*, Liverpool Hope Press, Liverpool, 157–74.

Edmondson, S.E., Gateley, P.S., Rooney, P.J. and Sturgess, P. (1993), 'Plant communi-ties and succession', in Atkinson and Houston, 1993, 65–84.

Erlebnis Wattengebeit (1997), *Perception of the Waddesea Area/beleving van de Wadden*, Dienst Landbouwkundig Onderzoek, Staring Centrum Instituut voor Onderzoek Landelijk Gebied, Waddenverenigig, Harlingen, The Netherlands.

Forman, R.T.T. (1995), *Landscape Mosaics: The Ecology of Landscapes and Regions*, Cambridge University Press, Cambridge.

Grime, J.P. (1979), *Plant Strategies and Vegetation Processes*, Wiley, Chichester.

Harrison, C., Limb, M. and Burgess, J. (1987), 'Nature in the city – popular values for a living world', *Journal of Environmental Management*, **25**, 347–62.

Houston, J.A. and Jones, C.R. (1987), 'The Sefton Coast Management Scheme', *Coastal Management*, **15**, 267–97.

Jackson, H.C. (1979), 'The decline of the sand lizard, *Lacerta agilis*, population on the sand dunes of the Merseyside coast, England', *Biological Conservation*, **16**, 177–93.

Jones, P.S., Kay, Q.O.N. and Jones, A. (1995), 'The decline of rare plant species in the sand dune systems of South Wales', in M.G. Healy and J.P. Doody (eds.), *Directions in European Coastal Management*, Samara Publishing, Cardigan, 547–55.

Meulen, P.D. van der, Jungerius, P.D. and Visser, J. (eds.) (1989), *Perspectives in Coastal Dune Management*, SPB Academic Publishing, The Hague.

Plater, A., Huddart, D., Innes, J.B., Pye, K., Smith, A.J and Tooley, M.J. (1993), 'Coastal and sea level changes', in Atkinson and Houston, 1993, 23–34.

Pye, K. (1990), 'Physical and human influences on coastal dune development between the Ribble and Mersey estuaries, northwest England', in K.F. Nordsrom, N. Psuty and B. Carter (eds.), *Coastal Dunes: Form and Process*, Wiley, Chichester.

UK Government (1993), *Sustainable Forestry: The UK Programme*, Command Paper 2429, HMSO, London.

Van Zoest, J. (1992), 'Gambling with nature', in R.W.G. Carter, T.G.F. Curtis and M.J. Sheehy-Skeffington (eds.), *Coastal Dunes*, Balkema, Rotterdam, 503–14.

Wheeler, D.J., Simpson, D.E. and Houston, J.A. (1993), 'Dune use and management', in Atkinson and Houston, 1993, 129–49.

DUNE MANAGEMENT AND COMMUNICATION WITH LOCAL INHABITANTS

FREEK ZWART

Staatsbosbeheer, Friesland, The Netherlands

Introduction

In 1899 the Staatsbosbeheer, the National Forest Service, was established in The Netherlands to manage the state forests and to plant what was then regarded as wasteland, the heathlands and shifting sands. Nowadays the National Forest Service is concerned with forestry, landscape conservation, recreation and nature conservation. This paper describes Staatsbosbeheer, the National Forest Service in the Netherlands, its tasks as site manager and its role as a manager of the Dutch coastal dunes, in particular on the Wadden islands of Terschelling and Vlieland. Dune management and the importance of communication with the local population are examined. The paper concludes with some remarks and warnings.

The National Forest Service

The tasks of the National Forest Service include nature management, outdoor recreation, landscape management and the harvesting of such natural products as timber and reed. It manages a total of 230,000ha of forest and nature areas in the Netherlands. The organization has a central unit and eight regional units in charge of the actual management of sites. Each regional unit is subdivided into districts. The province of Friesland has a 33,000ha management area comprising a wide range of nature types and within the Vlieland/Terschelling district the management area covers 11,000ha of (mainly) dunes (including 1000ha of dune afforestation), beach plains and mud-flats.

The Dutch Coastal Dunes

The Dutch coastal dunes cover a total of some 40,000ha, a substantial part of which (more than 17 per cent) lies in the Vlieland/Terschelling district. The dunes in Terschelling and Vlieland are vast continuous areas, largely unspoilt compared to the mainland dunes. They are of international significance for their natural value and are also popular recreational areas. The National Forest Service has an important role on both these islands. It manages more than 80 per cent of Terschelling and the entire open area of Vlieland. This creates a special bond with the local inhabitants, more so as 80–90 per cent of the islands' inhabitants rely on tourism for their income. Most visitors come for the islands' natural beauty and landscape.

Dune Management, Past and Present

Management priorities were very different during the first half of this century. Initially dunes were planted with trees and land was brought under cultivation where possible. Large stretches of dunes were drained. On the island of Terschelling 600ha of forest was planted and 300ha of moist dune valley was prepared for agriculture (arable land and grass). Many kilometres of ditches were dug, many of which are still in use, some in areas currently managed for nature. Over recent decades the ecological effects of acid rain have also been apparent.

These two factors, the large-scale human intervention of the past and the deposition of nitrogen, have been a threat to the characteristic flora and fauna and will eventually lead to a monotonous landscape. The forests, planted early this century, were pure forests, 80 per cent of which consisted of pine plantations (*Pinus nigra*) all in the same age class.

Management plan

The current management plan envisages a return to the original natural situation where possible, without damaging the interests of third parties. So far this has led to a number of drastic management measures, quite different from what had been done before. These include:

- the restoration of the hydrological system by filling in ditches;
- the large-scale reconstruction and restoration of wet dune valleys;
- the large-scale thinning of pine plantations and converting them into the more natural broadleaf forests.

All this was done in a landscape that appeared quite natural and unblemished, which was why it met with such storms of protests from the local inhabitants. Our management area in Vlieland/Terschelling is their backyard. The islands' inhabitants are clearly stakeholders.

Support

Without a broad base of support and good working relations with those involved it is impossible to get management objectives off the ground. This is especially true in our district where relations are so close.

Support is vital. We need the local population's support to carry out our regular management duties and preserve the quality of nature and recreation facilities. The organization also needs the government's support to survive.

Public Information, Communication and Participation

Relations between the National Forest Service and the islands' inhabitants were notoriously bad in the past. Conflicts and misunderstandings abounded. Local people vented their frustrations by vandalism, deliberate fires and occasional fist fights. Direct communication between the National Forest Service and the islands' inhabitants was virtually non-existent. For questions concerning management activities people turned to the mayor or the municipal executive. This did not do the National Forest Service any good.

Around 1980 the Service decided to improve communication systematically. A round of meetings was organized to explain the newly drawn-up management plan to the local inhabitants. These early meetings were not always easy. At last people were able to express their grievances. 'This is the first time in 75 years that we have heard from you,' a local farmer grumbled, and he was right. From then on we took care to let people know about management measures and the reason for them at an early stage through the local media. At this stage public information was still a rather one-sided affair.

For the present management plan (1992–2002) we handled things differently. After rounds of consultations with all the stakeholders involved, a draft management plan was drawn up. Opinions were invited and changes were made where necessary, after which the final plan was adopted.

This time all the stakeholders had had their say. Real communication had taken place; people felt they could make a genuine contribution. Meanwhile structural consultation groups sprang up all over the island with representatives of local authorities, other government bodies, tradespeople, farmers, sailors, fishermen, nature organizations and cultural clubs.

On the island of Vlieland the whole process has been taken one step further to what could be called *participation*. There, in an entirely open planning process, consultations with local inhabitants start even before pen is put to paper. This approach is still in its infancy. The framework is there but the details still have to be filled in.

Evaluation

On the island of Terschelling in 1995 anonymous door-to-door polls were held to get people's opinions about 16 different management activities carried out by the National Forest Service. Ratings could be given from 1 to 10 (1 = bad, 10 = excellent). Where ratings were insufficient people were asked reasons. The polls yielded the following results:

- 64 per cent said public information had improved over the past decade;
- Intensity of public information and good ratings seem related: the more information, the higher the ratings;
- The National Forest Service was given an overall rating of 7.3, which is more than sufficient;
- In the past large groups of islanders opposed the National Forest Service; now the general opinion is positive although some inhabitants are still negative about some of the activities or the organization.

On the island of Vlieland, prior to the introduction of the management plan, a similar evaluation was held, but here public information was something new. The local population had never before been informed about management plans; in the past consultations never took place. Although management activities on the island of Vlieland are very similar to those on Terschelling, the outcome of the polls was very different. Whereas on Terschelling the National Forest Service was given an overall rating of 7.3, Vlieland's population gave it an overall 4.4.

Conclusions

Lack of communication gives low ratings, incomprehension and a bad image. Communication can help to improve ratings and boost support. This makes it easier to get management objectives off the ground. Improving one's image calls for a systematic approach and cannot be done overnight.

Two final warnings:

- communication is a means, not an end – it does not automatically solve problems;
- communication is something that needs to be done in a professional manner.

CAN HUMAN EROSION BE ACCEPTED IN THE SEASIDE DUNES? FROM SAND-DYKE TO MOUNTAIN SCENERY: AN EXAMPLE FROM RØMØ, DENMARK

MARTIN REIMERS

The National Forest and Nature Agency, Denmark

ABSTRACT

In recent years there has been a growing awareness that frontal dunes should be allowed to develop more freely than before. At the same time, however, more and more effort is used to prevent human erosion. The dunes are polluted visually with signs, fences and boardwalks. With the Danish Waddensee island Rømø as an example, this paper recommends that human traffic here and there in the frontal dunes should be more accepted. It is a form of nature management which helps to preserve the natural dynamics of the dunes and reduces the need for expensive infrastructure.

Introduction

When visiting dune areas, for instance in Holland or Germany, I, as a Dane, am surprised to find how restricted access is to the dunes. The human traffic is regulated with fences, boardwalks and signs. When you are not used to all this infrastructure, you find it visually polluting. There is a strong feeling of walking around in a museum and not in wild nature, as though people have become spectators and are no longer part of nature (see Plate 1). Where there is heavy and increasing visitor pressure it is now an almost expected development to zone and separate people and nature.

The white dunes closest to the beach represent a dynamic natural formation which continuously changes in response to winds and storm surges. Mobility is the soul of the white dunes. The question should therefore be raised whether erosion by people in the dunes should be always seen as bad, or if in some places it could be accepted as a form of influence in line with the erosion caused by sea and wind.

Denmark is lucky in having large dune areas covering 127,000ha compared to its relatively small population of only 5 million. It means that the erosion caused by visitors in the Danish dunes has been small. There is, therefore, according to Danish nature conservancy law, free pedestrian access to the dunes almost everywhere.

However, in recent years, problems with erosion in the dunes along the West Jutland coast have risen considerably because of the steady increase in tourists to the area. The case for management and zoning is therefore growing, even in Denmark.

The importance of allowing the white dunes to develop in a natural way is recognized in Denmark. The dunes are only stabilized where it is necessary.

Rømø

Rømø is situated in the southwestern part of the Danish Waddensee close to the German border. The island is connected to the mainland by a 9km long dam, and has an area of 120km². The soil is mainly composed of sea- and wind-deposited sand. Eight hundred people live on the island all year, but in the summer season the population swells to 15,000. Dependent on the weather Rømø receives between 1 and 2 million day visitors a year.

The island is especially known for its beach, which consists of pure white sand for a length of 17km and a width of 1–3km. Almost all tourists visiting the island go to the beach. An important factor on this beach is that car access is generally permitted.

Along a length of the beach there is a frontal dune ridge which has a height of 5–10m. The

Plate 1. Nature as a museum; boardwalk on the island of Amrum, Germany.

human erosion on these frontal dunes is less intense than on similar beaches without car access. Trampling is more diffuse when parking is along the beach rather than in concentrated sites behind the dune ridge. These dunes have grown from beach level since a serious storm surge in 1981. They have developed in a natural way helped by the planting of marram *Ammophila arenaria*. Each winter the storm surges erode some 10m from the front of the dunes but this is then regained through the next summer. The wind erosion forms blowouts in the dunes.

In 1995 the active dune stabilization was reduced. Previously it included the planting of marram, removing projections (recontouring outcrops) and placing seaweed in blowouts. The reason for stopping the stabilization work was primarily to let the dunes

Plate 2. *The white dunes at Rømø as they looked before the dune stabilization stopped.*

Plate 3. *The white dunes at Rømø as they look after the dune stabilization has stopped.*

develop more freely. Since the stabilization of the dunes has stopped their appearance has changed radically due to the influence of the wind and the sea together with trampling caused by visitors. Until 1995 the dunes looked rather like a dyke heavily covered with grass (see Plate 2). Since then they have broken up into a geomorphologically more interesting landscape with peaks, crests and gullies. Here and there gaps have developed which have been flooded under storm surges.

Now there is much more white sand in the dunes than before, and the dunes have gained a beautiful appearance (see Plate 3). They appear today like a hilly mountain scene, free from defacing signs, fences and stakes. The variation in the dunes has increased considerably. There are now more micro-habitats for insect and plant species than before. The abundance of shortlived plants, for instance sea rocket *Cakile maritima,* has increased. Last but not least, State Forest District Lindet, which owns and administers the beach and the dunes, has cut its expenses by about £30,000 a year.

The local population has been dissatisfied to a certain degree with the fact that the wind, the sea and the human influence now have free rein in the dunes. This is probably because the population traditionally has great respect for the damage that the drifting sand, the moving dunes and the storm surges have caused the island in the past.

It will be interesting to follow the development of the dunes in the coming years. Will they grow or erode, will the dune ridge disappear, stay or drift inland, or will storm surges through the breaches in the dune ridge threaten the hinterland areas? The questions are many, and the answers are blowing in the wind!

It may become necessary to stabilize the dunes and perhaps regulate human traffic if drifting sand or storm surges threaten privately owned hinterland areas. But until that happens the development of the dune will be given back to the wind, the storm surges and the erosion from trampling feet.

Discussion

It is of course a happy combination of circumstances that gives us the chance to give the powers of nature and dune pedestrians free rein at Rømø. Only time will show if it can continue.

With increased visitor numbers and erosion in European dunes, it is generally becoming more necessary to implement dune stabilization and visitor management works. But let us still attempt, where it is possible, to let people walk freely in the frontal dunes. Let us try not to pollute the beautiful dunes visually with signs, fences and boardwalks. Let the dunes be *wild* as nature intends.

The author therefore hopes that in future people will have opportunities to wander through the white dunes rather than always being channelled into fenced footpaths. There should be areas where free walking in the dunes is allowed and even preferred – where people can be users, nature managers and part of nature itself. Let us do nature, and thereby ourselves, the favour of not turning the dunes into a museum.

COASTAL EROSION AND TOURISM IN SCOTLAND: A REVIEW OF PROTECTION MEASURES TO COMBAT COASTAL EROSION RELATED TO TOURISM ACTIVITIES AND FACILITIES

A.M. WOOD

33 East Claremont Street, Edinburgh, Scotland

ABSTRACT

Dune systems are particularly attractive to tourists and experience heavier visitor pressure than many other coastal environments. As part of a research report undertaken for the Scottish Tourism Coordinating Group, a series of site studies was carried out to provide a structured evaluation of a range of options available to combat mainly small-scale erosion effects, and to gauge their effectiveness in real situations. The report was published by Scottish Natural Heritage in 1994 as part of a series of Review Papers.

Context

The coastline of Scotland is over 10,000km long. The quality of the landscape, ranging from the highly indented, mountainous western seaboard to the estuaries and cliffs of the North Sea coast, contributes to an outstanding tourism image both nationally and internationally.

As one of Scotland's foremost tourism assets, the coast is a major focus for informal and formal recreation. Dunelands are a particular magnet within the spectrum of coastal landscapes. In addition to their landscape and natural heritage interest, they are also the site of many of the country's finest golf courses, which attract enthusiasts from all over the world (see Stubbs, this volume). While their attractiveness as an asset is highly valued, the fragility of the dunelands is not recognized to the same extent.

The beaches of Scotland have been extensively described (Ritchie and Mather, 1984) and some of the problems of erosion addressed by the Highland Beach Management Project in the late 1970s (CCS 1980a). Many Scottish beaches remain unspoilt, especially those on the islands, due to the limited number of visitors and generally limited access.

However, especially on the mainland, problems do occur and most of these are in some way related to human activity, including camping and caravanning, beach access and trampling, car parking and over-grazing. An extensive body of expertise has developed in the practical management of mainly small beach and dune areas, with well-tried techniques for access control, dune stabilization and revegetation. Publications such as those by the Countryside Commission for Scotland (CCS, 1980b), the British Trust for Conservation Volunteers (BTCV, 1986) and the Institute of Terrestrial Ecology (Ranwell and Boar, 1986) offer step-by-step guidance to some of the principles of dune management where recreation pressures are high.

Coastal erosion and its effects on coast protection, infrastructure and amenity is still an issue at many beaches. Many shorelines now have some form of protection against erosion. However, more recently, changes in environmental awareness are now being reflected in a move towards techniques which seek to accommodate the natural coastal processes; this approach is not only proving to have benefits in terms of protecting and enhancing the landscape and nature conservation value of the coastline, but is also leading to more enduring and cost-effective schemes (Ministry of Agriculture, Fisheries and Food, 1993).

In 1992, the Tourism and the Environment Task Force was set up in Scotland to increase environmental awareness in the tourism industry, encourage sensitive development and promote sustainable practice. Coastal erosion was identified as one of a number of key issues to be examined as a possible threat to sustainable tourism development, and a study was commissioned.

Terms of Reference and Content

In essence, the study aims were:

- To look at coastal erosion specifically from a tourism perspective, giving a basic practical introduction to the topic for those with a tourism remit, particularly at the local level;
- To concentrate on small-scale erosion problems at a level which impacted directly on discrete tourist activities, or where the causes of erosion problems were thought to be directly linked to tourist activities;
- To provide a specific Scottish dimension by reviewing the application of erosion control and management techniques at a range of case study sites around the coastline.

The content of the report is presented in three main sections. Section 1 provides a general introductory review, setting coastal erosion within the overall context of natural coastal processes. The review concludes that most Scottish coastal erosion problems are likely to have a local origin, probably brought about by human actions. Such activities might include dredging, coastal structures, land reclamation, over-stabilization of coastal dunes, inadequate access provision and alteration of beach water tables. It is important that such problems are correctly diagnosed otherwise treatment can be ineffective and costly. Solutions fall into two broad categories: those that allow the re-establishment of natural processes and those that hold the line. The decision on which is most appropriate is probably taken on a mix of financial, political, and increasingly environmental criteria.

Section 2 reviews the range of options available to combat small-scale coastal erosion, providing an evaluation of the appropriate techniques under headings including effectiveness, sustainability, aesthetics, and cost. For sand dune coasts the techniques described included beach recharge, sand fencing, planting dune grasses, sand stabilization, recontouring and visitor management.

Section 3, Site Case Studies, is the key section, providing an analysis of the application of techniques in real situations. Four examples from the 15 sites examined are outlined below.

Case Studies

Royal Troon Golf Club

The coastal edge of the southern section of Troon South Sands beach coincides with the boundary of the first and second holes of the championship course at Royal Troon Golf Club, which regularly hosts the Open Championship. While the beach complex is regarded as naturally very stable with regard to wind erosion, it is less so for marine attack, and the coastal edge was subject to severe storm erosion in two successive winters, 1989 and 1990, when retreat of up to 15m occurred.

In 1991, 470m of gabion mattresses were laid, topped by a gabion basket. The original ground profile to the rear of the slope was reinstated using sand washed onto the beach during the storms, and the resulting bund soiled and seeded.

The work was considered to have been partially successful. Substantial sand accretion and embryo foredune formation with colonization by lyme-grass *Leymus arenarius* was observed in some areas. In others the gabions were still exposed, and were thought likely to remain so as the sand supply was poor. It was recommended that building of a more extensive protective foredune should be considered, harnessing the favourable sediment movement in the area, reinstating the semi-natural appearance of the coastal edge.

Gullane Bay

Gullane beach is located on the East Lothian coast some 12 miles from Edinburgh and is subject to intensive informal recreational use with over 200,000 visitors per year. The dune system was badly damaged by military manoeuvres in the Second World War, which all but removed the foredune, and by the 1960s a bare sand plain extended from the beach across almost the entire former dune slack area.

Work began in 1962, and by the late 1970s a new foredune ridge had been established, through sand fencing and the planting of marram grass *Ammophila arenaria*. Extensive planting of sea buckthorn *Hippophae rhamnoides* was also carried out in the late 1960s and early 1970s. The new ridge continued to be eroded by destructive waves and further intervention has been considered necessary. A range of techniques have been used, including recontouring of a major section of the foredune, planting of marram, sand fencing, wave-barrier fencing, and visitor management in the form of paths, signage and the employment of site staff.

It was considered possible that the successful dune reconstruction work might have taken too much sand out of the system and left the site more vulnerable to marine erosion. Once a steep seaward face develops the dune may grow vertically and overwhelm any ability of the vegetation to trap sand, so that it dies back. The increasing instability, probably aided by trampling damage, will lead to more and more bare sand and may lead to a second phase of sand movement. The planting of sea buckthorn may not have helped as the plant has removed native dune species, impairing any self-healing capacity of the dune. Mechanical beach cleansing was also considered to be a problem.

Gairloch

Gairloch beach is reported to be the most visited beach on the northwest coast of Scotland, and is one of the very few sandy beaches in this part of Scotland easily accessible from the main coastal route. It is without doubt an important attraction in the Gairloch area.

The geomorphological status of the beach/dune complex is relatively stable, as its exposure to the west is counterbalanced by a sufficient quantity of sand within the local coastal cell. However, there is a localized tendency for erosion to occur at the northern end of the beach, with deposition at the southern, more sheltered end, a process known as *rotation*. The concentration of visitor pressure is also at the northern end, and the coincidence of natural and human forces has brought about the need for remedial action.

The original remedial strategy adopted was to channel visitor access along the most appropriate routes, restrict use of the most vulnerable areas, and attempt revegetation. A recent episode of substantial erosion of the main seaward face of the dunes prompted more radical experimental action. Sediment transfer has been carried out from the south to the north end of the beach. Sand was removed from a localized berm feature on the lower part of the south section of the beach, and part placed against the eroded dune face, with the remainder placed on the upper beach and allowed to blow onto the dunes. This small-scale beach nourishment in effect mimics, in reverse, the natural sediment movement process in an inexpensive way, and is regarded as the best alternative in reconciling conflicting requirements for optimal coastal defence and conservation of the aesthetic quality of the beach.

Gruinard Bay

The beach at Gruinard Bay South forms part of one of the most photogenic views on the entire western seaboard of Scotland. It enjoys a deservedly high profile in tourist literature, and its level of use places it in the top five in the Highlands.

The problem was primarily one of visitor management, in particular the management of car parking and provision of a safe route of access to the beach, in a manner which did not degrade the quality of the landscape setting.

The natural coastal processes here tend to exacerbate the impact of visitor pressure. The location and aspect of the beach mean that it is fully exposed to the funnelling of wind and wave forces from the northwest. This factor, in combination with a lack of sand supply to the local cell system from either offshore or river sources, has led to a net retreat of the coastal edge, evidenced by a steep bare sand scarp on the single large dune which backs the beach.

Environmental upgrading work, part of an integrated scheme first drawn up by Highland Regional Council in 1987, has been implemented, including dune stabilization, a timber access stairway to the beach, and reinstatement of road verges. Efforts to stabilize the sand scarp immediately adjacent to the access stairs met with only limited success, however. Work carried out in 1988 included marram planting and thatching. At the time of survey (1993), only a sparse covering of small marram plants was observed. Their failure to thrive may have been due to the very limited supply of windblown sand available from the beach and to the steepness of the dune scarp. Growth of marram was observed to be more vigorous lower down the face closer to the beach. It was considered that reprofiling the dune might be necessary if stabilization was ultimately to be successful. Regardless of whether or not this can be achieved, it was recommended that a hard engineering solution would be entirely inappropriate in this location.

Conclusions

Dune systems continue to be magnets for visitors. Visitor pressure may contribute to instability and erosion in some cases.

The effectiveness of current soft and hard engineering techniques in maintaining a healthy dune system can be variable; success is more likely where there is a fuller understanding of the wider coastal processes at work.

Although many of the areas and problems are on a small scale, they need to be kept under constant review to monitor both erosion and the effectiveness of management techniques.

In certain locations, both the results of erosion and the solution adopted have physical and visual impacts which may be perceived as detracting from tourism value.

The results of the study support the current recognition of the need for a co-ordinated approach to coastal erosion as part of an integrated strategy of coastal zone management.

References

ASH Consulting Group (1994), *Coastal Erosion and Tourism in Scotland,* SNH Review 12, Scottish Natural Heritage, Edinburgh.

British Trust for Conservation Volunteers (1986), *Sand Dunes: A Practical Conservation Handbook,* BTCV, Wallingford.

Countryside Commission for Scotland (1980a), *Highland Beach Management Project: Final Report 1977–79,* Countryside Commission for Scotland, Battleby.

Countryside Commission for Scotland (1980b), *Information Sheets on Beaches: Dune Grass Planting, Reseeding of Dune Pastures etc.,* Countryside Commission for Scotland, Battleby.

Ministry of Agriculture, Fisheries and Food (1993), *Coastal Defence and the Environment: A Guide to Good Practice,* Ministry of Agriculture, Fisheries and Food.

Ranwell, D.S. and Boar, R. (1986), *Coast Dune Management Guide,* Institute of Terrestrial Ecology, Huntingdon.

Ritchie, W. and Mather, A.S. (1984), *The Beaches of Scotland,* Countryside Commission for Scotland, Battleby.

THE MANAGEMENT OF THE SAND DUNE AREAS IN WALES: THE FINDINGS OF A MANAGEMENT INVENTORY

RICHARD H. DAVIES

Department of Geography, University of Wales, Swansea

ABSTRACT

During the summer and autumn of 1996 a questionnaire-based study was carried out targeting the 50 recognized sand dune areas along the Welsh coast. The aim of the study was to collect data on human impacts and management of sand dunes in Wales. This paper presents and discusses the results of the study.

The questionnaire approach identified the key uses of the dunes to be recreation, military uses, educational uses and wildlife protection. Most of the dune sites in Wales have facilities for general visitors; few have resources aimed at those visitors interested in wildlife and conservation. The key problems identified on the dune sites were visitor pressure, undergrazing and scrub encroachment. Management was carried out for five main purposes: wildlife protection, coastal conservation, recreation, military purposes and education.

Introduction

The sand dune areas along the Welsh coast have attracted relatively little research in terms of human impacts and associated management. Although the *Sand Dune Survey of Great Britain* compiles details of the dune vegetation in Wales (Dargie, 1995), the human impacts and management of the dunes are only touched upon. To improve understanding of the observed present dune vegetation, and to provide information useful for conservation management planning, information on human impact and present management is necessary.

Background

Davies et al. (1995) advocate that an effective management strategy depends on the availability of essential information which is objectively measured rather than anecdotal. The aim of this study was to investigate the current human impacts and management of the sand dunes of Wales in an objectively measured way. This was carried out by a postal questionnaire to establish:

- who and what are the key users of the sand dune sites in Wales;
- the key problems arising from these uses;
- whether the sites are managed and by whom;
- which aspects of the dunes are managed (notably the vegetation species and habitats) and what the management strategy is;
- whether the sites have a management plan.

The response rate to the questionnaire was good, with 43 sites returning fully completed questionnaires – a response rate representing 86 per cent of the sand dune sites in Wales.

Management

Not all of the sand dunes sites surveyed in Wales are managed; eight sites are not managed (see Fig. 1). The sites with the longest current management regime are (up to 1997) Newborough Warren on Anglesey (42 years) and Oxwich Bay (34 years). Information was not available for 14 sites. Overall, six have been managed for over 30 years, seven for between 20 and 30 years, seven for between 10 and 20 years and one for less than 10 years. Sites managed for the longest period will have had longer to get the management correct, with perhaps more experimentation in the early days. The vegetation and landscape present today is very much a product of past management techniques.

Figure 2 shows the types of site manager. Many sites are managed by the Countryside Council for Wales (CCW) or local authorities. Mixed management at 15 sites involves two or more managers; in seven cases one of these is CCW, managing with the National Trust at Whiteford and Stackpole, the local authority at Kenfig, the Ministry of Defence (MOD) at Laugharne and Pendine, the MOD and Tenby Golf Club at Tenby and the Forestry Commission at Newborough Warren. A further three are managed by the local authority and another body, a private owner at Hillend, the MOD at Pembrey, and the harbour-master at Conwy/Deganwy. The other three sites with mixed management are at Broomhill (National Trust and private owner), Caldey Island (National Trust and private owner) and Morfa Bychan (National Trust, Trustees of the estate and the golf course). Mixed management will involve greater discussion and possible conflict over the precise objectives of managing a site.

Of the 33 sites that are managed, 11 have formal management plans, a further seven have management plans that are being written or approved, and 15 are managed but do not have current management plans. Two of the sites that are currently not managed have had management plans prepared. These plans are generally only published within the organizations concerned (e.g. CCW) but 13 are available for consultation.

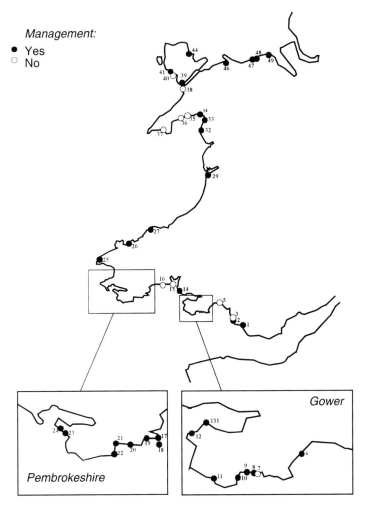

Figure 1. *Management status of the sand dune sites in Wales. (For key to sites see Table 2.) Box on left represents Pembrokeshire sites and box on right the Gower peninsula sites.*

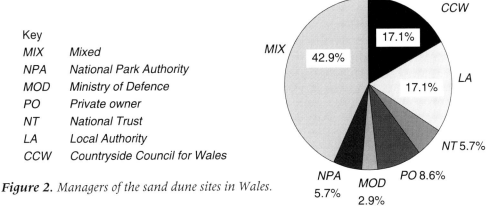

Key

MIX	Mixed
NPA	National Park Authority
MOD	Ministry of Defence
PO	Private owner
NT	National Trust
LA	Local Authority
CCW	Countryside Council for Wales

Figure 2. *Managers of the sand dune sites in Wales.*

Management Objectives

The management objectives collated were reordered under the six categories shown in Table 1. Table 2 shows which sites are managed, by whom and the length of time (to 1997) that they have been managed. The management objectives shown in Table 1, appropriate to each site, are also shown in Table 2 and reasons for management are shown in Table 3.

The first category in Table 1 concerns geomorphology and is a list of general state-

Table 1. Management objectives on the sand dunes of Wales

A. Objectives concerning general sand dune geomorphology

1. Maintain and enhance the current range of natural and semi-natural sand dune habitats and associated species.
2. Allow natural development through natural geomorphological processes.
3. Maintain a high landscape quality.
4. Safeguard all notable flora and fauna.
5. Maintain viable populations of rare species.

B. Objectives concerning specific sand dune habitats
– to maintain the status and diversity of...

6a. strandline, foreshore and mobile dunes
6b. semi-fixed and fixed dunes
6c. dune slacks
6d. heaths and dune heaths
6e. dune mire
6f. open water areas
6g. wooded areas
6h. dune grassland

C. Objectives concerning species
– to maintain populations of nationally rare and scarce species in a favourable condition

7a. birds
7b. bryophytes
7c. vascular plants
7d. invertebrates
7e. herptiles
7f. mammals

D. Objectives concerning human use of the sand dune areas

8. Stabilize the dunes and prevent sand dune encroachment.
9. Continue to accommodate visitor access to the site without compromising the nature conservation interest.
10. Encourage appropriate recreational use of the dunes.
11. Develop use of the dunes for appropriate formal and informal education.
12. Promote study and research.
13. Maintain and enhance the dune system as a natural coastal defence.
14. Prevent erosion and damage to the site by visitors.

E. Human land use objectives

15. Ministry of Defence objectives.
16. Meet all legal and other duties.
17. Manage visitors and improve tourist attractions.
18. Implement management.

F. Miscellaneous

Y. Site managed, but with no formal objectives.
U. Management plan under review.
S. Site-specific/species-specific objectives.

Table 2. Management of the sand dune sites in Wales. Italicized names represent no reply to questionnaire. Y = yes, N = no; CCW = Countryside Council for Wales, LA = local authority, NT = National Trust, PO = private owner, MOD = Ministry of Defence, G = golf course, T = estate trustees, FE = Forest Enterprise, H = harbour-master

Site	Site name	Managed	Time managed (years) to 1997	Manager	Objectives (see Table 1)
1	Merthyr Mawr	Y	23	CCW	1, 2, 3, 4, 6g, 9, 10, 11, 12, S
2	Kenfig Dunes	Y	20	LA/CCW	1, 2, 4, 9, 11, 13, 16
3	Maram Burrows	N			
4	*Baglan Bay*				
5	Crymlyn Burrows	N			
6	Black Pill to Brynmill	Y	?	LA	1, 7a, 9, 11, 12, 18
7	Pennard Burrows	N			
8	Penmaen Burrows	Y	30	NT	3, 7a, 9
9	Nicholaston Burrows	Y	13	NT	1, 9
10	Oxwich Burrows	Y	34	CCW	1, 5, 6a, 6c, 6h, 10, 11
11	Port-Eynon to Horton	Y	21	LA	8, 14
12	Hillend to Hills Tor	Y	21	LA/PO	14
13	Whiteford Burrows	Y	31	CCW/NT	1, 3, 5, 6a–h, 7a, 9, 11, 12, S
14	Pembrey Coast	Y	17	LA/MOD	U/MOD
15	Laughame Burrows	Y	?	CCW/MOD	4, 5, 6a–c, 7g, U
16	Pendine Burrows	Y	?	CCW/MOD	4, 5, 6a–c, 7g, U, MOD
17	Tenby Burrows	Y	?	G/MOD/CCW	15
18	Caldey Island	Y	?	PO/NPA	8
19	Lydstep Haven	Y	?	PO	17
20	Manorbier/Swanlake	Y	?	LA	9, 10/U
21	Freshwater Bay East	Y	11	PO/NPA	U
22	Stackpole Warren	Y	21	CCW/NT	U
23	Brownslade/Linney	Y	?	MOD	15
24	Broomhill Burrows	Y	?	NT/PO	1
25	Whitesand Bay	Y	?	PO/NPA	Y
26	The Bennet	Y	21	NPA	U
27	Poppit Sands	Y	21	NPA	U
28	*Towyn Warren*				

Table 2 continued

Site	Site name	Managed	Time managed (years) to 1997	Manager	Objectives (see Table 1)
29	Ynyslas	Y	18	CCW	U
30	*Tywyn to Aberdovey*				
31	*Fairbourne*				
32	Morfa Dyffryn	Y	16	CCW	1, 2, 4, 9, 11, 12, 16
33	Morfa Harlech	Y	11	CCW	9, 10
34	Morfa Bychan	Y	?	NT/T/G	Y
35	Pwllheli/Pen-Y-Chain	N			
36	Traeth Crugan	N			
37	Tai Morfa	N			
38	Morfa Dinlle	N			
39	Newborough Warren	Y	42	CCW/FE	1, 2, 6a–h, 7a–f
40	Penhrhynoedd-Llangadwaladr	N			
41	Aberffraw	Y	?	CCW	1, 2, 6a, b, c, f, h, 7b, c, d
42	*Valley*				
43	*Tywyn Gwyn*				
44A	Traeth Dulas	Y	31	PO	Y
44B	Traeth Lligwy	Y	31	PO	Y
45	*Red Wharf Bay*				
46	Conwy/Deganwy	Y	17	LA, H	Y
47	Kinmel Bay	Y	6.5	LA	Y
48	Rhyl to Prestatyn	Y	?	LA	1, 2, 4, 9, 10, 11, 12, 13
49	Gronant to Talacre	Y	?	LA	1

ments about the sand dune system which normally tend to pave the way for more specific objectives in the later sections. These general objectives encapsulate other ideas and notions, and are generally aimed at preserving the whole dune system, for wildlife value and coastal defence.

The second category, which is more specific than the first, concerns the objectives associated with sand dune habitats. The objectives cover all the main dune habitats, and various habitats are mentioned in site management plans depending on which habitats require attention. The objectives again emphasize the importance of maintaining the status of the dune habitats. All eight habitat types are mentioned in management plans for Whiteford Burrows and Newborough Warren NNRs.

The fourth category concerns visitors and users of the site, and deals with access to

Table 3. Reasons for management at the dune sites in Wales

	Coastal defence	Recreation	Wildlife protection	Education	Military
Sites	Kenfig Dunes	Merthyr Mawr	Merthyr Mawr	Merthyr Mawr	Pembrey Coast
	Rhyl to Prestatyn	Kenfig dunes	Kenfig dunes	Kenfig dunes	Laugharne Burrows
		Blackpill-Brynmill dunes	Blackpill-Brynmill dunes	Blackpill-Brynmill dunes	Pendine Burrows
		Penmaen Burrows	Penmaen Burrows	Oxwich Burrows	Tenby Burrows
		Nicholaston Burrows	Nicholaston Burrows	Morfa Dyffryn	Brownslade/Linney
		Oxwich Burrows	Oxwich Burrows	Rhyl to Prestatyn	
		Port-Eynon to Horton	Laugharne Burrows		
		Hillend to Hills Tor	Pendine Burrows		
		Lydstep Haven	Caldey Island		
		Manorbier/ Swanlake	Broomhill Burrows		
		Morfa Dyffryn	Morfa Dyffryn		
		Morfa Harlech	Newborough Warren		
		Rhyl to Prestatyn	Aberffraw		
			Rhyl to Prestatyn		
			Gronant to Talacre		
Total	2	13	15	6	5

the dunes without damage, encouraging recreation, education, research and the maintenance of the dune system for coastal defence. An important objective is accommodating visitor access without compromising the nature conservation interest of a site. This is related to the multiple use of many dune sites for conservation and recreation.

The sites which have the most comprehensive management plans and objectives are those where CCW is involved in their management. CCW makes time and provision for writing detailed plans of what needs management, why it needs management and how the management objectives will be achieved.

Conclusions

The sand dunes of Wales are managed for different purposes, broadly speaking *nature conservation* and *use*. In terms of nature conservation the dunes of Wales are managed to conserve the geomorphology, habitats, flora and fauna of the dunes. In terms of uses the dunes are managed to allow visitor access, education and research and coastal defence, and to prevent the nature conservation objectives from being compromised.

Acknowledgments

Thanks must be expressed to my supervisors Dr J.A. Edwards and Professor J.A. Matthews; and also to all the dune managers and representatives of CCW and other organisations who helped to make this study a success.

References

Dargie, T.C.D. (1995), *The Sand Dune Vegetation Survey of Great Britain: A National Inventory: Part 3 – Wales,* Joint Nature Conservation Committee, Peterborough.

Davies, P., Williams, A.T., and Curr, R.H.F. (1995), 'Decision making in dune management: theory and practice', *Journal of Coastal Conservation,* **1(1)**, 87–96.

Davies, R.H. (1997), 'The sand dunes of Wales: uses, problems and management responses', MPhil thesis, University of Wales, Swansea.

Section 4

THE SEFTON COAST LIFE PROJECT

THE SEFTON COAST LIFE PROJECT:
A CONSERVATION STRATEGY FOR THE SAND DUNES OF THE SEFTON COAST, NORTHWEST ENGLAND

P.J. ROONEY

Liverpool Hope University College, England
Project Officer, Sefton Coast Life Project 1996–1999

ABSTRACT

The sand dunes of the Sefton Coast, north west England, have suffered many of the problems typical of dune sites across Europe. A co-ordinated response to these problems began in 1978 through the Sefton Coast Management Scheme. Within the framework of the Scheme, funding was secured through the EC Life-Nature programme to assist with conservation efforts on the dune coast and in particular to assist with the preparation of a conservation strategy for the sand dunes of the Sefton Coast. The Sefton Coast Life Project was operational between 1995 and 1999.

Introduction to the Sefton Coast

The total area of dunes in Great Britain, not including blown sand deposits already lost to development, is about 56,000 hectares. The total dune resource in England is estimated at 12,000 hectares (Doody, 1991).

The Sefton Coast lies between the Mersey and Ribble estuaries in northwest England. Despite loss to development it is still the largest area of open dune landscape in England, and one of only six sites in Great Britain with more than 1000ha of dune habitat (Doody, 1991). The main dune system is 17km long and up to 5km wide, comprising a total area of approximately 2000ha. High, rolling dunes near the coast are backed by sandy plains and heathland and low-lying agricultural land overlying earlier deposits of sand.

The dunes of the Sefton Coast have suffered similar problems to many other dune sites across Europe. Until recently the dune area covered more than 3000ha, but residential development, afforestation, industry, military sites, roads, golf courses and holiday parks have considerably reduced this. The dune coast is now part of the coastline of the Metropolitan Borough of Sefton, which stretches from the commercial docks in the northern part of Liverpool in the south to the seaside town of Southport in the north.

Sefton is a local government unit with a population of approximately 300,000. It is estimated that the Sefton Coast lies within one hour's journey of over 5 million people and that over 500,000 people visit the sand dunes and beaches each year. Pressures from housing development and uncontrolled recreational access led to widespread environmental degradation during the 1960s and 1970s. In 1978 the Sefton Coast Management Scheme was established (Houston and Jones, 1987) to tackle the problems resulting from uncontrolled human use of the dunes.

The Sefton Coast Management Scheme is a working, voluntary partnership of land managers, co-ordinating their actions and agreeing objectives for the whole coastal area. It has been successful in tackling years of neglect, in repairing recreational damage, establishing nature reserves and developing skills in habitat management. Initial efforts were targeted on sites designated for nature conservation.

The Sefton dune coast today is managed mainly for nature conservation, coastal defence, sport and amenity. Sefton Council is the largest single landowner, with English Nature, the National Trust and the Territorial Army responsible for other substantial areas. Golf, with seven individual courses, is a significant land use within the dune area. Land ownership and land uses create a variety of habitats and a range of intensity of use often benefiting nature conservation, but also present a challenge for the integrated management of the whole dune system.

By the early 1990s the Coast Management Scheme partners realized that it was important to look beyond the safeguarded sites to the whole dune and coastal resource. Actions within the Sefton Coast Management Scheme area are now guided by a coastal zone management plan, revised in 1997 (Sefton MBC, 1997). Nature conservation is one element within this strategic plan.

Threats to the Dunes and the Need for Management

In spite of the work of the Sefton Coast Management Scheme, by the early 1990s threats to the dune system remained. The once extensive dune system had been fragmented into smaller units by development, and only about half of the remaining dune area was under active management by conservation land managers. Without continued efforts, there was the potential for conflicting conservation objectives because of the multiple land use and ownership of the Sefton Coast.

A general lack of grazing pressure had caused over-stabilization of the dunes, increasing dense grassland and scrub at the expense of open dunes and bare sand. This process reduced the area available for dune habitats and species with the effect of further fragmenting them. This recent vegetation trend was accelerated by dune stabilization works, introduced plant species and a reduction in rabbit *Oryctolagus cuniculus* populations following the introduction of myxomatosis.

Conifer planting has had a major impact on dune habitats. On the Sefton Coast conifer planting began in the late nineteenth century. As well as replacing the natural dune habitats, the plantations encourage scrub growth, reduce sand mobility, draw down the water table and acidify the soil (Atkinson, 1988).

Recreational visits at popular sites on the Sefton Coast are increasing by about 10 per cent each year. Management of recreation pressure continues to be necessary following early work by the Coast Management Scheme partners to restore damaged dune areas and provide a management infrastructure of footpaths and car parks. The problems of overuse are well known for dunes (Brooks and Agate, 1986). However, it is now also recognized that light recreation pressure can be beneficial in dune areas by supplementing rabbit grazing, so helping to maintain the open qualities of the dunes (Rooney and Houston, 1998).

Much of the remaining dune area of the Sefton Coast is now a candidate Special Area of Conservation. The designation brings with it particular obligations for the conservation of important habitats and species. Two of the dune habitats present are listed as 'priority' in the Habitats Directive. These are fixed (grey) dunes (Corine Code 16.221 to 16.227 Fixed dunes with herbaceous vegetation), and dune heath (Corine Code 16.23 Eu-Atlantic fixed dunes, *Calluno – Ulicetea*). Species occurring on the Sefton Coast and listed in either Annex II or IV of the Habitats Directive include petalwort *Petallophyllum ralfsii* (a bryophyte), great crested newt *Triturus cristatus*, natterjack toad *Bufo calamita* and sand lizard *Lacerta agilis*.

On the Sefton Coast it is believed that around half the area of priority fixed dune habitat and almost all the priority dune heath habitat are within golf courses and military sites, and outside of the cSAC.

The Sefton Coast Life Project

The urgent need for additional nature management resources led to an application by key partners in the Sefton Coast Management Scheme to the Life-Nature Fund, the EC's main source of funding for environmental projects. Within the framework of the Sefton Coast Management Scheme, funding was secured to assist conservation efforts on the dune coast and in particular the preparation of a conservation strategy for the sand dunes of the Sefton Coast. The bid was approved in July 1995. The Sefton Coast Life Project was operational between September 1995 and June 1999, with an active project team between February 1996 and March 1999. The Project was managed by a Project Steering Group of the main partners, under the auspices of the Sefton Coast Management Scheme. Work was guided by an Action Plan executed over a three-year period. The project operated with a total budget of over 1 million Euro, and was co-financed by the EC Life-Nature Fund.

The core partners
The core partners of the Sefton Coast Life Project were existing members of the Sefton Coast Management Scheme.

- Sefton Metropolitan Borough Council is a local government body, and co-ordinates the work of the Sefton Coast Management Scheme. The Council is the largest owner of coastal dune land in Sefton, and operates a Ranger Service. The Ranger Service is responsible for all aspects of site management including the protection of species and habitat management.
- English Nature is both the government statutory nature conservation agency and a coastal landowner. Two National Nature Reserves managed by English Nature, Ainsdale Sand Dunes and Cabin Hill, lie within the project area.
- The National Trust is a charity and the UK's largest land-holding non-govern-mental organization. The Trust owns a property at Formby on the Sefton Coast and actively manages the dune area in the interests of nature conservation.

Between them, the partners manage most of the dune nature reserves.

Aims of the project

The main aim of the project was 'to develop a strategic plan to manage the whole of the cSAC by consolidating management planning, improving conditions for key species and carrying out management actions to protect duneland habitats whilst also raising awareness and support amongst visitors and locals'. In particular, the project was expected to

- support the purchase of key sections of the coastal dunes and heath;
- establish nature reserves over the most sensitive locations;
- develop educational and information nature trails at appropriate sites;
- develop nature conservation management plans in co-operation with golf course managers and other landowners;
- undertake habitat restoration and species recovery actions;
- organize workshops to share the experiences of this programme with managers of similar dune SACs across Europe.

The aims of the Sefton Coast Life Project may be summarized as follows:

- To prepare a nature conservation strategy for the sand dunes of the Sefton Coast.
- To undertake emergency actions, purchase land at risk and carry out practical management work.
- To achieve sustainable habitat management through conservation management plans, species and habitat strategies and the development and dissemination of best practice in dune management.

The project concept

The actions of the Sefton Coast Life Project were within the context of the Sefton Coast Management Scheme's coastal zone management plan. They were therefore set within the reality of coastal management. One of the fundamental principles of the project was that the conservation experience and skills developed by the core partners be used for the benefit of others to encourage greater awareness of the need to conserve the whole dune system. In particular, therefore, the project involved the golf course and military site managers.

The core area for the Sefton Coast Life Project and the nature conservation strategy was the candidate Special Area of Conservation (cSAC). However, it was recognized by partners that to achieve effective and sustainable management of this core area, conservation must look beyond these artificial boundaries. Therefore a 'whole coast' approach was adopted. Although the conservation strategy for the sand dunes of the Sefton Coast has a decreasing sphere of influence beyond the core cSAC, it integrates with other initiatives such as Shoreline Management Plans and the Sefton Coast Management Scheme to benefit the core area of European interest. This approach effectively creates 'buffer zones' to the 'core' area and looks at the Sefton sand dunes in a wider geographical context. It recognizes that the Sefton dunes do not lie in isolation and that the natural linkages to the foreshore zone, to the adjacent estuaries of the Ribble, Alt and Mersey, to the inland low-lying mosslands and to the urban areas within Sefton must be recognized and understood.

Dissemination

Across Europe there is a general lack of public understanding of dune conservation. Dunes are seen as 'wasteland' and pine woods as 'improvement'. There is often strong opposition to conservation projects such as scrub clearance, grazing and control of access.

The Sefton Coast Life Project placed great emphasis on dissemination of information. In addition to organizing international workshops and a symposium, guidelines for dune management techniques and good practice case studies were published on the Internet, in general natural history publications and in specialized publications aimed at conservation and coastal managers.

The Life Project involved individuals and community groups in the preparation of site management plans and practical work as part of the Local Agenda 21 process. The preparation of the draft conservation strategy for the sand dunes of the Sefton Coast is an example of this approach. Through these processes the aims of the project were disseminated more widely and supplemented the more traditional media of newsletters, press articles, guided walks and illustrated lectures.

Preparing the Strategy

A small Working Group of project partners and advisors was convened between September 1996 and April 1997. This provided a focus for the Nature Conservation Strategy for the Sefton Coast SAC by defining its remit, the process of production, information required and relationships to other strategies and initiatives.

Consensus building

Drafting of the nature conservation strategy took a 'consensus-building' approach. This broke from the traditional consultative approach that is driven by 'experts' and does not actively involve key parties. Through consensus building common ground was sought between interest groups and land managers. It promoted agreement rather than conflict and provided an open-ended process in which new approaches and opinions could be brought forward.

The process of consensus building was as important as the outcome. It helped to build relationships and increase people's ownership of the outcomes. This was particularly important in the multiple-use environment of the Sefton Coast cSAC.

Issue groups

Thirteen 'issue groups' were the core of the consensus-building method used in drafting the nature conservation strategy. Groups were active between February 1997 and January 1998. Issue groups considered a single subject area:

- Pollution
- Invertebrates
- Economic issues/leisure/tourism
- Soils
- Herpetofauna
- Geomorphology
- Higher and lower plants
- Birds
- Mammals
- Foreshore
- Community and environmental groups
- Land managers

Through the issue groups the views and experience of a wide range of key players on the Sefton Coast were drawn upon. Participants in each of the issue groups were people well briefed in their subject area – naturalists, academics, land managers, community and environmental groups. In total over 140 individuals from 69 organizations were involved in the process.

Participants in each group varied according to the group issue. The numbers of participants from each organization and the number of times each issue group met were limited. This served to focus their discussion and product. The Life Project Officer acted as the secretariat and guided each meeting. All meetings followed a similar approach – conservation issues were identified, discussed, and action points (opportunities) suggested.

The results of each issue group were presented in a standard format as a draft report. Most of the draft reports were then subjected to targeted consultation determined by participants in the issue group. The revised reports were then made available for public comment in libraries in Sefton. The final report of the issue groups is a good record of perceptions of the key conservation issues and required actions for the sand dunes of the Sefton Coast. Its use extends beyond the compilation of the strategy; it is a document that will give general advice to the Sefton Coast Management Scheme.

Endorsement
The draft strategy was endorsed for consultation by the Steering Group of the Sefton Coast Management Scheme in February 1999.

Strategy Aims and Function

The strategy concentrates on the features and areas identified as being of European importance, while recognizing related and supporting features. It therefore takes the cSAC as the 'core' area for action with a 'buffer' area, defined as the boundary of the Sefton Coast Management Scheme plan, constituting a significant but lesser priority zone. The strategy provides guidance for the sustainable management of the natural habitats and species of Community interest, as defined in Article 1 of the Habitats Directive.

The strategy contributes to broader nature conservation and biodiversity initiatives for the local area and region by helping to identify some of the most important issues for nature conservation. It is not, however, a nature conservation strategy for the whole of the Sefton Coast area, nor all of the natural features present.

The aims of the strategy are:

- To develop conservation objectives, leading to the achievement of favourable condition, for the priority and other Annex I habitat types, and Annex II and Annex IV species listed in the Habitats Directive that occur on the Sefton Coast, as a contribution to favourable conservation status across their whole range within the European Community.
- To contribute to the strategic approach taken to the protection and management of the dune habitats and dune processes on the Sefton Coast, linked to shoreline management and planning.

Strategy Structure and Operation

The strategy operates at four levels:

- *Aims*
  The strategy aims are intended to promote a consistent and co-ordinated approach to the conservation management of features of European importance across the dune coast and especially within the area of the cSAC.

- *Goals*
  The goals elaborate the aims into statements of objectives.
- *Guiding Principles*
  The guiding principles are more specific statements. They provide more detailed guidance for partners and users of the strategy to encourage consistency and the implementation of management. Where relevant, some recommendations for action are stated. These may be further developed in the section on priority actions.
- *Priority Actions*
  Following the main body of the strategy is a list of priority actions. An annual review of progress with the strategy will be completed and reported to the Steering Group of the Sefton Coast Management Scheme.

The strategy embraces a number of key themes, including:

- the interpretation of the aims of the Habitats Directive into local goals;
- habitats and species;
- climate and coastal change;
- water conservation;
- use of the coast and its effect on nature conservation interests;
- monitoring and surveillance.

The strategy explains in some detail the implications of the Habitats Directive for the Sefton dune coast. It is aimed at the site managers and agencies who, collectively, have a responsibility towards the conservation of the special interest of the Sefton Coast cSAC.

The concept of 'favourable condition' is used as the local target for natural habitats and species listed in the Habitats Directive. This tailors the European-wide concept of 'favourable conservation status' in the Habitats Directive to meet the requirements of an individual site (Countryside Council for Wales, 1996). To assist the achievement of favourable condition, and to guide the use of resources, the strategy provides objectives for the priority natural habitats and species of Community importance occurring on the Sefton dune coast. The objectives, where possible, define target values and limits of acceptable change for key attributes of each feature.

The strategy is a realistic and forward-looking framework guiding management actions in the multiple-use environment of the Sefton dune coast. It contributes to the fulfilment of both the land managers' and local community's aspirations for the Sefton dune coast in the context of the Sefton Coast cSAC and the obligations of the Habitats Directive. The production process has helped to bring people closer together in their understanding and actions. Most importantly, the strategy supports the sustainable management and maintenance of the European conservation interests and further integrates nature conservation into the coastal zone.

Project Actions

Land purchase
A range of supporting actions were undertaken by the Life Project in the preparation and adoption of the strategy. The first, in September 1995, was to assist Sefton Council in the purchase of a large area of duneland at Lifeboat Road and Ravenmeols. This secured a piece of land where there had been longstanding proposals to develop a golf course. The land purchase was jointly funded by Sefton Council, English Nature, the Countryside Commission and the Life-Nature fund. It was the most important land purchase on the Sefton dune coast for over twenty years.

Practical action
Practical field staff formed part of the project team. Their work included extensive emergency habitat restoration such as scrub clearance, mowing and species recovery actions. Project field staff worked alongside partners' field staff, conservation volunteers and other site staff such as golf course greenkeepers. Contractors and grant incentives were also used to initiate practical conservation work.

   This approach was an important strength for the Life Project. It not only assisted in progress towards favourable condition for natural habitats and species of Community importance, but also, through the availability of real practical assistance, helped to create new partnerships, catalyse additional actions and put in place a lasting support more likely to endure beyond the project period. The golf clubs and military land are examples of sites on the Sefton dune coast which have benefited from this action-led approach and are consequently more actively involved in nature conservation activities.

   Where possible, the project supported and promoted the application of sustainable management, especially following habitat restoration or species recovery. For example, grazing is a sustainable habitat management tool, particularly important for fixed dunes, the natterjack toad and the maintenance of dynamic physical processes. The Life project supported the extension of an established sheep grazing project at Ainsdale Sand Dunes National Nature Reserve to areas cleared of scrub at Ainsdale Sandhills Local Nature Reserve.

Management planning
The preparation of conservation site management plans was a significant part of the Life Project. Given the complex land ownership of the Sefton Coast, site management plans have a prominent role in co-ordinating management between sites and enabling the efficient and effective management of sites, processes, habitats and species.

   Management plans were prepared for seven golf courses, Lifeboat Road and Ravenmeols Dunes (Sefton Council), Altcar Rifle Range (military) and a small dune area known as Kenilworth Road and Embankment (Sefton Council) surrounded by modern housing development. With the assistance of the Life Project, partners reviewed and updated several of their management plans for nature reserve areas. The Life Project also pioneered a innovative approach to the preparation of golf course conservation management plans. These plans are a hybrid between a standard nature conservation plan and the hole-by-hole plan usually preferred by golf clubs.

Some sites did not necessitate a full management plan. Instead, conservation guidelines were prepared and agreed with the landowners at sites including a municipal golf course, a private park, farmland and a churchyard.

Information and Monitoring

To assist strategy development, survey work was completed for the Annex II and Annex IV species (petalwort, great crested newt, natterjack toad and sand lizard). Habitat surveys using the National Vegetation Classification were digitized. To manage these and other data sets, a Geographical Information System (GIS) was established for the dune coast. This, together with a resource centre, presents survey data to land managers in an accessible form.

A monitoring programme is included in the conservation strategy for the sand dunes of the Sefton Coast. Life Project partners and other partners in the Sefton Coast Management Scheme have given a commitment to maintain and develop monitoring systems for the Sefton Coast.

Remaining Concerns

Concerns remain for the quality and use of the sand dunes of the Sefton Coast, even following the successful implementation of the Sefton Coast Life Project and the sustained success of the Sefton Coast Management Scheme partnership.

Although there are plans to increase the area grazed by domestic stock, currently a lack of extensive grazing over most of the dune system accelerates the process of seral succession, producing increased vegetation cover with a trend towards scrub and woodland on the fixed dunes. With this change there is evidence of decreased species diversity (Edmondson et al., 1993) and changes in soil type and structure (James, 1993). The reduction in grazing pressure in recent decades on most sites following the post-myxomatosis decline of the rabbit population has increased the rate at which these changes are occurring.

Without management intervention, the processes described above will produce a mix of dune scrub and woodland habitats of low species diversity, with the earlier stages of succession and open dunes being poorly represented. The priority habitats of fixed dunes and dune heath may be squeezed out by this change without a sustained and appropriate management response.

The effects of conifer plantations and coastal erosion may further reduce dune processes, the fragmentation of habitats and the loss of species at some sites. Small fragmented areas are more difficult to conserve than large complete areas.

The trend of increasing recreational use on the dune coast has the potential to put severe pressure on recreation management infrastructure, operation of natural processes, and sustained survival of habitats and species.

Public understanding of dune processes, coastal change and the need for management is still generally poor. As emergency habitat restoration schemes often necessitate a dramatic landscape change and use of large and heavy machines, they are the source of conflict between members of local communities and the dune managers. Conflict resolution diverts scarce resources away from conservation management.

Future Challenges

Site managers of the Sefton dune coast face challenges that are not unique to their site. Rather, they are held in common with many other European dune systems. They may be summarized as follows:

- *No further loss*
  A large proportion of the dune resource has already been lost to development and the fragmenting effects of these losses are sometimes acute. Further loss is not desirable, although all natural habitats and populations of species are subject to natural change.
- *Achievement of favourable condition for habitats and species*
  This is the primary responsibility of site managers at a local level. Favourable condition embraces local distinctiveness while contributing to the European goal of favourable conservation status.
- *Achieving wide public support*
  Managers can work towards this through effective education, interpretation and an open and inclusive management system. Wide public support is essential for sustained conservation management success.
- *Multiple-use management*
  The cSAC, as part of the Natura 2000 network, is about people and nature living together.
- *Impacts of climate-induced change*
  Surveillance is required to detect change, together with a management system that is both flexible and responsive to change.

Conclusions

The multiple land use, multiple ownership and international conservation value of the Sefton dune coast present a complex situation within which to develop shared conservation objectives. The potential for conflict is great.

Nature conservation actions are now within the context of the Sefton Coast Management Scheme's coastal zone management plan. This creates a realistic management situation, helps to focus nature conservation issues and provides a common coastal management vision. The voluntary working partnership of the Sefton Coast Management Scheme enables continuing progress.

The use of a consensus-building approach to prepare a nature conservation strategy for the Sefton Coast has demanded a higher level of resources extended over a longer time period than the traditional consultative approach. However, it has created a firm foundation for the strategy, produced more robust objectives, brought land managers and users closer together and increased their mutual understanding. The process has been as valuable as the product. Through this process, the production of the strategy and other project actions, nature conservation is being more closely integrated into actions within the coastal zone. However, the potential for conflict has only been reduced, not removed. The approach, together with the product, requires continuous implementation and review for sustained success.

Acknowledgments

The Sefton Coast Life Project was a partnership between Sefton Council (Leisure Services Department and Planning Division), English Nature and the National Trust. It was co-financed by the European Commission.

References

Atkinson, D. (1988), 'The effects of afforestation on a sand dune grassland', *British Ecological Society Bulletin*, **29(2)**, 99–101.

Brooks, A. and Agate, E. (1986), *Sand Dunes, A Practical Handbook*, British Trust for Conservation Volunteers, Wallingford.

Countryside Council for Wales (1996), *A Guide to the Production of Management Plans for Nature Reserves and Protected Areas*, Countryside Council for Wales, Bangor.

Doody, J.P. (1991), *Sand Dune Inventory of Europe*, JNCC/EUCC.

Edmondson, S.E., Gateley, P.S., Rooney, P.J. and Sturgess, P.W. (1993), 'Plant communities and succession', in D. Atkinson and J. Houston (eds.), *The Sand Dunes of the Sefton Coast*, National Museums and Galleries on Merseyside, 65–84.

Houston, J.A. and Jones, C.R. (1987), 'The Sefton Coast management scheme: project and process', *Coastal Management*, **15**, 267–97.

James, P.A. (1993), 'Soils and nutrient cycling', in D. Atkinson and J. Houston (eds.), *The Sand Dunes of the Sefton Coast*, National Museums and Galleries on Merseyside, 47–54.

Rooney, P.J. and Houston, J.A. (1998), 'Management of dunes and dune heaths; experience on the Sefton Coast, north west England', in Claus Helweg Ovesen (ed.), *Coastal Dunes – Management, Protection and Research. Report from a European Seminar, Skagen, Denmark, August 1997*, Danish National Forest and Nature Agency and the Geological Survey of Denmark and Greenland, Copenhagen, 121–29.

Sefton MBC (1997), *The Sefton Coast Management Plan, Second Review 1997–2006*, Sefton Metropolitan Borough Council.

TOWARDS BEST PRACTICE IN THE SUSTAINABLE MANAGEMENT OF SAND DUNE HABITATS: 1. THE RESTORATION OF OPEN DUNE COMMUNITIES AT AINSDALE SAND DUNES NATIONAL NATURE RESERVE

D.E. SIMPSON
Sefton Coast Life Project, England
and M. GEE
English Nature

Background

Ainsdale Sand Dunes National Nature Reserve (NNR) was purchased by English Nature in 1965 to protect a nationally outstanding sand dune site. These dunes were listed in 1915 by the eminent English naturalist Charles Rothschild in his schedule of 'areas worthy of preservation', with Ainsdale dunes being regarded as of 'primary importance' and of 'especial interest'. Subsequently a series of national and international designations have confirmed its importance as a dune system and wetland.

In the early twentieth century, part of what would become Ainsdale Sand Dunes NNR was being planted up with pine trees as the northern extension of a wider scheme across Formby Point (see Plate 1). Planting was instigated by the landowner, Charles Weld-Blundell, who had been inspired by the vast pine plantations created in Les Landes in southwest France. His aim was to stabilize the dunes and turn the 'wasteland' into a more productive estate providing a timber crop, woodland products and improved opportunities for agriculture and game. Shelterbelts of sea buckthorn *Hippophae rhamnoides* were planted to protect the young pines. The first pines were planted around 1900 and areas continued to be planted until the 1960s.

Until recently the woodlands covered 130ha, about a third of the Reserve area, mostly of Corsican pine *Pinus nigra laricio*. They were divided between a generally poor-quality 40ha seaward frontal woodland and a more healthy 90ha landward woodland. The woodlands today support an important population of the red squirrel *Sciurus vulgaris*, now rare in England (see Shuttleworth and Gurnell, this volume). The wet slacks were generally not planted, although drainage ditches were put in to lower the water table.

Threats

From the time of planting, the pines began to exert an ever stronger influence on the dunes. On the plantation sites, natural dune mobility slowed and stopped. The water table dropped, light levels were lowered, temperature extremes were reduced and soil character changed. Species characteristic of the open dunes were lost, to be replaced by relatively fewer woodland species. With time the maturing pines began to seed into surrounding dunes and drier slacks. In the areas of open dunes remaining between the plantations, conditions were altered by the shelterbelt effect of the trees. This reduced the influence of prevailing salt-laden winds that play an essential role in maintaining the character of the dune system. Myxomatosis in the rabbit population from the mid-1950s compounded this problem, so that by the late 1980s scrub woodland had colonized the unplanted slacks within the plantations and adjacent open dune habitat. Site managers were aware of the need to deal with this problem; a variety of techniques have been tested including mowing, cutting followed by herbicide treatment, turf-stripping, mechanical clearance and domestic grazing (Houston, 1997).

Water plays an important role in the ecology of the dune area. Habitat changes, as described, and drainage have led to a loss of wet slack habitat, an important habitat for protected species such as the natterjack toad *Bufo calamita* and petalwort *Petallophylum ralfsii* and other scarce wetland plants such as grass-of-Parnassus *Parnassia palustris* and marsh helleborine *Epipactis palustris*. Although it may not be possible to remove all of the constraints resulting from drainage works (e.g. adjacent

Plate 1. *Pine tree planting at Ainsdale circa 1936. Photograph by R.K. Gresswell, English Nature Archive, Ainsdale Sand Dunes NNR.*

road and farmland drainage) the effects of pine plantations can be tackled, providing an opportunity to restore the important open dune habitat. Given the level of importance, it was essential to develop techniques to restore open dune habitats in some of the plantation areas, while ensuring the conservation of all habitat types and landscape values. The frontal woodlands were chosen as most appropriate for the restoration of open dune habitats because of their seaward location. The rearward woodlands have been retained as forest and a management scheme has been developed to increase their diversity, their value for rare species and their landscape appeal.

The Open Dune Restoration Project

The project aims to restore open dune communities in the area of frontal woodlands and adjacent duneland. The project has been described by Gee (1998a) in English Nature's *Enact* magazine.

The consultation process began in September 1983 within English Nature and was extended to local coastal partners in 1988. The outline proposals were included in the first review of the Sefton Coast Management Plan (Sefton MBC, 1989).

Initial project planning brought in forestry, coastal management, nature conservation and cartographic advisors both from within English Nature and from organizations such as the Forestry Commission and the Environment Agency. Consultation was later extended to include the local planning authority and coastal protection authority through the mechanism of the Sefton Coast Management Scheme. All aspects of the project were considered including landscape change, erosion and sand blow.

Project plan

An overall plan was prepared to remove the frontal woodland in four phases (see Fig. 1) together with any associated scrub. The phased clearance was planned to take place at three-year intervals in accordance with the site management plan (now revised, Gee, 1998b).

It was thought that gradual tree removal would be less dramatic and the monitoring of the effects of the first two phases would inform later phases of the project. After tree removal, it was intended to burn the timber waste, bury fire sites and graze the reclaimed dunes with domestic stock by extending an adjacent grazing project initiated in 1990. In this way it was hoped that scrub regrowth would be controlled and the desired mosaic of low dune turf with bare sand patches would be maintained. Any regrowth not controlled by grazing stock could be treated later with a selective herbicide.

Project Operation

Restoration work

Tariffing (pricing up) of the timber was arranged and a felling licence obtained from the Forestry Commission. Initial publicity included press releases, some guided walks,

Figure 1. *Ainsdale Sand Dunes National Nature Reserve, four phases of woodland removal.*

leaflets to local residents and presentations to local councillors and members of the Sefton Coast Management Scheme. A combined timber and scrub removal contract was tendered and awarded. Access tracks, extraction routes, stacking areas and main road access for timber wagons were identified and warning signs erected. Sensitive nature conservation areas were identified to contractors. During the operations some indirect costs, such as the damage to tracks, were identified and as a result will be better addressed in later phases.

The first area was cleared in 1992 (see Plate 2). It covered 11ha with 4.5ha of pines, at the northern end of the frontal woodland. At this time monitoring of dune profiles was started to record any sand movement resulting from the clearfell. Cut scrub and pine brash was stacked in rows to be burnt, with fire sites buried and capped with clean sand. In some areas the burning of the pine-needle layer was tried, in other areas the needle layer was scraped off and the site tidied up. Stock fencing was erected around the cleared areas as an extension of the Reserve's grazing project. Herdwick sheep, a hardy breed from the Lake District, were introduced to the area in November 1993.

The second phase began in September 1995 and was completed by April 1996 with 16ha of scrub and woodland cleared, following a similar pattern to the first phase.

Plate 2. *First phase of the Open Dune Restoration Project, Ainsdale Sand Dunes NNR. Photograph by Sefton Coast Management Scheme.*

Project improvement

Several aspects of the project's operation have changed or improved through experience.

Separate specialist contractors for forestry and scrub work are now favoured. At first it was hoped that one contractor could deal with all aspects of the project, but it was realized that in future phases it would be advantageous to clear scrub a year before felling the pines. This would reduce the intensity and impact of the works and enable closer supervision of each stage. In the second phase any marketable timber from the scrub can be salvaged.

Appropriate machinery needs consideration. All cutting equipment suffers from rapid wear and tear in the sand. The most efficient method was found to be the use of a mechanical harvester on the easier areas with chainsaws for the rough timber and less accessible areas. Mechanical harvesters can be quicker but they leave a lot of small brash which may prove harder to collect, especially when driven over and compacted.

Herbicidal treatment of cut stumps and regrowth from deciduous species has been developed through the project. In the first phase stumps were not treated, in order to speed up operations and because it was hoped that stock would control regrowth. However, this was not sufficient, and in the second phase stumps were treated immediately with a selective herbicide (Triclopyr) and a foliar spray was used on any regrowth the following summer.

During the first phase there were some problems with a late start in the winter period. It was therefore decided that future contracts would start in early September to be completed by late spring. This would reduce disturbance to wildlife and leave staff free for other priority work during the summer. A specialist contractor should undertake each part of the project. The main timber contract and scrub contract both need to specify as a condition the collection and burning of cut materials. A separate contractor should undertake burial and capping of fire sites with clean sand and general tidying up. This will help to ensure quality work to timed deadlines. Experience has shown the importance of having one responsible officer, the site manager, managing the project with a forestry consultant carrying out the tariffing or pricing, together with some specialist supervision.

Some of the revenue from timber felling was used to fund a research project to investigate the population of the red squirrels on the reserve and to assess the impact of the project on the coastal population. This was one way of returning revenue from the felling to improve knowledge and understanding of woodland ecology and to improve future conservation management. Although considerable revenue was generated through the sale of timber, the costs of associated scrub clearance, final tidy-up work and fencing for grazing stock resulted in a relatively small net revenue.

At the end of the second phase of woodland removal, some local people raised concerns about the impact of the project. Indeed this has developed into an active pressure group opposed to further phases of the project. The group have lobbied local politicians, sought the support of local residents and visitors and carried their campaign to the media (see Edmondson and Velmans, this volume). Countering the complaints and attempting to increase understanding of the project has given the site manager an additional workload. A leaflet has been produced outlining the rationale for the work in the context of the overall management of the site.

For the third and fourth phases, English Nature plan to repeat their normal consultation procedures but there will be greater provision of information and more dialogue with community groups. This is expected to form part of a long-term Woodland Plan developed in association with the Forestry Commission. However, it is unlikely that this alone will go far enough to resolve the current difference of views between English Nature and some sectors of the local community.

Effects of Management

A monitoring programme has given valuable information about the effectiveness of this restoration project. Vegetation has been monitored by National Vegetation Classification methodology and species-presence surveys. Other forms of monitoring have included fixed-point photography, aerial photography, groundwater measurement by dip-wells, topographic profiles to monitor dune movement and observations on rare species and general wildlife.

Environmental changes have been recorded across the area cleared of pine trees. The project has restored higher groundwater levels. An overall increase in bare sand habitat is recorded without any large-scale destabilization. Dune wildlife has responded to the restoration of dune processes. Dune vegetation communities have re-established; there has been a particularly rapid response from wet slack communities, with the return of orchid species, the scarce yellow bartsia *Parentucellia viscosa* and skullcap *Scutellaria galericulata*. In some areas dune ridge communities have seen a flush of ruderals such as rose-bay willowherb *Chaemerion angustifolium*, but these are declining as nutrient levels drop and surfaces stabilize.

Large-scale flooding of the dune slacks in the former frontal woodland area has encouraged natterjack toads to spawn. For some slacks this is the first recorded instance and may represent the first breeding in fifty years or more. By the spring of 1998, natterjack toads were confirmed to be breeding in large numbers in slacks across the areas cleared in the first and second phases. The sand lizard *Lacerta agilis* has also been recorded in the restored areas. The rare tiger beetle *Cicindela hybrida*, together with nationally declining bird species grey partridge *Perdix perdix*, skylark *Alauda arvensis*, linnet *Carduelis cannabina* and reed bunting *Emberiza schoeniculus* have all been recorded. The ringed plover *Charadrius hiaticula*, a scarce breeding species on the Sefton Coast, has nested on the reclaimed duneland area in what must have been the first breeding attempt here for many decades. Wheatears *Oenanthe oenanthe* have been recorded singing and prospecting rabbit burrows for nest sites. All these are examples of the restoration of duneland species to a part of their former historical range at Ainsdale.

Eurosite Quality Award

English Nature received a Eurosite Quality Label in 1997 for the quality of its habitat management and restoration work on the open dune restoration project at Ainsdale. The project to restore open dune habitats at Ainsdale is one element of a wider initiative to protect and enhance the remaining areas of duneland on the Sefton Coast.

Acknowledgments

The work of David Wheeler, former site manager at Ainsdale Sand Dunes NNR, and the support of the EC Life-Nature Fund is acknowledged in the preparation of this article. Paul Rooney, former Project Officer for the Sefton Coast Life Project, is acknowledged for the preparation of the Eurosite Quality Award application.

References

Gee, M. (1998a), 'New life for old dunes', *Enact*, **6(1)**, 6–8.

Gee, M. (1998b), 'Ainsdale Sand Dunes National Nature Reserve Management Plan 1998–2003', unpublished report, English Nature, Wigan.

Houston, J. (1997), 'Conservation management practice on British dune systems', *British Wildlife*, **8(5)**, 297–307.

Sefton MBC (1989), *Sefton Coast Management Plan, First Review*, Sefton MBC, Southport.

TOWARDS BEST PRACTICE IN THE SUSTAINABLE MANAGEMENT OF SAND DUNE HABITATS: 2. MANAGEMENT OF THE AINSDALE DUNES ON THE SEFTON COAST

D.E. SIMPSON
Sefton Coast Life Project, England
J.A. HOUSTON
Sefton Coast Management Scheme, England
and P.J. ROONEY
Liverpool Hope University College, England (1996–1999 Sefton Coast Life Project)

Background

The Ainsdale dunes and foreshore including the Ainsdale Sand Dunes National Nature Reserve (NNR) and part of the Ainsdale and Birkdale Sandhills Local Nature Reserve (LNR) (see Fig. 1) have long been recognized as an outstanding area of wildlife interest. Since the time of the 1915 Rothschild list of proposed nature reserves, a series of designations has been attached to the area including Nature Conservation Review Grade 1* site, Geological Conservation Review site, NNR, LNR, Ramsar site, Special Protection Area (SPA) (the intertidal area), Site of Special Scientific Interest (SSSI) and, most recently, part of the Sefton Coast candidate Special Area of Conservation (cSAC). A high proportion of the area comprises EU fixed dune priority habitat. Key species include the sand lizard *Lacerta agilis*, the natterjack toad *Bufo calamita* and the liverwort petalwort *Petalophyllum ralfsii*.

The area covers approximately 7km² of the 22km long Sefton Coast dune system. Some 5 million people live within one hour's drive of the Sefton Coast, putting considerable pressure on this natural resource. Management of the NNR is undertaken by English Nature, the statutory wildlife agency for England, while that of the LNR is by Sefton Council, the local authority. This is co-ordinated across the whole Sefton Coast by the Sefton Coast Management Scheme. Both agencies have produced management plans for their sites.

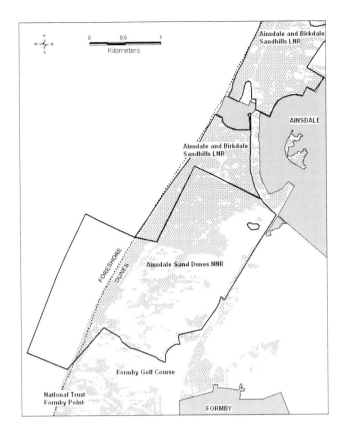

Figure 1.
Map of Ainsdale dunes.

Threats

The fixed dunes and associated species suffer from a number of factors threatening to reduce their nature conservation value. The spread of scrub and rank vegetation, in particular, leads to associated problems of soil development and the desiccation of slacks.

Scrub and growth of rank vegetation

Scrub invasion includes the establishment of tree and shrub species within the open dune landscape, particularly non-natives such as pine (predominantly *Pinus nigra laricio*), poplar *Populus* spp. and sycamore *Acer pseudoplatanus*, but also sea buckthorn *Hippophae rhamnoides* (native to the east coast of England) and birches *Betula* spp. This leads to the development of scrub woodland, a semi-natural habitat but one which results in a loss of valuable open dune habitat. In addition, low woody shrubs, such as creeping willow *Salix repens*, have grown tall and dense. Elsewhere an increase in rank grasses and herbs, particularly brambles *Rubus* spp., has produced a dense under-growth in areas not yet affected by scrub. These changes have led to a reduction of bare sand and short turf and therefore a reduction of the high ground-level temperature regimes which are a requirement of many dune species. Outbreaks of myxomatosis since the mid-1950s have significantly reduced rabbit numbers and have played a major role in these habitat changes.

Public pressure

Ainsdale beach is a popular tourist area and with large numbers of visitors to the beach, there are inevitably pressures put on habitats and species. The large number of visitors can lead to dune erosion and a loss of habitat quality. Associated problems such as litter, vandalism, fire and the collecting of rare species can cause considerable damage and take up staff time better spent on other nature management. In some circumstances, controlled or low trampling pressure can help to maintain short turf habitat and bare sand patches. Visitors come for a variety of reasons. Most come to the beach where they can park their car and make short forays into the dunes. Some come to walk in the dunes, often exercising a dog. Fewer come for quiet enjoyment, to experience the wild landscape and to observe the wildlife.

Aim of Management

Management aims to conserve the range of open sand dune habitats and species for which the area was designated, while enabling appropriate public access and informing and educating those visitors (Gee, 1998; Sefton MBC, 1993).

Vegetation management

Monitoring by fixed-point and aerial photography and casual observation has shown that important habitats and species would be lost without active intervention. At Ainsdale Sand Dunes NNR site managers have spent 30 years developing techniques to conserve the fixed dune habitats. These include scrub cutting and clearance, mowing, turf stripping, reprofiling and, more recently, grazing with domestic animals (Houston, 1997).

Scrub cutting and clearance

Initially site managers and volunteers using only hand tools carried out this work. Priority sites were targeted but only small areas could be cleared at a relatively high cost. Cut stumps required herbicide treatment but this was not always successful and was not deemed environmentally sound. The need for larger-scale operations and more sustainable techniques was clear. It was also found that rank ground vegetation was becoming a problem and needed special treatment. Mechanization of scrub clearance by using bulldozers, excavators or tractors with specialized rakes and grabs (see Plate 1) enabled large areas to be cleared and this was often more successful than scrub cutting because the roots were also pulled out (Rooney, 1998). However, such techniques are expensive, may cause some damage to dune topography and are not as sustainable and green as site managers may ideally desire. Monitoring studies have confirmed the value of scrub clearance as former dune habitats have returned, usually after a flush of ruderal plants (weeds) in the first few years. However, where other techniques cannot be used, and prior to grazing an area, this is still an important technique.

Mowing

Mowing was used on the NNR in the 1980s to control the height and density of creeping willow in dune slacks where it was rapidly overwhelming the short, botanically diverse dune turf. Trials began in 1979 and the technique has gradually been improved. Cut material has always been removed to avoid a build-up of nutrients and the development of thatch smothering the seedbank. In some areas a moss *Calliergon cuspidatum*, associated with creeping willow, has been raked out as it also smothers the underlying dune slack seedbank. Cuttings were hand-raked and removed off-site by tractor and trailer. Elsewhere on the Sefton Coast, forage-harvesters are now used to cut and collect vegetation on grassland and heathland. Both flail and rotary mowers have been used and both have been successful. Flails tend to produce a cleaner cut, though it is more difficult to rake and collect the chopped-up material. Rotary chains have been found to be more successful than blades, which tend to require more maintenance. Smaller tractors and mowers and low ground-pressure tyres cause less damage to slack topography and therefore microhabitat. Following a visit to Dutch sites in 1987, the mowing of creeping willow was changed on the NNR from autumn to spring, as this has been shown to reduce its vigour more effectively. However, this is not appropriate in wet slacks due to the presence of amphibians and damage caused to the soft turf.

Monitoring has shown an increased botanical diversity in mown slacks, but there

Plate 1. *Mechanized removal of sea buckthorn scrub, Ainsdale Sandhills LNR species recovery programme. Photograph by Sefton Coast Management Scheme.*

are problems associated with mowing. It is not a truly sustainable activity across all areas that require management. Some areas are just not accessible. Mowing can also cause soil compaction and reduce micro-topography, leading to greater uniformity of swards. Raking cut vegetation can be very labour intensive.

Turf-stripping and excavation

In the drought years of the mid-1970s many dune slack floors in the NNR were scraped (shallowly excavated) to provide breeding pools for the natterjack toad. It was found that the margins of these scrapes developed a diverse slack flora, which has been monitored over the years. Tracked excavators are the best equipment to use and most wheeled excavators should not be used in slacks. Small scrapes can even be hand-dug. However, vegetation succession dictates that without further management these scrapes will eventually turn to scrub. Continual intervention will be necessary to maintain early stages of succession unless the grazing pressure is relatively intense. Such regular intervention may cause damage to other site features and difficult decisions may have to be made. Where the off-site removal of the turf and spoil excavated is not practical, the materials should be buried, otherwise an area of ruderals (weeds) will develop around the scrape. A minimum depth of 1m of pure sand should cap the burial. Similar results were gained from the reprofiling of over-deepened scrapes, showing how resilient dune communities can be if provided with suitable conditions (Simpson, 1998).

Turf-stripping and excavation can certainly re-create early phases of dune succession and may be an important management technique for rare plants such as petalwort. However, both will cause disturbance which in the long term could lead to a significant percentage of the site being modified. It is costly and not a green approach. Ideally, in an actively accreting system, scrapes should not be necessary, as natural processes should result in the creation of embryo slacks.

Grazing

For centuries rabbits had been an important influence on the ecology of the Sefton dunes. Following the outbreaks of myxomatosis in the mid-1950s the balance was lost and the dunes became overgrown, with a loss of nature conservation value. Domestic grazing was reintroduced onto the Ainsdale dunes by English Nature in 1990. It was previously practised on the rearward dune areas along the coast until the late nineteenth century. Domestic grazing has proved to be the most successful and appropriate form of vegetation management. It has controlled target species such as creeping willow, and there has been an increase in species diversity with a corresponding return to a low structural mosaic of vegetation with bare sand patches. Domestic grazing has encouraged a resurgence in the rabbit population and successfully maintained early successional stages in scrapes. Herdwick sheep from the English Lake District have proved to be a particularly effective and appropriate type of grazing stock, with relatively low staff input and maintenance costs, allowing staff to deal with other key projects (Simpson and Gee, 1997). Hebridean sheep and Aberdeen Angus and Hereford cattle have proved their value, but presently for resource reasons cannot be accommodated within the grazing regime. With increasing knowledge and understanding, adjustments are being made to the grazing regime to ensure the best possible results are achieved for wildlife. Such is the success of domestic grazing on the Ainsdale Sand Dunes NNR that Sefton Council has approved a grazing trial on seven hectares of the adjacent LNR. This is an open access public area and the establishment of a grazing project will require the support of local people and visitors alike. Public relations work and the provision of access points along desire lines is planned. The initial survey work, design and site facilities for the project have been completed with assistance from the Sefton Coast Life Project.

Blowouts

Blowouts create new sandy habitats within the fixed dunes including incipient slacks, especially important for annual plants, specialized invertebrates, natterjack toads and sand lizards. They are an important element of the dynamic processes within the dune system. However, where recreation causes erosion, and particularly where property or infrastructure is threatened, dune restoration techniques are used. Dune stabilization works have been necessary in Ainsdale and Birkdale Sandhills LNR adjacent to a main trunk road, near a holiday village and to prevent sand encroachment onto a railway line.

Species management

Three species typical of open dune habitats present in the Ainsdale hills area are protected by both European and national legislation. These are the natterjack toad, sand lizard and petalwort. Ideally, management aims to conserve these species through broad-scale habitat or process management. However, to ensure their survival, specific measures are taken or considered for each species.

A series of natterjack toad *key pools* are maintained as breeding pools (see Plate 2). The key pool concept involves some intensive management, including occasional reprofiling to ensure open conditions, with a deepened section to provide water in drought years. Split paling fencing is used to protect pools in areas with high public pressure and interpretation signs and wardening help to explain the importance of this species (Simpson, 1992).

Sand lizards are most frequent in the transitional mobile/fixed dune habitat near the beach. Therefore there is the potential for disturbance from visitors. To reduce the risk, fencing is erected to protect some sensitive sites and minor paths may be re-routed. Where vegetation conditions are becoming overly fixed, for example due to a temporary decline in rabbit numbers, small bare sand patches may be maintained (EAU, 1992).

Petalwort is a liverwort which favours conditions found in the early phases of dune slack succession. Grazing, light trampling and disturbance appear to assist the conser-

Plate 2. An example of a managed key natterjack toad breeding pool. Photograph by Sefton Coast Management Scheme.

vation of this species. Presently (1999) a management strategy is being developed to address these issues.

Visitor management

Visitor management on the Ainsdale Sand Dunes NNR and Ainsdale and Birkdale Sandhills LNR is part of a wider zoning system used on the Sefton Coast as a whole. This identifies honeypot areas such as Ainsdale beach and closed or permit-only sanctuary areas such as the majority of the Ainsdale Sand Dunes NNR, together with a range of partial access and less intensively used sites in between such as the Ainsdale and Birkdale Sandhills LNR. This has the advantage of encouraging the majority of people who come to the coast for the beach and associated facilities to be concentrated in a few areas where intensive management can take place to accommodate them. For example, facilities include laid-out car parks, boardwalks to the beach, fencing to protect dune habitats, trails, information boards, toilets and food and drink outlets. In this way the integrity of the dune system is maintained and pressure is taken off important habitats and species elsewhere on the coast.

The quiet situation within Ainsdale Sand Dunes NNR provides a sanctuary for nature while allowing access by permit. The limited access also made the establishment of a sheep grazing project easier. With relatively low visitor numbers the sheep have not been troubled. Only small interpretation signs, together with articles in the local coastal magazine, guided walks and illustrated talks by the site manager have been required. Access by permit, or views from the three main permitted paths, have generally been sufficient for those wishing to see the NNR fixed dunes. The LNR has open access to public visitors and includes several permitted paths as well as many informal footpaths, particularly towards the beach. The waymarked paths encourage people to remain on them, leaving areas away from the path relatively undisturbed. In this way larger numbers of visitors can visit the dunes without undue disturbance to wildlife. However, they are able to experience the wild nature of the site, and to gain greater understanding of the need to conserve such areas. Nearer the beach access point, visitor access is more formal with beach car parking. Visitor management in the area includes information signs and facilities enabling inquisitive visitors to discover more about the wildlife interest of the area.

The Sefton Coast Management Scheme has worked for more than twenty years to ensure the best possible management of the coast. Through its guidance, the footpath network and associated signage and interpretation system has been developed and co-ordinated across the Sefton Coast. It has also enabled greater integration of management actions and encouraged developments such as the proposal for domestic grazing on the LNR. The scheme also assists with monitoring of dune condition across the whole coast and is working to develop management strategies for habitats and species across site boundaries.

Working towards Best Practice

The Ainsdale dune area has seen more than thirty years of conservation management which has been gradually modified and improved. As a result of the Sefton Coast Management Scheme, management is now more fully co-ordinated along the Sefton Coast. Strategies and plans inform site management and regular meetings are held for discussion and review.

The Sefton Coast needs to accommodate large numbers of visitors while effectively conserving the wildlife interest. At Ainsdale domestic grazing by sheep has proved to be the ideal management tool. However, scrub management and herbicide treatment are still needed in the early years of such a project. In the seaward dunes most popular with visitors and other areas inappropriate for domestic grazing, a variety of techniques are needed for vegetation management, such as mowing, scrub management with herbicide treatment and fencing of sensitive areas. Special management is required for a number of important species associated with the open dunes to ensure their survival. This has proved possible to achieve within annual work programmes. A wardening presence has proved to be important for visitor management and protection of rare species. In future it could be possible to manage sites better from a strategic viewpoint through, for example, the removal or variation of fencing boundaries and greater liaison between site staff across landownership boundaries.

With careful management large numbers of visitors can be accommodated while conserving dune wildlife and providing the opportunity for increased public understanding and support for dune conservation.

Acknowledgments

The authors would like to acknowledge the work of Mike Gee (Site Manager, Ainsdale Sand Dunes NNR), David McAleavy (Head of Coast and Countryside, Sefton MBC) and Peter Gahan (Site Ranger, Ainsdale Sandhills LNR), and the support of the EC Life-Nature Fund in the preparation of this article.

References

Environmental Advisory Unit (1992), 'Sand Lizard Conservation Strategy for the Merseyside Coast', unpublished report, English Nature, Peterborough.

Gee, M. (1998), 'Ainsdale Sand Dunes National Nature Reserve Management Plan 1998–2003', unpublished report, English Nature, Wigan.

Houston, J. (1997), 'Conservation management practice on British dune systems', *British Wildlife*, **8(5)**, 297–307.

Rooney, P. (1998), 'A thorny problem', *Enact*, **6(1)**, 12–13.

Sefton MBC (1993), 'Draft Ainsdale and Birkdale Sandhills Local Nature Reserve Management Plan (Parts 1 and 2)', unpublished report, Sefton MBC, Southport.

Simpson, D. (1992), 'Conservation of the Natterjack Toad (*Bufo Calamita*) on the Sefton Coast', unpublished report, English Nature, Wigan.

Simpson, D. (1998), 'Bringing back the slacks', *Enact*, **6(1)**, 9–11.

Simpson, D. and Gee, M. (1997), 'Setting up a grazing project', *Enact*, **5(4)**, 23–26.

TOWARDS BEST PRACTICE IN THE SUSTAINABLE MANAGEMENT OF SAND DUNE HABITATS: 3. MANAGEMENT FOR GOLF AND NATURE ON THE SEFTON COAST

D.E. SIMPSON
Sefton Coast Life Project, England
P.J. ROONEY
Liverpool Hope University College, England (1996–1999 Sefton Coast Life Project)
and J.A. HOUSTON
Sefton Coast Management Scheme, England

Background

Golf courses have been a feature of the landscape of the Sefton Coast for over a century. The vast extent of sand dunes provided excellent terrain with marvellous views and was judged to be as good as the sandy links of the east coast of Scotland where the game first became popular. The West Lancashire Golf Club at Crosby developed the first course on the Sefton Coast in 1873, towards the southern end of the dune system. Since then, clubs have established courses along the length and breadth of the coast, with a present-day total of seven duneland (links) courses. Foster (1997) has written an excellent account of the history and development of golf courses on the Sefton Coast.

Golf courses today occupy over a quarter of the dune area, 550ha out of a total of some 2000ha, and much of this area includes the EU priority habitats of fixed dune and dune heath, with associated species such as the protected sand lizard *Lacerta agilis*. It is therefore essential, for the overall conservation of the dune system, that golf course management is sympathetic to nature and that clubs help to conserve their semi-natural duneland habitats (see Stubbs, this volume).

Protection and Development

Golf courses on the Sefton Coast have undoubtedly protected large areas of dunes from other forms of development, particularly housing. This has helped to maintain the integrity of the dune system as a whole and limited the degree of fragmentation.

The development of playing areas has obviously modified the natural duneland, although on the Sefton Coast only about 25 per cent of any course area is managed turf. Access is restricted to club members and visitors, and most players try to keep out of the rough! The golf courses can therefore act as real sanctuary areas for wildlife.

Traditional golfing management has included mowing, some grazing, scrub clearance, ditching, pond and woodland management. Considerable expertise in habitat management has been gained indirectly by green staff and Green Committee members (most golf courses are run by members with a Green Committee responsible for estate management). The nature conservation importance of the golf courses on the Sefton Coast is confirmed by designations such as Site of Special Scientific Interest (SSSI) and Site of Local Biological Importance (SLBI: a non-statutory designation by Sefton Council). Some sites are additionally of international (Ramsar) or European (cSAC) importance.

Wildlife Features

The golf courses of the Sefton Coast support a magnificent diversity of duneland habitats and species. In addition to fixed dune and dune heath habitats are mobile dunes, dune slacks, dune grasslands, ponds, scrub, semi-natural woodlands and conifer plantations.

Sand lizards are found on three of the courses, including the most northerly naturally occurring population in the UK at Hesketh Golf Club. The natterjack toad *Bufo calamita* is also found on three courses, including two with recent successful re-introductions. The nationally rare grey hairgrass *Corynephorus canescens* is found in profusion at Southport and Ainsdale Golf Club. The dune heath at Formby Ladies Golf Club includes two of only three plants of crowberry *Empetrum nigrum* known on the Sefton Coast. West Lancashire Golf Club is notable as having one of the highest population densities in the country of the declining skylark *Alauda arvensis*. Royal Birkdale Golf Club contains exceptional dune slack habitats, Hillside Golf Club has perhaps the most dramatic high dunes, and the Formby Golf Clubs contain a great variety of dune and heath habitats.

Management Planning

Golf course management has had significant benefits for nature. However, as on the nature reserves it became obvious in the 1970s and 1980s that site quality was declining. This was particularly due to invasion by scrub and rank grass which began to dominate the open dunes following the outbreaks of myxomatosis in the mid-1950s. English Nature responded to this problem by preparing site management plans for those courses which included SSSIs (Royal Birkdale and Hesketh). These gave a description of the site, an analysis of its importance, a rationale for management and objectives with related prescriptions. Five-year work programmes were also prepared. These plans were partially successful.

In 1996 a seminar entitled 'Life in the Rough' was organized by the Sefton Coast Life Project at the Royal Birkdale Golf Club to launch a golf and nature initiative on the Sefton Coast. This was the first time that representatives of all the Sefton Coast clubs had assembled to discuss environmental management. Following this meeting all the clubs on the Sefton Coast worked with the Sefton Coast Life Project to prepare management plans and to take action to conserve important habitats. The initiative was successful and has helped to renew a partnership between golf and nature and raise awareness about the importance of managing the dune system as a whole.

The new management plans follow the general format of the earlier English Nature plans but with some significant differences. The plans have been produced in partnership with the clubs and have been made more golfer-friendly, by including a *hole-by-hole* section in the rationale. An important prescription in these plans is to produce rare features maps to advise wildlife management and other activities on the course. The plans also recommend that a conservation officer be appointed by each club to oversee the works detailed in the plan and to provide continuity. Ideally the conservation officer would be part of a group including the Chair of Greens, the head greenkeeper, an English Nature officer where appropriate, a representative of the Sefton Coast Management Scheme and any other appropriate representatives from the club or environmental organizations.

Practical Conservation Work

Quite often conservation management on golf courses and nature reserves is a form of gardening, using fairly costly, but small-scale, projects to conserve specific features. A good example would be the selective cutting of scrub on south-facing slopes to improve habitat for the sand lizard. Such actions have been necessary to compensate for the restriction of natural dune processes in golf course areas, but management practices should, wherever possible, work more with natural processes and at the habitat level. Overall, conservation management needs to be considered more strategically in relation to the whole dune system. For example, habitat links could provide wildlife corridors within and beyond the golf courses.

Scrub management

Scrub is a natural component of fixed dunes but its spread needs to be controlled to conserve open dune habitats. Grazing is the natural way in which the growth of scrub is checked. On some golf courses rabbits will help to control scrub but, considered as a pest species, they are seldom tolerated in sufficient numbers to be effective. Scrub clearance on golf courses should aim to link up the sometimes fragmented open dune habitats. It may also be necessary to improve the habitat for rare species, such as the sand lizard. In some areas dune habitat has survived beneath the scrub (especially under white poplar *Populus alba*) and restoration is relatively simple. As a general principle fixed dune areas should be cleared of non-native species such as white poplar, and the more aggressive native scrub species including sea buckthorn *Hippophae rhamnoides* and birches *Betula* spp. should be reduced to an acceptable minimum. Some scrub, however, should be managed to grade into woodland and could be coppiced to enhance its value for wildlife. Other areas have been identified in management plans where the succession from scrub to woodland will be left to nature.

Scrub can be cleared in several ways, but all work is best done in the winter months for least disturbance to wildlife. Cutting with chainsaws or handsaws can be followed by stump treatment with an approved selective herbicide. Stump treatment is most effective before the end of the year and is less effective if there is rainfall soon after application. In the following growing season regrowth sometimes needs re-spraying with herbicide. Any treatment in areas with creeping willow and heather undergrowth will require care and spraying may not be appropriate.

On some sites with large areas of scrub, machinery such as excavators and tractors with specialized grabs can be used (Rooney, 1998). Fire sites should be chosen carefully to minimize damage to habitats and should be buried under at least a metre of pure sand.

Gorse requires coppicing to maintain its vigour. The central portion of old stands should be cut first and when this has regenerated sufficiently the outer region can be cut. If gorse cover is required it can be introduced to areas either as young plants grown in a nursery or by rotavating in gorse litter and seed.

Mowing

Mowing is an essential feature of golf course management that helps to retain open dune grasslands in the semi-rough and rough. By slight modifications to mowing regimes great benefits can be gained for wildlife. For example, by varying the time of cutting or relaxing the mowing regime, flowers can be allowed to set seed. Other areas can be mown in alternate years or less frequently to open up the sward and to provide invertebrates and small mammals with a suitable habitat mosaic.

Some courses may need to consider whether they are mowing too great an area and too often. The area in front of the tee, or carry, may offer opportunities for reduced mowing, while islands of rough on the fairway could provide interesting hazards as well as valuable wildlife habitat. A curving boundary to the rough is more visually pleasing than a straight one and can improve conditions for wildlife.

Forage harvesters are one of the most popular pieces of mowing equipment for nature conservation as they can both cut and collect vegetation (see Plate 1). They combine a flail mower with a large box for cuttings. Cuttings should always be collected to reduce nutrient enrichment and the build-up of thatch. Simple rotary and flail mowers are equally effective but flail machines produce more finely chopped cuttings, which are less easy to rake up and collect. Rotary mowers are probably more efficient when equipped with chains rather than blades, which require more maintenance. All cuttings should be removed off-site. The removal of cuttings from fairways, greens and tees is also important for golf management as it helps reduce the build-up of nutrients, which in turn helps to maintain finer grasses requiring less management.

Plate 1. *Forage harvester in dune slack. Photograph by Sefton Coast Life Project.*

Mowing in slacks needs special consideration. In wet slacks, vehicles need to have low ground-pressure tyres to reduce the likelihood of compaction and disturbance. If creeping willow is a target species to control, the cut should be made in spring where water table conditions allow. Mowing at this time of year more effectively reduces the vigour of creeping willow than an autumn cut.

The mowing of selected areas of heathland can ensure the maintenance of a mosaic of habitat structure, providing niches for different species. The semi-rough is usually more regularly mown, thus maintaining heather in pioneer and vigorous phases. Firebreaks can provide another opportunity to increase structural diversity in a heathland habitat. Sandy firebreaks created on south-facing slopes can be very valuable for rare reptiles and invertebrates. In the rough, however, areas should be identified for less regular cutting of heather to allow building and mature phases to be present, with areas cut on, for example, a ten- or twenty-year rotation. This should ensure that these areas also stay vigorous and free from infestation by the heather beetle

Lochmaea suturalis. In other areas heather could be allowed to senesce and die back. These areas could subsequently be turf-stripped to initiate the cycle once again. It is advisable to mow in early spring as then the heather is put under less stress than with an autumn cut. However, if seed is being harvested for restoration projects a late autumn cut is required.

Turf-stripping on dune ridges

Turf-stripping is used by greenkeepers to provide turf for repair works. Such stripped areas, usually on drier rough ground, create valuable bare ground habitats for wildlife. Turf-stripping is usually carried out on a small scale, often with hand tools, although it can be mechanized. Bare ground habitats are valuable in slacks and also on ridges. They provide basking and egg-laying habitats for the sand lizard, burrowing habitats for the natterjack toad, and a range of habitat requirements for rare invertebrates such as the tiger beetle *Cicindela hybrida*, solitary bees and wasps. Bare areas are also important for the survival of annual dune grasses and flowers. With some planning, turf-stripping could be developed on many golf courses to be even more valuable for dune wildlife. Repair works on golf courses should use local turf rather than importing turf that could introduce non-native species.

Dune restoration

Where a blowout has become a management problem or works have disturbed sand, it may become necessary to restore or stabilize the area. Techniques are well known (Brooks and Agate, 1986) and include the planting of marram *Ammophila arenaria*, reprofiling, fencing and access control. In this way valuable dune grassland habitat can be created.

Heather restoration

Heathland restoration has been successfully undertaken on a number of golf courses on the Sefton Coast. This has included a project to eradicate a problem with the heather beetle, which had killed areas of older heather. At Formby Golf Club, with advice from the Sports Turf Research Institute (Taylor, 1995), the old dead heather was mown short and the ground scarified using a forage harvester. As a result the regeneration of heather was very successful.

Water management

Water management is a vital issue on links golf courses. Best practice to conserve water includes techniques such as wetting agents, drought-tolerant grass seed mixes and efficient irrigation systems. Links fairways were never intended to be bright green throughout the year!

Effective water resource management requires a strategic approach that considers a whole dune system and groundwater unit. The Environment Agency are taking a lead role in this matter (see Elliot and Birchall, this volume).

Wetland habitats

Wetland habitats are often at a premium on sand dune golf courses. Because of drainage requirements to maintain the course in a playable condition all year round, wetland habitats may be restricted to reservoirs, feature ponds and ditches. These features can be managed to improve their conservation value by increasing the area of shallows, lengthening the edge and reducing shading by trees.

Natural slacks on Sefton Coast golf courses are usually dry (or drained) and are often incorporated into the design of the course as fairways. However, there may also be opportunities to recreate dune slack communities on many golf courses. Particularly appropriate locations would be scrub-invaded and dried-out slacks in the rough and out-of-play areas. It may also be possible to create slack habitats as part of any reconstruction works such as the creation of new tees. Works in wet areas should be undertaken with a tracked vehicle such as an excavator or bulldozer, with little more than the surface turf and roots being scraped off. Enriched spoil should preferably be removed from the site or, if this is impractical, buried and capped with at least a metre of pure sand. Care should be taken with surface levels to ensure that these areas only flood in the winter months, creating species-rich and colourful dune slacks, which could also be used as breeding pools by the natterjack toads.

Creating natterjack toad breeding pools

Golf courses provide an opportunity to restore the rare natterjack toads to large areas of their former range, even where drainage and scrub encroachment have denied them natural breeding sites. Small lined pools can be installed in suitable out-of-play areas (see Plate 2). Butyl rubber liners are ideal and long lasting. These pools should be drained in late summer to control invertebrate predators. This can be through a drainage system incorporated into the pool, as developed by the Sefton Coast Life Project and the Royal Birkdale Golf Club, or by using a pump. They can be left to refill with winter rainfall. If there is a drain, refilling the pool can be left until late spring to deter natterjack toad competitors such as the common toad *Bufo bufo*. A few pieces of wood debris can be scattered in marginal areas to provide emerging toadlets with protection from desiccation and predators. Care should be taken to ensure that water does not wick out of the pond through sandy margins to the surrounding dunes. A concrete liner may help to buffer water acidity on heathland courses.

Woodland management

There can be opportunities for the selective removal of trees on golf courses to improve conditions for golf and nature. The removal of trees can open up and connect areas of dune and heathland habitat, allowing species to move more freely between areas and thus reducing the danger of localized extinctions. It can also restore views that have been gradually lost over the years and re-establish true links playing conditions. Valuable woodland edge habitat can be increased by the selective removal of trees along otherwise straight edges. This could be of particular benefit to species such as

Plate 2. *A lined natterjack toad breeding pool at the Royal Birkdale Golf Club. Construction of the pool was co-financed by the Pond Life Project, Sefton Coast Life Project and the Royal Birkdale Golf Club. Photograph by Sefton Coast Life Project.*

bats. Cutting back trees from the edge of dune slacks can improve their value by re-exposing them to the elements, reducing leaf-fall into them, restoring water levels and generally slowing down the pace of vegetation succession. Around greens and tees, tree removal helps to increase air flow and light levels, thereby improving the quality of the turf and reducing playing surface problems such as disease and frost.

Informing management

The extent, distribution and condition of habitats and species need to be monitored to check whether management objectives are being met.

Fixed-point photography can give valuable information about the condition of the course from a ground-level perspective. On the Sefton Coast a photographic record of each golf course has been completed through the work of the Sefton Coast Life Project and these have been transferred to a Geographic Information System (GIS). Photographs are taken from each tee with a view looking down the fairway. Additional shots are taken of important habitats or areas of concern. A guide to fixed-point photography has been given to each club so that the photography can be repeated at appropriate intervals. Aerial photography is also available.

Rare species require more detailed monitoring, usually as part of wider coastwide studies co-ordinated by specialists. On the Sefton Coast, for example, work on the rare natterjack toad population includes data collection concerning adult assembly, spawning and toadlet emergence which is summarized in an annual report. All actions for this species are co-ordinated through an agreed local species strategy. Increasingly, through this strategy, golf courses are being identified as potential areas for the re-establishment of the population to its former range.

A coastwide network for monitoring water levels is to be established in 1999 by the Environment Agency, with particular reference to the licensed abstraction of water by golf courses for irrigation (see Elliot and Birchall, this volume). The golf courses are increasingly being asked to demonstrate that their licensed abstractions are sustainable and not detrimental to adjacent wetland features.

Interpretation

Nature conservation needs to become more central to the management of golf courses. On the Sefton Coast, the Life Project developed several interpretation ideas, including a regular newsletter *Linkslines,* the use of interpretation boards and guided walks and talks for club members. The process of producing a management plan with its rounds of consultation helps to increase knowledge and understanding within the club of the course's wildlife importance.

Training

Training events for green staff should incorporate wildlife identification, ecology, conservation, management techniques, equipment and the value of the management of the rough for golf and nature. Training events give opportunities for golf course consultants, head greenkeepers, greenkeepers and members to share their knowledge and experience to the benefit of the conservation of the course.

Golfing management

As part of normal course management, tees, greens and even fairways may occasionally be relocated. On any Site of Special Scientific Interest this would require a consent from English Nature. If such plans do go ahead there may be opportunities to improve duneland habitat considerably. For example, scrub could be cleared, wetlands created and dune and heathland habitats restored.

Pest control is a routine activity on most golf courses. Rabbit control is often necessary, for example, to prevent damage to the turf on greens. Rabbits as grazers, however, are a valuable component of a duneland ecosystem and should not be controlled if they are not causing a problem. In particular, rabbit control should not be by gassing. Burrowing animals such as the natterjack toad and sand lizard will also succumb to gassing control methods.

The ecological management of golf courses has been promoted through the European Golf Association Ecology Unit's 'Committed to Green' initiative (Stubbs, 1997 and this volume). This provides a procedure for golf clubs to develop an environmental approach to course management, validated by a quality control certificate.

Acknowledgments

Ian Kippax and Kevin Gould – Hesketh Golf Club
Brian Kenyon, Derek Turner and Martin Twist – Hillside Golf Club
Peter Rostron and Chris Whittle – Royal Birkdale Golf Club
Bob Hutt, Roy Williams and Mike Mercer – Southport and Ainsdale Golf Club
Keith Wilcox, Austin Cartmell and Derek Postlethwaite – Formby Golf Club
Val Moran and John Bourhill – Formby Ladies Golf Club
Graham Dudley and John Muir – West Lancashire Golf Club
Ian Macmillan – Hankley Common Golf Club
Mickey Dodd – Burnham and Berrow Golf Club
Bob Corns – English Nature, Taunton
Bob Taylor – Sports Turf Research Institute
    The support of the EC Life-Nature Fund is acknowledged in the preparation of this article.

References

Brooks, A. and Agate, E. (1986), *Sand Dunes: A Practical Handbook*, British Trust for Conservation Volunteers, Wallingford.

Foster, H. (1997), *Links along the Line*, Ainsdale and Birkdale Historical Association, Birkdale.

Rooney, P. (1998), 'A thorny problem', *Enact*, **6(1)**, 12–13.

Stubbs, D. (1997), *The Committed to Green Handbook for Golf Courses*, European Golf Association Ecology Unit, Information Press, Oxford.

Taylor, R.S. (1995), *A Practical Guide to Ecological Management of the Golf Course*, British and International Golf Greenkeepers Association and the Sports Turf Research Institute, Bingley.

Section 5

INTERNATIONAL AND NATIONAL PRIORITIES: STRATEGY AND IMPLEMENTATION

SAND DUNES AND THE HABITATS DIRECTIVE: PREPARATION OF THE UK NATIONAL LIST

JOHN HOPKINS and GEOFF RADLEY

Joint Nature Conservation Committee UK and English Nature

Introduction

One of the most important recent developments in the field of international wildlife legislation was the adoption in May 1992 by member states of the European Union of *Council Directive 92/43/EEC on the Conservation of Natural Habitats and of Wild Fauna and Flora*, known more or less universally as the Habitats Directive.

Intended to provide member states of the European Union with a mechanism to meet their obligations under the 1979 Bern Convention and to compliment the provisions of the 1979 Birds Directive, the main aim of the Directive is 'to contribute towards ensuring biodiversity through the conservation of natural habitats and of wild fauna and flora in the European territory of the member states to which the treaty applies' (Article 2).

The 24 Articles of the Directive (Table 1) include a range of measures including conservation of features in the landscape which are important for wildlife, the protection of listed species from damage, destruction, or over-exploitation and the surveillance of species and habitats. However, the force of member states' legal obligations under the various Articles varies considerably, ranging from the strictly obligatory through to the discretionary. Member states' most stringent obligations relate to the selection, designation and protection of a series of sites, to be called Special Areas of Conservation (SACs). This paper deals with the selection of this list of sites in the UK. Along with the Special Protection Areas (SPAs) classified under the 1979 Birds Directive these sites are intended to form a coherent ecological network, Natura 2000.

Special Areas of Conservation

The selection of SACs is intended to provide protection for the most threatened and important habitat types and species found in the European Union and the procedure to be followed is set out in Article 4. Annex I of the Directive lists 169 *habitat types of community interest*, whilst Annex II contains a list of 623 *species of community interest* which must be protected within the SAC series.

Table 1. Summary of the main Articles of the Habitats Directive (Articles not listed are largely of an administrative character)

Article 1. Definition of terms used in the Directive (e.g. Natural Habitat Type of Community Interest; Site of Community Importance).

Article 2. Aims of the Directive.

Article 3. Objectives of the Natura 2000 site series, which comprises protected sites under both the Birds Directive and Habitats Directive.

Article 4. Procedures for selecting Special Areas of Conservation (SACs) for habitats listed at Annex I and species listed at Annex II, using criteria listed at Annex III.

Article 6. Site protection obligations.

Article 7. Harmonization of site protection mechanisms under the Birds Directive with those under the Habitats Directive.

Article 10. Measures to encourage the conservation of habitat features outside the SAC series, which are important to the coherence of the Natura 2000 site series.

Article 11. Requirement to carry out surveillance, particularly of priority habitats and priority species.

Article 12. measures to strictly protect animals listed in Annex IV(a)

Article 13. Measures to strictly protect plants listed in Annex IV(b)

Article 14. Measures to encourage maintenance of the favourable conservation status of exploited species in Annex V.

Article 20. Establishment of a committee of member states to assist in implementation of the Directive.

Article 22. Measures concerning reintroductions, introductions and education/publicity.

In addition, a number of these habitat types and species are afforded *priority* status, indicated by an asterisk in the relevant Annex, and are afforded preferential treatment in the selection of sites and also enhanced legal protection. Article 4.1 requires that member states should employ criteria listed at Annex III of the Directive in the selection of sites. It is not therefore intended that all occurrences of the Annex I habitat type or Annex II species should be included within the SAC series but rather that a selection of the best examples is made, although for some rare types the list is likely to be exhaustive.

Article 4 of the Directive also sets out an administrative procedure and timescale for the establishment of SACs:

- Member states to prepare a 'national list' of sites of importance for Annex I habitat types and Annex II species, based upon criteria in Annex III Stage I and to submit this national list to the European Commission by June 1995.
- The national list to be considered in the light of the criteria listed in Annex III Stage 2 and within the context of the six biogeographical regions described in Article 1(c) (i.e. Alpine, Atlantic, Boreal, Continental, Macaronesian, Mediterranean) and the EC as a whole. Member states and the European Commission to adopt sites on member states' lists, plus any agreed additions as *Sites of Conservation Importance* (SCIs) by June 1998.
- SCIs to be designated by member states as SACs within six years of adoption by the Commission.

At the time of writing the UK has submitted a national list of 315 sites of which 27 are selected primarily due to the presence of dune habitat (see Fig. 1). However, due to the slow submission of sites from other member states, particularly with regard to the Atlantic Biogeographical Region in which the UK occurs, negotiations have yet to take place concerning which of the sites on the UK list should be formally adopted as SCIs and designated as SACs. To date only the site list for the Macaronesian Biogeographical Region has been agreed.

Annex I Dune Types

Sand dune habitats are well represented in Annex 1. Ten of the 76 Annex I habitat types found in the UK are specific to sand dunes (Table 2). These ten habitats encompass a very large part of all of the variation exhibited by dune habitats in the UK. They are, however, a heterogeneous assemblage of habitats and exhibit the following pattern of variation which needs to be taken into account in site selection:

a) **Ecological character.** The official reference for the identification of Annex I habitats is *Interpretation Manual of European Union Habitats* (European Commission DGXI, 1996). In the UK the National Vegetation Classification

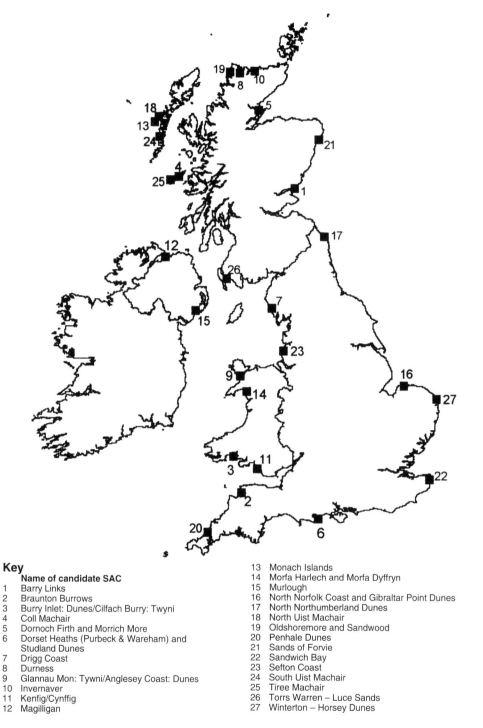

Key

Name of candidate SAC
1   Barry Links
2   Braunton Burrows
3   Burry Inlet: Dunes/Cilfach Burry: Twyni
4   Coll Machair
5   Dornoch Firth and Morrich More
6   Dorset Heaths (Purbeck & Wareham) and
    Studland Dunes
7   Drigg Coast
8   Durness
9   Glannau Mon: Tywni/Anglesey Coast: Dunes
10 Invernaver
11 Kenfig/Cynffig
12 Magilligan

13 Monach Islands
14 Morfa Harlech and Morfa Dyffryn
15 Murlough
16 North Norfolk Coast and Gibraltar Point Dunes
17 North Northumberland Dunes
18 North Uist Machair
19 Oldshoremore and Sandwood
20 Penhale Dunes
21 Sands of Forvie
22 Sandwich Bay
23 Sefton Coast
24 South Uist Machair
25 Tiree Machair
26 Torrs Warren – Luce Sands
27 Winterton – Horsey Dunes

Figure 1. UK national list of sites selected to protect Annex I coastal sand dune habitats.

(NVC) (Rodwell, in press) has been used to characterize the Annex I habitats where possible (although not all types have clearly defined relationships to the NVC). The habitats listed in Annex I vary considerably in the range of ecological variation they encompass. For example, in the case of *Embryonic shifting dunes* the habitat can be described in terms of a single NVC community, and exhibits a limited range of floristic variation around the coasts of Britain. In contrast *Humid dune slacks* is equivalent to five NVC communities and * *Fixed dunes with herbaceous vegetation (grey dunes)* is equivalent to six NVC communities, with both habitat types hosting rich assemblages of plant and animal species strongly differentiated within and between sites in terms of successional stage, pH and biogeography. The most extreme case is provided by machair, which corresponds to an assemblage of dune, wetland and cultivated habitats found on the sand plains of northwest Scotland (Ritchie, 1975 and see Angus, this volume) and may be more sensibly seen as a landscape type or *habitat complex*.

b) **Extent.** Dune habitat is comparatively rare in the UK; no Annex I habitat covers more than 20,000ha in contrast to the more than 1.5 million ha of blanket bog which occurs. However, the individual Annex 1 habitats vary considerably in abundance, from *Dune juniper thickets*, which are estimated to cover less than 20ha in total, through to *Fixed dune with herbaceous vegetation (grey dunes)* and *Machair* which each cover more than 10,000ha, an approximately 500-fold difference in abundance.

c) **Distribution.** Most of the dune habitats are widely scattered around the UK's coast, although there is an emphasis on the west coast, where physical conditions favour dune development. However, some are geographically restricted, notably *Dune juniper thickets*, found at only two sites in Scotland, and *Machair* and *Decalcified fixed dunes with Empetrum nigrum* which are similarly confined to Scotland.

d) **Structure and function.** The Annex I types cover the classic geomorphological sequence of vegetation found on UK dunes, ranging from *Embryonic shifting dunes*, very open communities developed in unstable upper shore situations, through to various fixed dune habitats supporting a range of closed grassland, heath and scrub types on stable sands. Although for the most part habitat types are characterized by freely draining substrates, in the case of *Humid dune slacks* maintenance of suitable hydrological condition is also essential to the survival of the habitat.

One habitat deserves special mention at this stage, because although it occurs in the UK, the decision was taken not to select sites for it in the UK. This is *Dunes with sea buckthorn (Hippophae rhamnoides)*. The decision was taken because sea buckthorn has been widely introduced within the UK to sites outside its traditional range. It normally behaves as an introduced species, spreading rapidly at the expense of other Annex I habitat types, whose conservation often requires its control or even eradica-

Table 2. Annex I sand dune types of the United Kingdom

Annex I dune type	Description	Extent (ha)	Distribution	Comments
*Dune juniper thicket	Stands of *Juniperus communis* on stable sand dunes	<100	Scotland	Found at only two sites in the UK, Morrich More and Invernaver
Embryonic shifting dunes	Corresponds to SD4 *Elymus farctus* ssp. *boreali – atlanticus* foredune community of the NVC	101–1000	All UK coasts	Typically found as a narrow zone. Extent at any site may vary from year to year according to storm frequency and intensity. Mainly a feature of a limited number of accreting dune systems
*Eu-Atlantic decalcified fixed dunes (Calluno–Ulicetea)	Corresponds to H11a *Calluna vulgaris – Carex arenaria* heath *Erica cinerea* sub-community, H11c *Calluna vulgaris – Carex arenaria* heath species-poor sub-community and H1 d *Calluna vulgaris – Festuca ovina* heath *Carex arenaria* sub-community.	101–1000	All UK coasts	Covers most of the fixed dune vegetation of acidic dune systems. Intergrades with **Decalcified fixed dunes with *Empetrum nigrum*** at some Scottish sites
*Decalcified fixed dunes with *Empetrum nigrum*	Corresponds to H11b *Calluna vulgaris – Carex arenaria* heath community *Empetrum nigrum* ssp. *nigrum* sub-community	101–1,000	Scotland	Often found as complex mosaics with ***Eu-Atlantic decalcified dunes Calluno–Ulicetea***, and other Annex I types
Humid dune slacks	Corresponds to SD13 *Salix repens – Bryum pseudotriquetrum* dune slack community, SD14 *Salix repens – Campylium stellatum* dune slack community, SD15 *Salix repens – Calliergon cuspidatum* dune slack community, SD16 *Salix repens – Holcus lanatus* dune slack community, SD17 *Potentilla anserina – Carex nigra* dune slack community	101–1000	All coasts	Covers most of the spectrum of dune slack vegetation identified in the NVC, though it grades into **Dunes with *Salix arenaria***. Also covers a range of other vegetation types found in geomorphological dune slacks

Table 2 continued

Annex I dune type	Description	Extent (ha.)	Distribution	Comments
Dunes with sea buckthorn (*Hippophae rhamnoides*)	Stable dune and dune slack colonized by *Hippophae rhamnoides*	101–1000	All coasts	Widely introduced in the UK and often spreading at the expense of other Annex I habitat types
Dunes with *Salix arenaria*	Corresponds to SD16 *Salix repens – Holcus lanatus* dune slack	101–1000	All coasts	Most of UK resource found on Sefton Coast, elsewhere found as small stands on the fringes of mature dune slacks. Very closely related to humid dune slacks, both floristically and functionally
Shifting dunes along the shoreline	SD6 *Ammophila arenaria* mobile dune community	1001–10,000	All coasts	Dependent upon maintenance of sand movement within the dune system
***Fixed dunes with herbaceous vegetation**	SD7 *Ammophila arenaria – Festuca rubra* semi-fixed dune community, SD8 *Festuca rubra – Gallium verum* fixed dune grassland, SD9b *Ammophila arenaria – Arrhenatherum elatius* dune grassland *Geranium sanguineum* sub-community, SD11 *Carex arenaria – Cornicularia aculeata* dune community, SD12 *Carex arenaria – Festuca ovina – Agrostis capillaris* dune grassland	>10,000	All coasts	Covers wide range of dune pasture types which are maintained by grazing
Machair	Coastal sand plains supporting a wide range of dune and other vegetation which do not fit easily into the NVC.	>10,000	northwest Scotland, especially on Hebridean Isles	Distinctive formation of dune, pasture and wetland vegetation forming on calcareous blown sand under the influence of heavy grazing, a cool moist climate and often rotational cultivation (cf. Ritchie, 1975)

tion. The long history of pastoral management in the UK means that sea buckthorn does not have the settled niche that it has in some continental dune systems.

Annex II Species

By comparison with other member states in the Atlantic Biogeographic Region, the UK has limited responsibilities for the conservation of Annex II species (Hopkins and Buck, 1995). Sites are proposed on the UK national list relating to 43 species, of which only five have significant occurrences on sand dunes such that sites have been selected for the UK list:

- great crested newt *Triturus cristatus*;
- narrow-mouthed whorl snail *Vertigo angustior*;
- slender naiad *Najas flexilis*;
- fen orchid *Liparis loeslii*;
- shore dock *Rumex rupestris*;
- petalwort *Petallophylum ralfsii*.

Of these species only one, petalwort *P. ralfsii*, a bryophyte characteristic of open moist dune sands, is found only on sand dunes in the UK, although in the case of the so-called fen orchid *L. loeslii* decline in management of fen habitats in the UK means this is now an extremely rare species found at only a few sites, with most of the UK population found in open dune slacks at Kenfig Dunes and Burry Inlet, South Wales. Shore dock *R. rupestris* is a globally rare species with most of the world population found on sandy beaches and at the foot of sea cliffs on the British coast, with one occurrence on sand dunes at Newborough Warren in North Wales. The great crested newt *T. cristatus* and slender naiad *N. flexilis* are species dependent upon freshwater bodies in dunes, which forms only a part of the UK range of ecological situations in which they occur, while the narrow-mouthed whorl snail *V. angustior* occurs in a range of flushes in the UK including dune slacks in South Wales.

Although this group of Annex II species are taxonomically heterogeneous they are all characteristic of wet sands or aquatic dune habitats and their listing in Annex II reflects the particular conservation importance of wet dune systems.

The EC Process and Criteria for Site Selection

The process to be followed for site selection is set out in Article 4 and Annex III of the Directive. It is a two-stage process:

> *Stage 1* Assessment of the relative importance of sites containing examples of the individual Annex I habitat types and Annex II species in each member state;

Stage 2 Assessment of the overall importance of the sites in the context of the appropriate biogeographical region and the EC as a whole.

It is important to realize that in Stage 1 the focus is on individual Annex I Habitats and Annex II Species, with each initially having an independently selected national site series.

The criteria to be employed at Stage I are set out in Annex III. They can be summarized as follows:

Habitats
- Degree of representativeness, i.e. the degree to which the habitat example conforms to the described type, including its main variations, as defined in the *Interpretation Manual of European Union Habitats.*
- Area. The sites which contain a high proportion of the national resource of a particular type have been favoured.
- Degree of conservation of structure and function and restoration possibilities, i.e. whether the typical physical and biological characteristics of the habitat are present in a sustainable ecosystem.
- Global assessment, i.e. overall assessment based upon the above.

Species
- Proportion of the national population at the site; those sites hosting a high proportion of the resource have been favoured.
- Degree of conservation and restoration possibilities of features of the habitat that are important for the species.
- Degree of isolation of the population.
- Global assessment, i.e. overall assessment based upon the above.

In addition, member states are required to classify sites on their national lists according to their relative value for each Annex I habitat type and Annex II species and to identify which of the sites in their national lists are selected for priority habitat types and species.

The criteria in Stage 2 are intended to be used to assess the sites at the level of the six biogeographical regions and the EC as a whole. The Stage 2 criteria may be summarized as:

- Relative value of the site at national level;
- Relationship of the site to migration routes or its role as part of an ecosystem on both sides of one or more community frontiers;
- Total area of the site;
- Number of Annex I habitats and Annex II species present;
- Global ecological value (i.e. overall assessment based on the above) on the site at the level of the biogeographical region and/or the EC as whole.

The Stage 1 and Stage 2 criteria must be read alongside other site selection requirements or qualifications set out in the Directive, which include:

- Restriction on site selection obligations relating to widely dispersed and aquatic species (Article 4.1);
- The requirement to contribute towards the maintenance of *favourable conservation status,* a term defined in Article 1(e) for habitats and 1(i) for species (Article 2.2 and Article 3.1);
- The obligation each member state has to select a series of sites which reflect the proportion of the EC resource of a given habitat or species within their national territory (Article 3.2).

There are numerous difficulties in the interpretation of the criteria for the selection of SACs. In 1994 Directorate General XI of the European Commission, the UK Environment Departments and the Joint Nature Conservation Committee (JNCC) jointly organized a meeting of member states with territory within the Atlantic Biogeographical Region in order to discuss the interpretation of criteria in the selection of national lists. Important conclusions of this meeting relevant to the selection of sites for dune habitats and species are as follows:

a. Balancing the national lists.
b. Acknowledging that outstanding single-interest sites in terms of quality, extent or range make an important contribution to the Natura 2000 network. Special emphasis will be given to identifying and delimiting sites containing complexes of interests on Annexes I and II as valuable ecological functional units.
c. Member states will give significant additional emphasis in number and area to sites containing priority habitat types and species.
d. In considering the degree of representativeness of Annex I habitat types on individual sites, member states will take account of the best examples in extent and quality of the main type (i.e., that which is most characteristic of the member state) and its main variants, having regard to geographical range.
e. Acknowledging that sites containing Annex I habitat types and Annex II species at the centre of their range will make an important contribution to Natura 2000. Member states will take responsibility for proposing sites containing habitats and species that are particularly rare in the member state, with a view to preserving the range.
f. Acknowledging that certain habitat types and species listed in Annexes I and II are relatively common and extensive in certain member states. These member states will have particular responsibility for proposing a proportion of the resource that is sufficient to contribute significantly to the maintenance of the habitat types and species at a favourable conservation status (Hopkins and Buck, 1995)

Table 3. Summary of selection principles in the UK

Main factors	Reference
Habitats	
i) Representativeness	Annex III Stage 1A. Conclusions of Atlantic Biogeographical Region Meeting (para. d) Article 1e
ii) Area of habitat	Annex III Stage 1A. Article 1e
iii) Conservation of structure and function	Annex III Stage 1A. Article 1e
Species	
iv) Proportion of UK population	Annex III Stage 1B. Article 1i
v) Conservation of features important for species survival	Annex III Stage 1B. Article 1i
vi) Isolation of species populations	Annex III Stage 1B. Conclusions of Atlantic Biogeographical Region Meeting (para. g)
General	
vii) Priority/non-priority status	Annex I Stage 1D. Conclusions of Atlantic Biogeographical Region Meeting (para. c)
viii) Rarity	Conclusions of Atlantic Biogeographical Region Meeting (para. e)
ix) Geographical range	Article 1e. Article 1i
x) Special UK responsibilities	Article 3.2. Conclusions of Atlantic Biogeographical Region (para. f)
xi) Multiple interest	Annex II Stage 2. Conclusions of Atlantic Biogeographical Region Meeting (para. b)

Table 3 summarizes the principles applied to the selection of the UK national list and sets out the derivation of each principle from the Directive and/or the Atlantic Biogeographical Region meeting.

Considerations in the UK Approach to Site Selection

There is a long-established protected site series under UK domestic legislation, the Site of Special Scientific Interest/Area of Special Scientific Interest (SSSI/ASSI) system. It is clear, however, that simply to list all of these sites on the UK national list would not meet the requirements of the Directive. Not all SSSIs/ASSIs are selected for one or more of the Annex I habitat/Annex II species, and the criteria applied to their selection do not conform to those listed in the Directive. A clean-sheet approach to the listing of sites has therefore been carried out in order to ensure that the selected sites conform as closely as possible to the requirements of the Directive.

Previous experience of site selection exercises (Ratcliffe, 1977; Nature Conservancy Council, 1989) has however been drawn upon in planning the compilation of the UK national list. On the basis of this experience, the UK decided against adopting a quantitative approach to the evaluation of sites using the Annex III criteria (a decision also reached by the other member states of the EU). The problem of adopting such an approach is that scaling or quantifying assessments would be needed for each of the criteria. For example, it is not possible to attach simple numeric values to attributes such as 'degree of representativeness', or 'degree of conservation of structure and function' or 'degree of isolation of population'.

Moreover, as member states are required to apply multiple criteria in the selection of sites, there would be a need to transform the values attached to single attributes into a composite score. There is currently no widely agreed method for determining such weightings and of then integrating data into a single overall index (Margules, 1986). Further, in any attempt to produce single indicator assessments of a number of criteria there is the problem that intercorrelations are likely to introduce bias (Usher, 1980). For example, in the Annex III criteria there will always be a degree of positive correlation between the area of a site and the number of Annex I habitat types and Annex II species present on the site. As yet there is no consensus as to how these problems should be resolved and therefore informed specialist judgment has played the main role in the selection of sites.

Process of Site Selection in the UK

In the selection of sites in the UK, therefore, the process adopted has been one of extensive discussion and peer review of the site list, which has been adapted to take account of new knowledge arising from surveys or as interpretative documents became available.

Responsibility for implementation of the Directive rests with the UK government, who have commissioned the UK statutory conservation agencies (Countryside Council for Wales, English Nature and Scottish Natural Heritage along with the Environment Heritage Service for Northern Ireland) to advise on which sites should be included on the UK national list. The co-ordination of the work by the conservation agencies was carried out by a committee of senior staff from each of the conservation agencies under the chairmanship of the Chief Officer of the UK Joint Nature Conservation Committee, which acts on behalf of the conservation agencies with regard to international conservation matters.

A range of professional groups and statutory committees have scrutinized the list in its various iterations as follows:

- Specialist working groups (covering woodlands, coastlands, marine, freshwater, other lowland habitats and species) drew up the initial list of sites for evaluation and quality-assured subsequent changes to the list recommended by other groups.
- Local staff of the county conservation agencies were required to ensure that sites selected were the ones in their area of geographical responsibility that best satisfied the selection criteria.
- The management boards or project boards of the individual conservation agencies have taken an overview of the representation of sites in their country to ensure an evenness of response and proper representation of habitats and species.
- The governing bodies of the individual agencies have approved the proposals within their geographical areas of responsibility.
- The Joint Nature Conservation Committee has formally approved the proposals prior to providing formal advice to government.
- Government departments have scrutinized the selection process and the resulting list to satisfy themselves that the Agencies have acted in accordance with the requirements of the Directive.

Public consultation, including consultation with national and regional bodies with an interest in land use issues and nature conservation, along with consultation with owners and occupiers of individual sites, has also played a significant part in the process, in order that government can further assess the appropriateness of the conservation agency recommendations. A first public consultation in relation to 280 sites was published in March 1998. A second consultation relating to 42 new sites and 48 amended sites was carried out in October 1997.

The overall process of selecting Special Areas of Conservation in the UK is described more fully in Brown et. al., 1997, and Hopkins and Buck, 1995.

Interpretation of Site Selection Principles in Relation to Dunes

Selection principles relating to Annex I habitats

Representativeness

A first requirement of evaluation of any habitat is that it conforms sufficiently to the general habitat type and its main variations as described in the *Interpretation Manual of European Union Habitats* (European Commission DGXI, 1996).

For some habitats, such as *Embryonic shifting dunes*, there is a limited range of ecological variation. In this case the principal axis of variation is geographical and runs from north east to south west, with lyme grass *Leymus arenarius* most abundant in the north east of Britain and sand couch *Elytrigia juncea* most abundant in the south west. Sites for this habitat have been chosen to encompass this range.

Other habitats, such as *Shifting dunes along the shoreline with Ammophila arenaria (white dunes)*, have more than one axis of variation. There is a geographical axis of variation: *Shifting dunes* in southern Britain are marked by the occurrence of species such as Portland spurge *Euphorbia portlandica* and sea holly *Eryngium maritimum*. However, the most marked axis of variation relates to the degree of instability. This may occur within a site, reflecting the varying rate of sand accretion through the shifting dune area. Sites selected for this habitat have again been chosen to cover the range, but across this range sites showing good internal zonation within this habitat have been favoured.

For other types such as **Fixed dunes with herbaceous vegetation (grey dunes)* the type is varied and selection of sites has needed to take account of a complex pattern of variation. The most widespread type is Atlantic dune (*Mesobromium*) grassland. This consists of a short sward characterized by red fescue *Festuca rubra* and lady's bedstraw *Galium verum*, and typically rich in the species of calcareous substrates. This type exhibits geographical variation and variation due to different levels of soil moisture, which are taken account of in the site series. However, this Annex I habitat type also encompasses other distinctive communities. There is a taller type of dune grassland, characteristic of northeast England, in which bloody crane's-bill *Geranium sanguineum* is prominent. In eastern England, where the climate is drier and more continental, there are areas of fixed dune vegetation dominated by lichens and bryophytes.

Area of habitat type

For the most part the habitat examples selected are those which are the largest in the UK. In general this selection has been reinforced by other criteria, as representativeness and degree of conservation of structure and function are generally best developed in larger habitat examples.

Conservation of structure and function

Habitat structure and function involve a number of interrelated components. Structure can relate to a variety of biotic and abiotic features. Function relates to the way in which the biotic and abiotic features of the habitat interact over time. In the case of dune habitat, maintenance of function has been a vitally important consideration, and those habitat examples where function remains intact have been positively selected.

Central to function within dune systems are physical processes. Dune systems are formed by the wind-borne transport of sand from drying beach plains, and the continued link between beach and dune is vital to the continued health of the system. Continued internal redistribution of sand within a dune system is also essential for the maintenance of many dune habitats.

For dune wetlands, the dune aquifer and the hydrological processes that maintain it are prerequisites for successful habitat conservation. For the vast majority of UK dune habitats grazing is essential to their continued existence in order to halt successional change.

Selection principles relating to Annex II species
Proportion of UK population
For the most part, the sites chosen are those which appear to host the largest populations. Particular attention has been paid to those sites which host more than 10 per cent of the population of a given species, as is the case with fen orchid *L. loeslii* at two South Wales dune sites.

Conservation of features important for species survival
The selection of species sites has paid particular attention to the features thought to be important for species survival. Often these coincide with features of importance for habitat conservation, but not always. Petalwort *P. ralfsii*, for example, appears to be favoured by the kind of compaction associated with vehicle usage (Brown et al., 1997).

Isolation of species populations
This factor was felt to be relevant to only a small number of species populations in the UK, none of which occurs on dunes.

General selection principles
Priority and non-priority
The Directive requires that special attention is given to sites containing priority habitats and therefore special attention has been given to *Fixed dunes with herbaceous vegetation (grey dunes), Decalcified fixed dunes with Empetrum nigrum, Eu-Atlantic decalcified fixed dunes (Calluno–Ulicetea)* and *Dune juniper thickets (Juniperus spp.)*, for all of which a high proportion of the UK resource is therefore selected.

Rarity
The habitats listed in Annex I vary considerably in their abundance. For purposes of implementing the Directive special emphasis has been given to those habitats which cover less than 1000ha, which includes a high proportion of dune habitats.

Geographical range
Sites have been selected to reflect, in broad terms, the geographical ranges of each of the Annex I habitats and Annex II species within the UK. Where a habitat is widespread, as is the case with *Fixed dunes with herbaceous vegetation,* the site series has aimed to cover the full range, but selection has been more rigorous in the central part of its range where it is most widespread and abundant.

Special UK responsibilities
Comprehensive data on the distribution of dune habitats within the EU is hard to come by. However, on the basis of our existing knowledge we have assumed that we have a special responsibility, along with the Irish Republic, for machair. We also feel it likely that we have a special responsibility for the Atlantic dune (*Mesobromium*) grassland variant of **Fixed dunes with herbaceous vegetation.*

Multiple interest

Strictly speaking multiple interest only becomes a criterion during Stage 2 of the site selection process. However, the UK has considered it in preparing its national list. This is mainly because the UK has decided to submit to the European Commission only high-quality sites that it believes meet the Stage 2 criteria (so far as we are able to judge) and are worthy to be designated as SACs.

In the case of the dune habitats and species, although the starting point for selection of sites has been the independent evaluation of the most important example of each Annex I habitat and Annex II species, a striking feature of the outcome of this exercise was that many sites have been independently nominated as hosting high-quality examples of many different dune habitats (Table 4). The species sites have also overlapped with those selected on habitat grounds.

This reflects a fundamental ecological truth concerning dune systems, which is that the most valuable examples of individual habitat types, and the best populations of most species, nearly always occur as part of sites where there are a variety of different habitats. Variety is related to the size of the system – bigger sites generally have more habitats – but it also tends to reflect the continued operation of the physical and ecological processes that maintain diversity.

The UK dune site list and its future management

As can be seen from Figure 1 the 27 sites selected for the UK national list occur on all the coasts of the UK where dune systems are developed and provide representation of a high proportion of the total UK dune resource.

Fitting the long-term conservation of this series of dune sites into the framework provided by the Habitats Directive is not as straightforward as might be imagined. The dynamic nature of dune systems lies at the heart of this difficulty.

While the Habitats Directive recognizes the importance of function for the conservation of habitats and species, the model of conservation that it envisages is essentially static. Sites are selected within defined boundaries for a defined series of features. This is illustrated by Article 6.2 of the Habitats Directive, which obliges member states to 'avoid the deterioration of natural habitats and the habitats of species for which the areas have been designated'. This model cannot be applied simplistically to dune systems, where continued change is an essential part of successful conservation. The different habitats within a dune system have a dynamic relationship with each other and change within sites is inevitable.

Mobile dune communities rely upon instability of the dune system, which may well destroy areas of fixed dune vegetation. Conversely, stable dune habitats such as dune heath, grassland and scrub communities can normally only expand at the expense of mobile communities. Dune slacks are also dynamic features. Continued creation of new slacks, often from blowouts, is highly desirable in order to maintain the communities characteristic of the early successional stages of dune development, notwithstanding the fact that such events inevitably destroy existing habitats.

Table 4. Annex I sand dunes types for which coastal dune sites on the UK national list are selected

Name of Candidate SAC	Embryonic shifting dunes	Shifting dunes along the shoreline with Ammophila arenaria (white dunes)	*Fixed dunes with herbaceous vegetation (grey dunes)	*Decalcified fixed dunes with Empetrum nigrum	*Eu-Atlantic decalcified fixed dunes (Calluno-Ulicetea)	Dunes with Salix arenaria	Humid dune slacks	Machair	*Dune juniper thickets (Juniperus spp.)	Dunes with sea buckthorn (Hippophae rhamnoides)	Total number of sand dune Annex 1 types
1 Barry Links	1	1	1			1	1				5
2 Braunton Burrows	1	1	1				1				4
3 Burry Inlet: Dunes/Cilfach Burry: Twyni	1	1	1				1				4
4 Coll Machair								1			1
5 Dornoch Firth and Morrich More	1	1	1		1		1		1		6
6 Dorest Heaths (Purbeck & Wareham) & Studland dunes	1		1				1				3
7 Drigg Coast		1	1								2
8 Durness			1								1
9 Glannau Mon: Twyni/Anglesey Coast: Dunes	1	1	1			1	1				5
10 Invernaver		1	1	1	1				1		5
11 Kenfig/Cynffig		1	1				1				3
12 Magilligan		1	1				1				3
13 Monach Islands								1			1
14 Morfa Harlech and Morfa Dyffryn	1	1	1				1				4
15 Murlough					1						1
16 North Norfolk Coast and Gibraltar Point Dunes	1	1	1				1				4
17 North Northumberland Dunes	1	1	1				1				4
18 North Uist Machair								1			1
19 Oldshoremoren and Sandwood							1				1
20 Penhale Dunes	1		1								2
21 Sands of Forvie	1	1	1				1				4
22 Sandwich Bay	1	1	1				1				4
23 Sefton Coast	1	1	1				1				4
24 South Uist Machair								1			1
25 Tiree Machair								1			1
26 Torrs Warren – Luce Sands	1	1	1				1				4
27 Winterton – Horsey Dunes	1	1	1								3

It is not therefore realistic or desirable to expect strict conservation targets based on the status quo at the time of SAC designation to be met. A more flexible approach will be required. The definition of Favourable Conservation Status in Articles 1(e) and 1(i) of the Directive is helpful in this respect as it makes it clear that it has to be assessed for both habitats and species across the whole territory of the member state. Local losses of individual habitats can therefore be offset by gains elsewhere.

However, even whole dune systems are bound to change. Dunes exist in a delicate balance with the beaches that front them. Geomorphologically they act as reservoirs of 'spare' sand. They absorb sand from the beach in times of accretion and they release it back at times of erosion. Accretion is not difficult to accommodate within the framework of the Habitats Directive, but erosion is more difficult. Erosion is of course a creative as well as a destructive force. At one level it releases sand that may be fuelling accretion elsewhere, at another it maintains mobile dune habitats within systems where they would otherwise be absent. The end result is, however, a net loss of habitat area.

The concept of Favourable Conservation Status could also absorb change to whole dune systems if accretion and erosion were roughly in balance. However, this is not the case. Out of 121 English dune systems, 67 were recorded as undergoing net marine erosion when visited between 1987 and 1990 (Radley, 1994), whereas only 21 were recorded as undergoing net progradation. Relative sea level rise, increased storminess and increasing tidal range mean that this trend will continue and probably intensify.

This is perhaps the greatest challenge facing dune conservation. Several responses are possible and some are already being tried in the UK. One answer is to compensate for the losses due to erosion by habitat management measures to reclaim former areas of dune lost to other land uses such as forestry and agriculture. A good example of such management is the removal of the frontal plantations within Ainsdale Sand Dunes National Nature Reserve (see Simpson and Gee, this volume). Beach feeding may be an answer at the local level. This technique is being used to maintain the Gibraltar Point dune system, but sustainability has to be an issue, as does the potential impact on other Annex I habitats such as *Sandbanks which are slightly covered by the sea at all times*. In the longer term we may wish to facilitate managed retreat of whole dune systems, and it would be prudent to develop planning policies to protect the immediate hinterland of dune systems from irreversible development.

It remains to be seen, however, whether all of these measures will be capable of maintaining either the current area of each Annex I dune habitat or the total area of dune. It is likely that these have fluctuated in the past for entirely natural reasons. Stability, even at the level of the member state, may not necessarily be achievable in future.

Dunes are of course not alone in being highly dynamic features. All the habitats characteristic of soft coasts, the majority of marine habitats and many of the species that depend on them are naturally dynamic. Devising conservation objectives, management measures and reporting targets for these features which fit the spirit of the Directive without contravening its letter is a major challenge for the UK and other EU member states.

References

Brown, A.E., Burn, A.J., Hopkins, J.J., and Way, S.F. (1997), *The Habitats Directive: Selection of Special Areas of Conservation in the UK*, Joint Nature Conservation Committee Report No. 270, Peterborough, JNCC.

European Commission DGXI (1996), *Interpretation Manual of European Union Habitats; Version Eur 15*, Brussels, European Commission DGXI.

Hopkins, J.J. and Buck, A. (1995), *The Habitats Directive Atlantic Biogeographical Region*, Joint Nature Conservation Committee Report No. 247, Peterborough, JNCC.

Margules, C.R. (1986), 'Conservation evaluation in practice', in M.B. Usher (ed.), *Wildlife Conservation Evaluation*, London, Chapman and Hall, 297–314.

Nature Conservancy Council (1989), *Guidelines for the Selection of Biological SSSIs*, Peterborough, NCC.

Radley, G.P. (1994), *Sand Dune Vegetation Survey of Great Britain, Part 1: England*, Peterborough, Joint Nature Conservation Committee.

Ratcliffe, D.A. (1977), *A Nature Conservation Review: The Selection of Biological Sites of National Importance to Nature Conservation in Great Britain*, 2 vols., Cambridge, Cambridge University Press.

Ritchie, W. (1975), 'The meaning and definition of machair', *Transactions of the Botanical Society of Edinburgh*, **42**, 431–40.

Rodwell, J.S. (in press), *British Plant Communities. Volume 5: Maritime and Open Vegetation*, Cambridge, Cambridge University Press.

Usher, M.B. (1980), 'An assessment of conservation values within a large Site of Special Scientific Interest in North Yorkshire', *Field Studies*, **5**, 323–48.

MANAGEMENT OF DANISH DUNES TODAY: THEORY AND PRACTICE

CLAUS HELWEG OVESEN

National Forest and Nature Agency, Denmark

Introduction

Denmark is rich in valuable dune areas along its 7500km of coastline. The total area of dunes according to recent mapping (1994) is 1283km². Free public access to the dunes is an old tradition in Denmark, and the effect of wear has arisen in recent decades. In certain cases this brings interesting dynamics into the dunes by replacing the wear from grazing animals in earlier times.

The management of dunes now follows Chapter 2 paragraphs 8–11 in the *Nature Conservation Act* of 1992. The paragraphs on dune protection areas stress the obligations to prevent sand drift where it seriously interferes with interests of housing or agriculture, but the conservation interests and public access in the dunes are also secured in the first paragraph of the Act and in Chapter 4.

Principles of Dune Management

Some of the most important guiding rules for modern dune administration in Denmark are listed below, followed by some comments on how they are realized in practical management.

1. A protection zone of a belt of the outermost 100m of dunes along the coast of Jutland

Here activities like building, digging etc., which can cause damage to dunes, are prohibited, and there is an obligation for the managing authorities to prevent sand drift where necessary. The belt is marked in the terrain by solid concrete poles.

Prevention of sand drift is of course still an important element in Danish dune management. It is often done by covering blowouts with pine branches and hay, planting marram and by cutting (reprofiling) 'dune-edges' where wind erosion is especially marked. New conifer plantations in open dune areas no longer occur, but the majority of dune woodland plantations are maintained through forestry practice (see 3 and 4 below).

2. Public access to dune areas with the possibility of channelling traffic where necessary

Public access is secured according to rules in the *Nature Conservation Act* (Chapter 4). A certain number of roads allow cars to pass through dunes, thus giving vehicular access to the broad sandy beaches along the North Sea. Elsewhere there is free access by foot to dune areas as long as one stays more than 50m from habitation. The managing authorities often channel traffic in heavily used sites by establishing pathways covered by wood chips, straw or other stabilizing material. In areas with large summerhouse hinterlands, like the island of Fanø at the north end of the Danish Wadden Sea, this is well established with many numbered pathways to the bathing beaches. This reduces the often heavy wear from traffic through the dunes between summerhouses, other hinterland areas and the beach.

3. Management of dune plantations for nature conservation interests as well as protection from sand drift

Dune plantations were typically planted about 100 years ago with mountain pine *Pinus mugo* as a prominent species. Other species such as lodgepole pine *Pinus contorta*, Scots pine *P. sylvestris* and sitka spruce *Picea sitchensis* have been much used in later plantations providing potential for some timber production. The plantations come under the new initiative for 'richer forest' started in the mid-1990s by the National Forest and Nature Agency. An attempt is now made to replace extensive use of sitka spruce and lodgepole and mountain pine partly by broadleaf species such as oak *Quercus* spp. and partly by other conifer species such as Scots pine, noble fir *Abies nobilis*, and others.

4. Clearing of dune plantations and restoration of dune heathland in certain areas

A number of plantations have little economic or conservation value and can be removed by felling to restore dune heathland. There is a proposal to clear a belt for the conservation of Atlantic dune heathland immediately east of the dune belt along the North Sea coast. Until now it has only been realized on an area of some 120ha per year in the period 1992–99, and for the next 5–8-year planning period the area of land cleared of plantations to restore dune heathland will be c. 90ha per year. Altogether some 850ha of open dune landscape has been restored in the period 1992–99. These figures are not high, but in Denmark, as in other European countries, initiatives to remove planted forest in the windy coastal landscapes are often met with rather strong local resistance, and successful projects typically need a lot of public persuasion.

5. Rabbits do not occur commonly in Danish dune areas and the traditional (sheep) grazing has almost ceased

Until about 1950 there was a tradition of having Danish dune sheep (a specific breed) grazing the dune areas. This interesting national breed of sheep has now nearly disappeared (one flock is left at Hulsig in northernmost Jutland). Trampling by wild and domestic animals (sheep, cattle, horse etc.) thus has little or no importance in Danish dunes – in most places the wear from human traffic is the dominating mechanical factor besides the wind. This, together with a stronger growth of grasses like *Deschampsia flexuosa* and *Corynephorus canescens*, means that the conditions for lichen growth have ameliorated, and today new colonization of *Cladonia*-dominated vegetation often occurs at the edge of pathways where the sand is kept open by a moderate amount of wear. This development has been observed and described on the fine dune heathlands of Rømø.

6. Nitrate deposition from precipitation has an effect on dune heathland, especially in the *Cladonia*-dominated lichen-heathland

The stronger grass growth mentioned above is probably an effect of an increase in nitrate deposition. This strongly influences inland dunes and sandy heathlands, but also has effects nearer to the coast. Direct effects on *Cladonia* species are studied in northern Jutland by Johnsen and Heide Jørgensen.

7. Management plans exist for the state-owned heathlands (c. half of the total area) and an overall mapping and evaluation of conservation interests (based on geomorphology and vegetation) is part of the background for plan revisions

State-owned dune areas (open as well as tree-covered) are managed according to 15-year plans, taking into account both sand drift prevention and conservation interests – recently also including the need for dynamics in dune areas. Most municipal and privately owned areas are managed in the same way and under supervision by the State Forest Districts. Conflicts between the interests of conservation and dynamics and landowners do, of course, occur.

8. A large proportion of the most valuable dune areas has been protected and managed as nature reserves for many years

Many valuable dune areas are protected as reserves. Some are very old and famous like Raabjerg Mile, Skallingen and Grenen (Scaw Spit). The best known is Raabjerg Mile, where protection was begun in 1902. This initiative was taken by the famous Danish botanist Eugen Warming (founder of plant ecology) on the basis of his pioneering studies on the ecology of coastal plant communities. An important but complicated goal since then has been to allow for the continuous movement of drifting dunes. Raabjerg Mile is also a recent example of a case where a conservation easement is sought for the areas just east of the dune over which it will move in the coming years, the speed being about 5–15m every year.

9. Dunes with heath and associated wetlands are managed in nature restoration plans, which are being realized in Denmark in the 1990s

The clearing of plantations and restoration of areas of grey dunes and dune heathland (both priority habitats according to the Habitats Directive) are important management tools, but so also is stopping the drainage of former wetlands in the dune hinterland. An important project, Hulsig Hede, is being implemented by the County of North Jutland and the State Forest District with support from the EU LIFE fund.

Conclusions

In modern dune management the aim is to let natural processes take their course as far as possible within the obligations on dune protection in the Nature Conservation Act. The activities of people in the dunes are a more or less natural substitute for the wear of wild and domestic animals. The Danish tradition of free access to the dunes is kept as long as the situation can be said to be sustainable, i.e. the dunes move to some extent, but do not disappear. The prevention of sand drift according to the law follows a differential scale according to the conservation and other land-use interests. The tendency is clearly to reduce the amount of dune stabilization work in less inhabited areas like Thy in northwest Jutland, where only special areas around car parks are stabilized, whereas landowners typically demand a higher level of stabilization, for example in areas with summerhouses.

References

Anthonsen, K.L., Clemmensen, L.B. and Jensen, J.H. (1996), 'Evolution of a dune from crescentic to parabolic form in response to short-term climatic changes: Råbjerg Mile, Skagen Odde, Denmark', *Geomorphology*, **17**, 63–77.

Christensen, S.N. and Brandt, E. (1994), *Danish Dunes, Mapping 1992–1994*, report (in Danish) edited by the National Forest and Nature Agency, Copenhagen.

Favennec, J. and Barrère, P. (1997), *Biodiversité et protection dunaire*, Life; Office National des Forêts; Lavosier tec and doc, Paris.

Jensen, F. (1994), 'Dune management in Denmark', *Journal of Coastal Research*, **10(2)**, 263–69.

Ovesen, C.H. (1994), *Danish Dunes: Monitoring, Management and Research*, report from a seminar, May 1992 (in Danish with English summaries), National Forest and Nature Agency, Copenhagen.

Ovesen, C.H. (ed.) (1998), *Coastal Dunes – Management, Protection and Research*, report from a European Seminar, Skagen, Denmark, August 1997, National Forest and Nature Agency and the Geological Survey of Denmark and Greenland.

Williams, A.T. and Bennett, R. (1996), *Dune Vulnerability and Management in England: Partnership in Coastal Zone Management*, Samara Publishing, Cardigan.

THE UK HABITAT ACTION PLAN FOR SAND DUNES

KATHY DUNCAN

Scottish Natural Heritage, Scotland

ABSTRACT

Action plans will be produced for 38 habitats in the UK. Their production is a direct consequence of the *UK Biodiversity Action Plan*, published by the British government in 1994, and the subsequent report produced by the UK Biodiversity Steering Group in 1995. Sand dunes are one of five coastal habitats that have been included within the list. The habitat action plan will describe the current status of sand dunes in the UK, the main factors affecting the habitat, current action, objectives and proposed targets for maintaining or increasing its extent and a list of actions needed to achieve the targets. After formal consultation by the Targets Group the final version will be sanctioned by the UK Biodiversity Group. This paper discusses the content of the sand dune habitat action plan and in particular the proposed targets and actions. The successful implementation of all the habitat and species action plans provides a significant challenge to the government and its agencies.

Introduction

The publication of the *UK Biodiversity Action Plan* (UKBAP) in 1994 (Department of the Environment, 1994) set out the government's commitments to the Convention on Biological Diversity, signed at the Earth Summit in 1992. The overall goal of the UKBAP is 'to conserve and enhance biological diversity within the UK and to contribute to the conservation of global biodiversity through all appropriate mechanisms'. To achieve this goal four objectives for conserving biodiversity were identified. These were to conserve and where practicable to enhance

1. overall populations and natural ranges of native species and the quality and range of wildlife habitats and ecosystems;
2. internationally important and threatened species, habitats and ecosystems;
3. species, habitats and natural and managed ecosystems that are characteristic of local areas;
4. the biodiversity of natural and semi-natural habitats where this has been diminished over recent decades.

One of the first outcomes of the UKBAP was the establishment of a Biodiversity Steering Group which would, among other things, develop costed action plans for key species and habitats. The intention was that each action plan, with its set of objectives, targets and agreed actions, would direct conservation activity of the key players to ensure consensus and the conservation of priority habitats and species within the UK. The Steering Group was made up of central and local government, the nature conservation agencies, the collections, business, farming and land management representatives, academic bodies and voluntary conservation organizations.

In 1995 the Biodiversity Steering Group reported back to government, identifying 400 species and 40 habitats which required action plans (Department of the Environment, 1995). As part of their work, the Steering Group had already prepared a first tranche of action plans for 116 species and 14 habitats; the remainder were to be written over the next three years. Responsibility for implementing the recommendations of the Steering Group report was assigned to the UK Biodiversity Group. Specific responsibility for the development of species and habitat action plans was given to the Targets Group, a sub-group of the UK Biodiversity Group.

In selecting key species and habitats in need of action plans, the UK Steering Group identified a number of criteria. 'Key species' were defined as those

- whose number or range has declined substantially in recent years; or
- which are endemic; or
- which are under a high degree of international threat; or
- which are covered by relevant Conventions, Directives or legislation.

'Key habitats' were defined as those:

- for which the UK has international obligations; or
- which are at risk, such as those with a high rate of decline especially over the last 20 years; or
- which are rare; or
- may be functionally critical; or
- are important for key species.

In 1998 a further 56 species action plans were published for priority vertebrates and vascular plants (UK Biodiversity Group, 1998). It is intended that the remaining 250 species action plans and 25 habitat action plans will be published by the end of 1999.

Production of Coastal Habitat Action Plans

Five key coastal habitats were identified as requiring action plans. In addition, 13 habitat action plans are being produced for marine habitats. These habitats are listed in Table 1.

The majority of the coastal habitat action plans have been prepared by a coastal

Table 1. Coastal and marine Habitat Action Plans
(*costed HAP produced in first tranche by UK Biodiversity Group)

Coastal habitats	Marine habitats
Coastal sand dune	Saline lagoons*
Machair	Sea grass beds*
Coastal vegetated shingle	*Sabellaria alveolata* reefs
Saltmarsh	*Sabellaria spinulosa* reefs
Maritime cliff and slope	Mud habitats in deep water
	Modiolus modiolus beds
	Sublittoral sands and gravels
	Littoral and sublittoral chalk reefs
	Mudflats
	Serpula vermicularis beds
	Lophelia pertusa reefs
	Maerl beds
	Tidal rapids

specialist group, made up of representatives from the Environment Agency, the Department of the Environment, Transport and the Regions (DETR), English Nature, the Countryside Council for Wales, Scottish Natural Heritage and Wildlife and Countryside Link. In addition, early drafts were circulated to a number of other coastal experts and those with an interest in the coast. It was hoped that by involving the key players in the preparation of the plans, any major problems could be identified and resolved early on.

In producing the action plans, it was recognized that there is often considerable overlap between the different habitats. Coastal habitats are often found in close proximity to one another and, given the dynamic nature of the environment, are often closely interrelated. In particular, there are obvious similarities between sand dunes and machair in Scotland and this has been recognized in these plans. In addition, a number of BAP priority species for which species action plans are being produced may also be found in coastal habitats. For example, dune gentian *Gentianella uliginosa*, petalwort *Petalophyllym ralfsii* and the tiger beetle *Cincindela hybrida* all have significant populations on sand dunes. It is therefore important that the different action plans (which in some cases are being produced by different specialist groups) support one another and that the targets and actions do not contradict.

The Sand Dune Habitat Action Plan

The sand dune HAP follows the agreed format established by the UK Steering Group and used in all the plans. Although it describes the current status and factors affecting the habitat, it is not intended to be a detailed description of its ecology. Rather, the emphasis within the plan is on the objectives and the actions required to achieve set targets.

The action plan is divided into 5 sections:

1. Current status

The current status describes the basic characteristics of sand dunes, their structure and function, geographical variation and the extent of the resource in the UK. It is known that sand dunes cover 11,897ha in England, 8101ha in Wales and 1576ha in Northern Ireland. Survey work on sand dunes is continuing in Scotland, where the current best estimate indicates there are 48,000ha of coastal blown sand, of which 15,000ha is machair. As machair is covered by a separate habitat action plan, the total area of sand dune in the UK is recorded as 54,500ha.

2. Current factors affecting the habitat

The action plan attempts to highlight the main factors causing a loss or decline in extent of the sand dune habitat in the UK. It is not an exhaustive list and there may well be other factors which are considered to affect, sometimes significantly, individual sites. However, the list is thought to cover all the major factors causing an impact on sand dunes in a UK-wide context. These are considered to be:

Erosion and progradation

Sand dunes, particularly on the seaward edge, are naturally highly mobile systems. The majority of dunes in the UK show net erosion rather than net progradation, often as a result of insufficient sand supply. Overall, however, the net loss of dune habitat as a result of erosion is thought to be low. In England Pye and French (1992) estimated it to be at a level of no more than 2 per cent of the total resource over 20 years.

Falling water tables

Dune slack communities are dependent on a seasonally high water table. In some cases, however, local extraction of water and drainage of adjacent land for agriculture or housing are causing a fall in the water table. This can have a subsequent detrimental effect on specialist slack flora and can encourage the invasion of coarse vegetation and scrub.

Grazing

The occurrence of grassland and heath vegetation on sand dunes in the UK is a result of a long history of grazing by sheep, cattle and rabbits. Both under-grazing and over-grazing will affect the typical fixed dune communities. The former leads to invasion by coarse grasses and the development of scrub and eventually woodland; the latter to loss of sensitive dune species and, in extreme cases, erosion of the vegetation cover. An appropriate grazing regime, however, will help maintain a species-rich dune grassland.

Recreation

Recreation is now a major land use on sand dunes. Moderate pressure may cause little damage, but excessive pedestrian and off-road vehicle use can cause significant destabilization and erosion of vegetation. A notable proportion of sand dunes have also been developed as golf courses. Although this rarely results in total loss of the dune communities, management can result in modification by mowing, fertilizing and reseeding of tees, greens and fairways.

Sea defences and stabilization

The construction of sea defences or other artificial stabilization measures can result, if inappropriately planned, in habitat loss and inhibition of dynamic coastal processes. They may also cause sediment starvation down-drift, thus extending the impact beyond the confines of the sites. Many sand dune systems, particularly in developed, highly populated and low-lying coastal areas, have been modified in this way. In England some form of sea defence was recorded on 52 out of the 121 dune sites surveyed, over 40 per cent of sites (Radley, 1994).

Beach management

Increasing demand for litter-free beaches for recreational pursuits has resulted in an increase in mechanical beach cleaning. As well as clearing litter, these methods also remove strandline and organic material at the top of the beach, thus hindering the process of embryonic dune development. As a result the dune system may start to show signs of erosion, even if the physical conditions would appear to favour accretion.

Forestry

Although not widespread in the UK, large-scale afforestation of dunes has occurred at a few locations, for example at Culbin in Moray. This results in loss of dune vegetation communities and lowering of the water table. However, management at Ainsdale Sand Dunes National Nature Reserve in Merseyside, and to a lesser extent Tentsmuir National Nature Reserve in Fife, has shown that dune communities can be restored in a relatively short time following felling.

Military use

During the Second World War many dune systems around the UK were used for construction of defensive installations and military training. Much of the damage caused then has now been reversed and a significant number of dune sites (particularly in Scotland) are still owned and used by the Ministry of Defence (see Baker, this volume).

Other uses

In addition, housing development, industrial development, waste tips, fly-tipping and sand extraction are all considered to be important factors. Indirect factors are considered to be atmospheric nutrient deposition, coastal squeeze due to rising sea levels and increased storminess, and disruption of coastal processes by marine aggregate extraction.

3. Current action

In this section of the action plan the statutory basis for conservation of the habitat and the relevant legislation, including planning and policy guidance, is summarized. It also describes the management initiatives and research carried out by statutory and voluntary bodies to promote the conservation of the habitat. This section, therefore, sets the context within which the habitat is currently being managed and protected and provides the baseline from which to prescribe future action.

It is recognized that a relatively large proportion of the resource is already protected in one way or another, through the Sites of Special Scientific Interest (SSSI) (Area of Special Scientific Interest, ASSI, in Northern Ireland) network, and more recently the Natura 2000 network. In addition, a number of initiatives from government are assisting in the protection and management of all coastal habitats. Of significance are the planning policy guidance notes produced by DETR (PPG 20 Coastal Planning) and the Scottish Office (NPPG 13 Coastal Planning). In addition, national Coastal Fora have now been set up in England, Scotland and Wales, and one is due to be set up in Northern Ireland. Sand dunes also fall within countryside management initiatives (such as the Countryside Stewardship in England) and individual sites are managed as, for example, National Nature Reserves and SSSIs.

4. Action plan objectives and proposed targets

The action plan identifies five broad objectives to ensure the long-term habitat status for sand dunes. Specific targets for the time period up to 2010 have been set wherever possible. These targets are specific, measurable, achievable, realistic and time-limited. As in all the action plans, targets are generally expressed as maintaining the status of the habitat or improving it where possible. For sand dunes the emphasis has been on maintaining or enhancing the natural functions of the system and encouraging the natural coastal processes to determine the extent of the dune resource.

The five broad objectives are:

1. The existing sand dune resource of about 54,500ha should be protected from further losses to anthropogenic factors, whether caused directly or indirectly;
2. The expected net losses due to natural causes of about 2 per cent of the dune habitat resource over 20 years should be offset by encouraging new dunes to accrete and where possible by allowing mobile dune systems to move inland;
3. Opportunities should be sought for restoration of sand dune habitat lost to forestry, agriculture or other human uses;
4. The biodiversity of the UK sand dune resource should be enhanced by encour-

aging natural movement and development of dune systems and by controlling natural succession to scrub and woodland;

5. Dune grassland, heath and lichen communities should be maintained on the majority of dune systems, but development of Atlantic woodland should be allowed on up to five carefully selected sites.

5. Proposed action with lead agencies

Essentially the most important component of the Habitat Action Plan, this section identifies the actions that are needed to ensure that the agreed objectives and targets are met. For each action, the statutory body or bodies deemed responsible for undertaking the action are identified in the action plan. The principle bodies involved are the government departments: the Department of Environment, Transport and the Regions (DETR), the Welsh Office, the Scottish Office and the Department of the Environment (Northern Ireland) (DoE(NI)); and the statutory nature conservation agencies, English Nature, Scottish Natural Heritage, the Countryside Council for Wales, Environment and Heritage Services and the Joint Nature Conservation Committee. In addition, specific actions will fall to other bodies such as the Environment Agency, Scottish Environment Protection Agency and Forest Enterprise. The areas of action that have been identified, categorized into six sub-sections, are listed below.

Policy and legislation

These actions relate to policy measures that would have beneficial conservation implications for sand dunes. Responsibility for these falls principally to the DETR, the Scottish Office, the Welsh Office and the DoE(NI). The actions are:

- To develop and promote planning policies and procedures which will aim to prevent further losses of sand dune habitat to development and exploitation and minimize them where they are unavoidable;
- To develop and promote agri-environment schemes which will encourage restoration and sustainable management of dune habitats;
- To develop and promote incentives to encourage the management and restoration of landward transitional dune habitats and, where appropriate, to allow landward movement of dunes;
- To develop and promote coastal zone management policies which allow the maximum possible free movement of coastal sediment and pay full regard to the conservation of sand dunes.

Site safeguard and management

These actions relate to the promotion of appropriate land management through site designation and management schemes, including habitat recreation and restoration. Responsibility for these falls principally to the statutory nature conservation agencies, although certain actions will also require input from the DETR, the Scottish Office, the Welsh Office, the DoE(NI), the Ministry of Agriculture, Fisheries and Food (MAFF) and Forest Enterprise. The actions are:

- To give notification of any remaining areas of sand dune habitat which meet national criteria as SSSIs and ensure appropriate management of designated sites;
- To use positive management agreements where appropriate to encourage sustainable grazing of sand dune SSSIs and ASSIs and other dunes where possible;
- To encourage golf course management policies and practices which are sympathetic to the flora and fauna of sand dune systems;
- To promote and encourage the restoration of open dune vegetation on afforested dune systems;
- To promote and encourage the restoration of dune vegetation on dune systems used for arable farming or agriculturally improved grassland;
- To monitor and regulate water abstraction and land drainage schemes which might affect water tables in sand dune systems, and promote remedial action where necessary;
- To discourage unnecessary stabilization of all dunes, and where appropriate promote managed destabilization measures on over-stabilized dunes;
- To support beach management strategies which encourage the protection of the seaward fronts of dune systems from unsustainable pressure by pedestrian or vehicular traffic, and discourage the use of mechanical beach cleaning close to dune fronts.

Advisory

These actions describe key areas where the provision of advice on the conservation of sand dune habitat is needed. Responsibility falls principally to the statutory nature conservation agencies, although action relating to agriculture and forestry will require input from the Ministry of Agriculture, Fisheries and Food (MAFF), the Scottish Office Agriculture, Environment and Fisheries Department (SOAEFD) and the Forestry Commission. The actions are:

- Where appropriate, to promote and develop demonstration sites for the restoration of dune vegetation on dune systems which have been converted to forestry or agriculture;
- To encourage the appropriate management of sand dunes by preparing and disseminating updated guidance material;
- To ensure all relevant agri-environment project officers and members of regional agri-environmental conservation groups are advised of the location of existing examples of this habitat, its importance and the management requirements for its conservation.

International

These actions describe what is needed at an international level to help conserve the habitat. Responsibility falls to the statutory nature conservation agencies. The actions are:

- To promote the exchange of information on sand dune ecology and management among European maritime states through organizations such as the European Union for Coastal Conservation (EUCC) and Eurosite;
- To ensure lessons from the Sefton Coast Life Project are widely disseminated and incorporated into good practice.

Future research and monitoring

These actions identify suitable areas for future research and monitoring to help the conservation of sand dunes. This includes research on the status of the habitat and on the main impacts on sand dunes. Responsibility falls to the statutory nature conservation agencies. The actions are:

- To co-ordinate information on changes in the extent and quality of the sand dune resource in the UK in order to enable effective monitoring of the objectives of the habitat action plan;
- To continue research into the use of remote sensing for monitoring soft coast habitats;
- To promote research into the causes of falling water tables in sand dune systems;
- To promote research on the effects on sand dunes of indirect influences such as nitrogen deposition, climate change and sea level rise.

Communications and publicity

These actions identify appropriate opportunities to raise awareness about the habitat. Responsibility for the first action falls to the statutory nature conservation agencies. However, responsibility for the second falls to the government departments and all those involved in the implementation of the plan. The actions are:

- To raise public awareness of the essential mobility of soft coasts and the value of maintaining unrestricted coastal processes;
- To promote awareness among decision-makers of the implications of the policies outlined in the habitat action plan.

The Development of Habitat Action Plans

The sand dune HAP, along with all other remaining 25 habitat and 250 species action plans, should be published by the end of 1999. Costs are currently being allocated to each action prior to a final consultation being carried out by the Targets Group. In some cases costs are assumed to be part of ongoing costs of existing organizations; for example the designation of further sites as SSSIs comes under the statutory nature conservation agencies' remit, and the promotion of appropriate planning policies comes under the remit of the DETR, the Scottish Office, the Welsh Office and the DoE(NI). In other cases additional funds will be needed, for example to allow research to be carried out, to encourage the restoration of dune habitats, to develop demonstration sites or to raise awareness.

The production of a costed habitat action plan for sand dunes will complete the first step in fulfilling UK domestic objectives under the Biodiversity Convention. Following its publication, the emphasis will naturally switch to the implementation of the plan. In the recent publication listing the second tranche of species action plans, the Biodiversity Steering Group regards the action plans as providing 'a robust scientific case for conservation action and building consensus between the key bodies involved about the best means of achieving biological objectives'. However, they go on to say that 'action planning is not an end in itself; the value of the work lies in providing a clear focus and platform for subsequent action' (UK Biodiversity Group, 1998). Lead partners are now being identified to take forward and co-ordinate the implementation of individual action plans. They will be responsible for co-ordinating and developing the work programme for a steering group of key players. However, responsibility for delivering the UK targets, provision of resources and the gathering of information for progress reporting will be shared between all members of the steering group. There is, therefore, still a considerable challenge, not only for the conservation agencies, but for all the other key partners, to work together to ensure the successful implementation of the action plans and the conservation of the most threatened and declining habitats and species in the UK.

References

Department of the Environment (1994), *Biodiversity: The UK Action Plan,* HMSO, London.

Department of the Environment (1995), *Biodiversity: The UK Steering Group Report*, HMSO, London.

Pye, K. and French, P.W. (1992), *Targets for Coastal Habitat Recreation,* Cambridge Environmental Research Consultants, Cambridge.

Radley, G.P. (1994), *Sand Dune Vegetation Survey of Great Britain: A National Inventory, Part 1: England,* Joint Nature Conservation Committee, Peterborough.

UK Biodiversity Group (1998), *Tranche 2 Action Plans, Volume 1: Vertebrates and Vascular Plants,* English Nature, Peterborough.

SYNOPSIS OF THE FLEMISH COASTAL DUNE CONSERVATION POLICY

JEAN-LOUIS HERRIER and ILSE KILLEMAES
Ministry of the Flemish Community, Department of Environment and Infrastructure, Belgium

Description of the Flemish Coastal Dunes

Flanders has a nearly straight coastline with a length of 65km. The Flemish coastal dune belt has a maximum width of 2–3km over a 14km length between the French border and Nieuwpoort, and between Knokke and the Dutch border over a length of 6km. Between Nieuwpoort and Knokke the dunes have a width of only 60–600m. The Flemish coastal dunes lie between sandy beaches on the seaward side and an alluvial plain of clay-like polders on the landward side. Between Nieuwpoort and the French border, the Flemish coast is characterized by huge parabolic dunes.

The sandy soils of the southwestern Flemish coastal dunes have a lime content of 8–20 per cent, and those of the northeastern part of the coastline about 4 per cent. These lime-rich 'young' dunes were formed mainly during the last 1000 years and are covered by essentially calcareous vegetation types.

Four fossil dune formations rise above the surrounding flat polder plain 2–3.5km inland at Adinkerke, Lombardsijde, Bredene and Knokke. These 'old' dunes were formed 1400–2000 years ago, in the case of the fossil dunes of Adinkerke 4500 years ago. They present a slightly undulating flat aspect and have decalcified acid soils.

Around the turn of the century the total area of the Flemish coastal dunes, including the saltmarshes along the Yzer river mouth at Nieuwpoort and the sea-inlet of the Zwin at Knokke, was estimated at up to 6000ha. The heavy urban development that took place during the twentieth century led to a decrease of this coastal dune area from 6000ha to 3600ha. The remaining dune areas are now heavily fragmented by urban development and roads.

The Legal Protection Framework

Legal instruments for the whole Flemish territory
Landscape protection
The first legal instrument available for the protection of dune and other nature sites was the 'Law of 7th August 1931 on the Protection of Monuments and Landscapes'.

Although aesthetically, culturally and historically orientated, it did enable landscape protection for sites based upon their natural landscape features. The first areas that were protected as 'landscapes' along the Flemish coast were the Westhoek (590ha, 1935) near the French border, and the Zwin (150ha, 1939) near the Dutch border. This law was recently replaced by the 'Decree of 16th April 1996 on the Protection of Landscapes'.

The following dune areas are protected as landscapes by Ministerial Order.

Along the western part of the coast:
- the Cabourg-domain (100ha, 1964) at Adinkerke;
- the Houtsaegerduinen (80ha, 1981) at De Panne;
- the Ter Yde-duinen at Oostduinkerke (200ha, 1992);
- the Simliduinen at Nieuwpoort (40ha, 1978);
- the Yzer river mouth (100ha, 1993);
- the Sint-Laureinsduinen (6ha, 1982);
- the Schuddebeurze (50ha, 1982).

Along the eastern part of the coast:
- the Groenpleinduinen (9ha, 1982) at Knokke;
- the Zwinbosjes (220ha, 1983) at Knokke;
- the 'Oude Hazegraspolder and inner dunes' (110ha, 1995) at Knokke.

The 'Decree of 16th April 1996 on the Protection of Landscapes' offers private and public owners of lands in a protected landscape area support to draw up a management plan and the chance to obtain subsidies for landscape improvements. Unfortunately, these opportunities are little used by private landowners as many of them are opposed to the landscape protection designation given to their properties.

Town and country planning
The 'Law of 29th March 1962 on the Organization of Town and Country Planning' enabled zoning plans to be drawn up and approved by Royal Decree during the 1970s. These zoning plans attribute to every part of the Flemish territory a clearly defined land utilization, such as green area, agricultural area, housing area, industrial area or military area. Over the whole Flemish region 152,000ha were designated as 'green area'.

There are three zoning maps for the Flemish coast: Brugge–eastern coast, Oostende–central coast and Veurne–western coast. On these maps 2500ha of dunes were attributed a green designation with at least 1400ha of dunes and dune-polder transition zones designated as housing development, military training ground, recreational area or agricultural area. Of these 'non-green' areas, about 250ha were developed, and thus lost for conservation purposes, between 1976 and 1993.

In 1997 a 'structure plan' for the whole Flemish territory was agreed upon in the Flemish regional government. This 'Structure Plan for Flanders' imposes on the

Flemish regional government the requirement to designate an additional 38,000ha of 'green areas' and 10,000ha of 'forest extension areas' across the whole of the Flemish territory by revision of the zoning plans.

Forestry

The 'Forest Decree of 13th June 1990' describes forest as 'all surfaces of land on which trees and woody scrubs are the main constituents of the vegetation'. In all wooded areas that conform to this legal description, whatever their legal status on the Town and Country Planning zoning maps, any felling of trees requires permission from the Forest Division of the Ministry of the Flemish Community. The Forest Decree also imposes the obligation to produce a management plan for each publicly owned forest and for privately owned forests with an area exceeding 5ha. From a conservationist point of view this protection of wooded areas should be viewed positively. However, disagreements sometimes arise between the forestry and conservation sectors.

Although the Forest Decree certainly promotes more nature-oriented forestry activity, some officers of the Forest Division regard areas of sea buckthorn *Hippophae rhamnoides* and even creeping willow *Salix repens* as forests and argue that management plans for the scrub-encroached dune areas should be enacted in the frame of the Forest Decree. The main disagreement is caused by the fact that some foresters have a devotion to any woody growth and are rather opposed to the restoration of herbaceous natural habitats by removal of shrubs and trees.

Nature conservation

The 'Law of 12th July 1973 on Nature Conservation' and its replacement, the 'Decree of 21st October 1997 concerning Nature Conservation and the Natural Environment', allow the statutory designation of 'nature reserve' to a natural area that is owned or rented by the Flemish regional government or non-governmental nature conservation organizations. The nature reserves managed by the Nature Division of the Ministry of the Flemish Community are called 'Flemish Nature Reserves' (replacing the 'State Nature Reserves') and those managed by non-governmental organizations are called 'Acknowledged Nature Reserves'. These designations are made by ministerial order and each nature reserve must have a management plan, also approved by ministerial Order.

For each Flemish Nature Reserve or group of Flemish Nature Reserves an advisory commission must be created. They consist mainly of scientific specialists, local conservationists and representatives of the provincial and municipal authorities. The constitution of these commissions is approved by ministerial order. The purpose of the advisory commissions is to assist the Nature Division Officer responsible for the management of the nature reserve.

Along the Flemish coast the following areas have the legal status of Flemish Nature Reserve:

- De Westhoek (340ha), at De Panne;
- De Houtsaegerduinen (80ha), at De Panne;
- Hannecartbos (34ha), at Koksijde (Oostduinkerke);
- D' Heye (13ha), at Bredene;
- De Baai van Heist (40ha), at Knokke-Heist (Heist).

In addition, the following areas were recently acquired by the Flemish Region and will be designated as Flemish Nature Reserves in the near future:

- Ter Yde (22ha), at Koksijde (Oostduinkerke);
- De Yzermonding (60ha), at Nieuwpoort;
- De Kleiputten van Heist (9ha), at Knokke-Heist (Heist).

Only one coastal dune site managed by a non-governmental organization, De Fonteintjes (12ha), between Zeebrugge and Blankenberge, has the legal status of 'Acknowledged Nature Reserve'.

The 1997 'Decree on Nature Conservation' requires the adoption and implementation of a 'Flemish Ecological Network' (Dutch abbreviation: VEN) and of an 'Interweaving and Supporting Network' (Dutch abbreviation: IVON). This 'Nature Framework' should provide a spatially and functionally coherent system of large managed natural areas (VEN) connected by a series of linear landscape elements, for instance hedgerows, and nature-oriented agricultural areas (IVON). Areas that are designated as 'agricultural areas with scenic value' on the zoning plans can be included only in the IVON, and not in the VEN. However, areas without scenic value that have the legal status of 'agricultural area' cannot be included in either the VEN or the IVON.

European Directives
Only limited areas of the Flemish coastal dunes were designated as Special Protection Areas in the frame of the European Bird Directive, 79/409/EEC, by decree of the Flemish government, 17 October 1988.

The greater part of the Flemish coastal dunes, saltmarshes and some adjacent beaches are included in the proposal of designation of Special Area of Conservation in the frame of the Habitats Directive by decree of the Flemish government, 14 February 1996.

Legal instruments unique to coastal dunes

In the frame of the 'Decrees of the 14th July 1993, 21st December 1994 and 29th November 1995 on the Protection of the Coastal Dunes' all the remaining ecologically important dune and dune-polder transition areas (about 1100ha) that were not yet protected as 'green areas' received legal protection by the creation of two designations – 'protected dune area' and 'agricultural area of importance for the dunes'.

The first designation was given to all dune sites that were identified on the zoning plans as housing development, military area or recreation area. Some 'green areas' that were threatened by development permissions granted before the zoning plans became effective were also designated as 'protected dune areas'.

The second designation of 'agricultural area of importance for the dunes' was attributed to dune and dune-polder transition sites identified in the zoning plans as 'agricultural area' or 'agricultural area with scenic value'. This was the result of strong opposition by the agricultural sector to a change in the planning status of agricultural land close to nature areas, even though the agricultural activity in these areas was already in strong decline because of bad drainage, poor sandy soils or spatial fragmentation caused by built-up areas and roads.

In 'agricultural areas of importance for the dunes' a number of exceptions are made, mainly in favour of farmers, to permit development rights. The most dramatic consequence of this is that the 'agricultural areas of importance for the dunes' cannot be included in the network of the VEN. This means that before a substantial part (about 800ha) of the coastal dune area can be included in the ecological framework of the VEN, its status on the zoning plans must first to be changed from 'agricultural area (with or without scenic value)' to 'green area'.

Management Policy

The 'Ecosystem Perspective for the Flemish Coast' that was produced between 1995 and 1996 by the University of Ghent and the Institute of Nature Conservation under the supervision of the Nature Division of the Ministry of the Flemish Community was approved by the Flemish Minister of the Environment in 1998. This document contains an inventory of the scientific knowledge of the Flemish coast and provides a scientific basis for the management and sustainable use by all concerned sectors (conservation, coastal defence, forestry, drinking water supply, recreation, agriculture etc.) of the beach, coastal dune and saltmarsh ecosystems.

Nature reserves

The Flemish Nature Reserves and other 'Regional Nature Domains' along the coast that are managed by the Nature Division of the Ministry of the Flemish Community amount to 600ha.

Although the Westhoek Nature Reserve was declared in 1957 and the Law of 12 July 1973 on Nature Conservation required the production of management plans for all legally designated nature reserves, it was not until 1995 that a management plan for this, the oldest of Flanders' coastal nature reserves, was produced. At the same time a management plan for the Flemish Nature Reserve of Houtsaegerduinen was also drafted. In 1996 both management plans were approved by ministerial order.

Before 1995 the Forestry Service had been responsible for the management of the Flemish (State) Nature Reserves. Under this regime the management policy for publicly owned land was to encourage vegetation succession towards scrub and woodland.

Only on 6ha, out of a total area of 420ha for both the Westhoek and the Houtsaegerduinen, was mowing carried out to maintain some scattered remnants of herbaceous vegetation. The change of policy occurred in 1995 when the management responsibility for the Flemish Nature Reserves was given to the Nature Division. Since then, management plans have been or are being worked out for all coastal Flemish Nature Reserves by the University of Ghent and other specialized research consultants, under the supervision of the Nature Division.

An essential element in the management plans are 'Target Habitat Types', described by the 'Ecosystem Perspective for the Flemish Coast'. The Target Habitat Types are based upon the habitat types listed in Annexes of the Habitats Directive. In this way, the management plans for the Flemish Nature Reserves are a contribution towards the implementation of the Habitats Directive.

A consequence of former management policies to encourage successional change is that most of the Flemish Nature Reserves along the coast are heavily overgrown by scrub, mainly sea buckthorn *Hippophae rhamnoides*, wild privet *Ligustrum vulgare*, exotic trees, and cultivars of poplars *Populus* spp. The main objective in the management plans is to restore herbaceous habitats through the large-scale removal of scrub, followed by grazing by horses and cattle ('pattern-oriented' management, as for example at Westhoek) or management of spontaneous development of woodland through the removal of exotic tree species and grazing by donkeys ('process-oriented' management, as for example at Houtsaegerduinen). The practical execution of the management plans for the Flemish Nature Reserves along the western part of the Flemish coast is a part of the LIFE-Nature project 'Integral Coastal Conservation Initiative'.

Coastal defence

The Flemish region owns nearly all dunes that are of importance for coastal defence from Westende to Knokke, with a total area of 350ha. The dune ridges adjacent to the sea between De Panne and Westende and between Knokke and the Dutch border are, on the whole, still privately owned. The management of the coastal defence dunes owned by the Flemish region is the responsibility of the Waterways and Coast Division of the Ministry of the Flemish Community. In the past their management was almost exclusively oriented towards ensuring dune stability for coastal defence by erecting fences to prevent trampling and planting marram and some exotic shrub and tree species such as Japanese rose *Rosa rugosa*, oleaster *Elaeagnus angustifolia*, tamarisk *Tamarix gallica* and white poplar *Populus alba*.

Marine erosion was sometimes countered by the building of concrete dykes in front of the dune belt, for instance along Westhoek, at De Panne and in front of the dunes near the Yzer river mouth at Lombardsijde. At other places, such as De Haan and Knokke-Heist, beach nourishment was used as a coastal defence technique.

In 1994, regular consultation was started between the Waterways and Coast Division, the Nature Division and the Institute of Nature Conservation. This has led to a greater mutual understanding between the coastal defence engineers and conservationists. Since then the planting of exotic sand-fixing shrub and tree species has ceased and a more dynamic approach to coastal defence has been promoted. Where it does not present a risk to public safety, the seaward dune belt is now allowed to have some measure of natural dynamic movement again. For all publicly owned coastal defence dune areas 'Site Perspectives' are now being worked out by specialized research consultants under the supervision of the Waterways and Coast Division. These Site Perspectives are inspired by the Ecosystem Perspective for the Flemish Coast, and are provided for coastal defence where there is no legal basis for management plans.

Regional forests

The Forest Division of the Ministry of the Flemish Community is responsible for the management of the regional forests. The Flemish region owns a stretch of 150ha of afforested dunes at De Haan. These dunes were afforested between the First and the Second World Wars, mostly with Austrian *Pinus nigra* ssp. *nigra* and Corsican pines *P. nigra* ssp. *laricio*, white and other poplars. Since the 1980s the management objective in these forests has been to transform the monotonous pine and poplar plantations into a more diverse deciduous woodland type. Although the Forest Decree clearly promotes a forestry policy that is more nature oriented, some old habits are maintained by the managers in the field: unafforested spaces with herbaceous vegetation are only reluctantly tolerated and planting of exotic tree species continues even where natural regeneration by indigenous trees is occurring.

In spite of these difficulties, foresters and conservationists have joined forces to encourage the afforestation of some ecologically degraded agricultural zones along the transition from dunes to polders and in the polder. The aim of this is to improve the natural quality of these areas and to increase the area available to the local population and tourists for recreation. This increased recreational area close to the seaside resorts may decrease the recreational pressure on the coastal dunes. One project has already been completed between Blankenberge and Zeebrugge where a 'Zeebos' of 20ha has been planted. Other projects, at Oostende and Nieuwpoort, are encountering tough opposition from private landowners as they wish to develop golf courses and urban areas on this land.

Water collection areas

Around 450ha of the coastal dune area is used for water abstraction. This activity lowers groundwater levels and reduces wet dune slack habitat, not only within the abstraction zones, but also across the whole dune system. One of the water supply companies, the IWVA, started a project of infiltrating purified surface water into one of its dune estates at Koksijde. The production of drinking water by infiltration should permit a significant reduction in the abstraction of natural groundwater and therefore restore more favourable levels.

The IWVA has started to produce management plans for its water collection dune areas. One of these sites, the Doornpanne (160ha) in Koksijde, is now being managed by extensive grazing by horses. Management planning for the water collection dune area is taking place in the context of the Forest Decree even though the site is not a real forest, but just a mosaic of herbaceous and shrubby vegetation types. The preference of the water supply companies for using the Forest Decree as the legal basis for management planning may be explained by the fact that it imposes fewer restrictions than the Decree concerning Nature Conservation and the Natural Environment.

Golf courses

There are two golf courses on the Flemish coastal dunes, one between Bredene and De Haan (40ha) and another in Knokke (80ha).

Although designated on the zoning plans as 'green areas', these golf courses do not benefit from any kind of conscious nature management. In the past the roughs were only mown a couple of times a year, resulting in very valuable dune grasslands. Since the 1980s golf course maintenance has been intensified, with an increased use of chemical fertilizers and pesticides on the fairways affecting the neighbouring roughs. In addition to this, the roughs now seem to be mown more frequently. The increased planting of exotic and other ornamental trees and shrubs also has a negative impact on the natural vegetation of the roughs.

It is suggested that the best long-term solution to the problems caused by golf courses is to relocate them from the dunes to ecologically less valuable land in the polder close to towns and to restore the natural dune habitats. However, in the short term an attempt should be made to make golf course owners and managers conscious of the ecological importance of the roughs and to promote more nature-oriented management.

Acquisition Policy

Problems

As described above, legal spatial protection of the remaining dune areas is not sufficient to maintain or restore their valuable natural habitats such as calcareous grassland, wet dune slack or dune heath. The main threats are scrub invasion by sea buckthorn, wild privet and exotic trees, high recreational pressure and the lowering of groundwater levels by drinking water abstraction. An appropriate habitat-oriented management strategy is required.

Only 1400ha of the remaining 3600ha area of coastal dunes are in public ownership by the Flemish region, the drinking water companies or some municipalities. The remaining 2200ha are legally protected, but privately owned by real estate companies, private individuals or families. These private dunes do not benefit from any kind of management and real estate speculation persists strongly. Most private owners hope that sooner or later the legal protection for these areas will be abolished. This speculation, together with the importance of the coastal region for tourism and the fact that a relatively large part of the dune area is situated next to good road networks, creates abnormally high land prices in comparison with legally protected wasteland in the interior. High land prices and the limited readiness of private landowners to sell their dunes to public authorities are major obstacles for the acquisition of coastal dunes by the Flemish region.

Since the availability of financial resources for the acquisition of nature areas in Flanders is relatively limited and land prices at the coast are extremely high, the creation of a specific financial instrument seems essential for the maintenance and protection of the dunes.

Measures

An 'Acquisition Plan for the Coastal Dunes and Adjacent Areas along the Flemish Coast' was produced in 1996 by the University of Antwerp and the Institute for Nature Conservation under the supervision of the AMINAL-Nature Division. This Acquisition Plan contains a priority classification of the dune areas, based on scientific criteria, and a number of guidelines to inform the dune acquisition policy of the public authorities.

A very important legal consequence of the incorporation of a natural area into the VEN is that it implies in that area the right of pre-emption to the Flemish region and to nature conservation NGOs. This measure should facilitate the acquisition of land for conservation purposes. The Flemish Minister of the Environment has already instructed the administration to incorporate into the VEN all dune sites that are designated as 'green areas' on the Town and Country Planning Maps or as 'Protected Dune Area' in the frame of the Decree on the Protection of the Coastal Dunes.

On 1 January 1998 a four-year project called 'Protection and Acquisition of Coastal Dunes' was started by the Nature Division of the Ministry of the Flemish Community to handle the indemnity claims of the private owners of dunes that are protected by the decrees on the Protection of Coastal Dunes, but are also subject to the administrative procedures for the acquisition of dune sites for conservation purposes by the Flemish Region. The Decree of 21 October 1997 concerning Nature Conservation and the Natural Environment also provides the Flemish Region and the municipalities with the power to expropriate land for nature conservation purposes.

During April 1997 a symposium was organized by the NGOs Belgian Nature and Bird Reserves and the World Wide Fund for Nature within the context of the LIFE Project 'Integral Coastal Conservation Initiative'. As a result, two propositions for a decree were introduced in the Flemish parliament, attempting to create a specific instrument for the acquisition of coastal dunes. This would have created a specific dune fund, organized as a regional service. However, the Flemish government did not approve this, but did agree to an annual budget for the Ministry of the Flemish Community of 150,000,000 BEF (3,750,000 Euro – 1998). This should enable the annual acquisition of at least 100ha of dunes.

Legal contract

The Flemish government has decided to pursue a legally binding agreement between the Flemish Region, the Province of West Flanders and Municipalities. This would regulate the acquisition and management of natural coastal dune sites. Its most important consideration will be to achieve co-operation between the various parties, realizing that the coastal zone can only be protected in a sustainable way by scientifically founded management.

This agreement will also provide for the establishment of an advisory body on the acquisition and management of coastal dunes. The advisory body will be composed of representatives of the Flemish administration responsible for nature conservation and coastal defence, the Institute of Nature Conservation, the Province of West Flanders and eventually the Municipalities who are party to the agreement. The advisory body will meet twice per year.

After consulting the advisory body, working groups on acquisition and management will deliver an annual report to the Flemish government, the Provincial Executive and the Municipal Executives. These reports will also be communicated to the Flemish parliament, the Provincial Council of the Province of Western Flanders and the relevant Municipal Councils.

GREEN ISLANDS AND NATURA 2000: NETWORKING AREAS, PEOPLE AND INFORMATION

ALBERT SALMAN

European Union for Coastal Conservation, The Netherlands

ABSTRACT

Ten years after its establishment EUCC is preparing an initiative which should provide new impetus to its mission and mobilize its membership: the Green Islands Network. We aim to bring together 2000 organizations each of which has responsibility for the conservation or management of an internationally important area. There are at least 2000 such areas in Europe and the Mediterranean. Site managers, local authorities, NGOs and expert organizations are invited to join and collaborate actively in this network.

Introduction

In September 1987 representatives of governments, NGOs and research institutes came together at the first European dune conference to consider the future of Europe's coastal dunes. Out of this was born the European Union for Coastal Conservation (EUCC) which is concerned with the integrity and natural diversity of the coastal heritage. EUCC has developed into the largest coastal network in Europe, with 750 members and member organizations in 40 countries and seven professional offices. Active National Branches were formed in 11 countries, in central and eastern Europe field projects were set up with the support of local communities, and some EUCC branches are even buying islands and coastal wetlands. Under the auspices of the Council of Europe and United Nations Environment Programme (UNEP), EUCC has developed the European Code of Conduct for Coastal Zones and is actively contributing to the development of a European Coastal and Marine Ecological Network (ECMEN).

The question, is how effective has this all been? This paper reviews some of the progress over the last ten years and makes a number of suggestions as to how we can improve our way of working and our effectiveness.

Dune Conservation in the 1990s

The various European dune conferences (Leiden, Seville, Sefton etc.) have enormously stimulated the process of networking, especially between dune managers and scientific experts. This has led to positive developments and local success stories in dune management in, for instance, Denmark, Germany, The Netherlands, England and France.

However, the decline and loss of sand dunes in Europe has not come to a halt. We continue to build golf courses and level dunes for tourist facilities, and many dunes are eroding due to the presence of artificial structures.

The question is: how can we reverse this trend? How can our lessons and success stories be understood by the politicians dealing with the 2000 coastal areas which form our European coastal heritage? How can we effectively communicate the lessons from Leiden, from Galway, from Sefton and elsewhere?

I would like to make a few observations.

With regard to ecological networks:
* local planning schemes do not always reflect the international importance of natural areas;
* the implementation of the EU Birds and Habitats Directives is slow and often insufficient.

With regard to networks of people:
* both local authorities and site managers tend to develop their own networks, especially at the international level;
* EUCC's membership includes both groups, but local authorities are a small minority;
* local politicians tend to focus on spatial planning, economic development and financial aspects, and not on conservation site management.

With regard to exchange of information on coastal management in Europe:
* reports, bilateral contacts, regional networks, workshops and conferences do not entirely meet the comprehensive and continuous need for information from the thousands of actors involved in coastal management;
* many local authorities and site managers lack the time and the capacity to select the specific information from the information overload;
* the European demonstration programme for coastal management is expected to provide us with interesting conclusions with regard to information needs.

The Green Islands Philosophy

EUCC believes that the ecological integrity and the natural diversity of the coastal environment can be preserved through better international communication and co-operation between local communities, site managers, environmental experts, NGOs and regional planners. All these actors are invited to make the connections between each other, and each to accept responsibility for one of the 'green islands' as an element of the European Ecological Network. The name of this partnership is the Green Islands Network.

The EUCC Secretariat has listed 2000 coastal and marine areas in Europe and the Mediterranean which need special attention. In February 1998 the EUCC Board approved a five-year strategy, Green Islands 2000, in an effort to contribute to their conservation through building partnerships, communication and local projects.

Ten years after its establishment EUCC is developing the Green Islands initiative to provide new impetus to its mission and to mobilize its membership in a partnership approach. The Green Islands Network should bring together 2000 organizations in an effort to contribute to the conservation and management of 2000 internationally important coastal areas. Conservation site managers, local authorities, NGOs and expert organizations are invited to join and collaborate in this network.

The name 'Green Islands' reflects the fact that many natural areas have become isolated islands of nature, surrounded by cities, industry and agricultural areas. With the Green Islands initiative EUCC aims to reinforce the connections between these coastal areas through networking in three different ways:

1. supporting for the development of ecological networks;
2. connecting people and their (local) networks;
3. contributing to the development of information and communication networks, including databases and modern communication mechanisms.

EUCC hopes to implement these objectives through joint projects involving professionals concerned with coastal planning, management and research. A large number of projects are already running or under preparation.

With regard to ecological networks:
* European Coastal and Marine Ecological Network (ECMEN) project;
* Green Islands for Natura 2000 project;
* Green Islands Demonstration Projects.

With regard to networks of people:
* development of the Green Islands Network.

With regard to exchange of information on coastal management in Europe:

- development of the European Coastal Guide.

European Coastal and Marine Ecological Network (ECMEN)

The concept of a coastal ecological network was first developed by the EUCC in 1991, as part of the European Ecological Network (EECONET). In September 1998 EUCC published the second report of the ECMEN project. This report:

- presents a new typology of European coastlines (landscapes, formations and habitats);
- presents a new map, *Coastal Systems of Europe* (see below);
- analyses a series of coastal and marine species and their patterns of migration;
- discusses the implications of current approaches to conservation;
- identifies the need for a wider approach to conservation, not only at a European level but also at a more local site and landscape level.

The potential value in developing an ecological network is compelling from the point of view of individual migratory species. Many mobile species rely on a sequence of sites, corridors and stepping stones for migration, feeding, resting and breeding. These often cross the boundaries of nation states and may cross continents. At the same time environmental influences such as water pollution may extend far beyond their original source. The fact that most of these influences affecting the conservation of species and habitats are ultimately related to policy formulation and management decisions suggests that understanding the nature of the biological network is only the first stage in developing a European approach to networks. Equally important are the institutional structures and legislative mechanisms, e.g. those following from the Habitats Directive.

The EUCC map *Coastal Systems of Europe* illustrates the diversity of coastal landscapes, formations and habitats as well as some important physical characteristics to be taken into account in ecological network-building. The complete map encompasses Europe from Georgia to Iceland, and can be made available in scale 1:4,000,000 (original) and 1:7,000,000. A first draft was published in September 1998. The EUCC Secretariat will continually improve and add to the map.

Green Islands for Natura 2000

In May 1998 EUCC launched its public awareness project Green Islands for Natura 2000 at various European locations. This project contributes to the implementation of the EU Habitats Directive. The implementation of the Directive will lead to a European network of protected sites, the Natura 2000 network. The success of Natura 2000 depends on the full participation of all member states.

The project consists of:

- a study of the progress in the implementation of the Habitats Directive in various member states;
- an evaluation of opportunities to enhance public and political support for Natura 2000;
- measures to enhance public and political support for Natura 2000;
- publicity about Natura 2000;
- the development and distribution of information material, such as leaflets, posters, a campaign video and an exhibition;
- the organization of Green Islands days, international events with opportunities for public participation and bilateral exchange (at a local level).

Green Islands demonstration projects

All over Europe field-based projects contribute to an integration of nature conservation and economic sectors, in site management or local development. It is an important task for us to stimulate the exchange of their results and their lessons between our members.

EUCC has already initiated nine field projects, in Russia (St Petersburg; Kingisepp), Estonia (Matsalu Bay; Virtsu), Latvia (Kemeri National Park; Randu Meadows), Lithuania (Nemunas Delta), Poland (Oder Delta) and Ukraine (Dnestr Delta). In these projects local people are involved in nature management and wise use of the areas. Local NGOs are trained on the job.

These EUCC projects serve as the first Green Islands demonstration projects. These can be defined as local projects that combine sustainable development with the conservation of an important natural area. In such projects the principles of ECMEN and the European Coastal Code can be brought into practice.

EUCC aims to bring together best practice examples and success stories in a database and to make them accessible via publications and the Internet (e.g. via the European Coastal Guide, see below). Conservation site managers and local authorities are invited to identify successful projects in their area, which will then be given international attention as Green Islands demonstration projects.

The Green Islands Network

We aim to develop the Green Islands Network with members and partners. Partners are expert organizations or companies which support the programme. Green Islands members are local authorities, site managers and NGOs which, first, accept a certain responsibility for the conservation or management of a coastal area as an element of the European Ecological Network (this responsibility could for instance include the application of the European Code of Conduct for Coastal Zones in the area), and, secondly, participate in an exchange of knowledge and experience between the members. Members will be asked for a contribution depending on their financial situation. The minimum contribution will provisionally be the EUCC membership fee for organizations. In various projects we already collaborate with project partners in

10 countries. We are aiming for the participation of 2000 members and partners within five years.

The European Coastal Guide

EUCC is preparing a European Coastal Guide (ECG) as an interactive, demand-driven facility to assist Green Islands members in an effective exchange of information.

The Guide will include information on:

- ecological relationships between coastal areas (e.g. ECMEN);
- opportunities for sustainable development in coastal management (including the Coastal Code and its supporting documents);
- best practice information: the Green Islands demonstration projects;
- international co-operation between coastal areas;
- international environmental policies and conventions;
- international funding and subsidy programmes;
- management of sand dunes and beaches.

There is often a need for assistance in the financing and implementation of environmental projects with local authorities and site managers. Green Islands members can call upon the Coastal Guide Helpdesk for such assistance. Sometimes it will be possible to bring members together who can jointly find a solution for a similar problem. The helpdesk can also assist in finding financial support programmes, for example EU funding. The ECG project will involve a database, an Internet site, e-mail, e-mail listservers, fax and a helpdesk facility.

Through the ECG we will pay special attention to the reinforcement of the coastal dune network, since this has been a core group within EUCC's membership from the beginning. If there is sufficient interest we will develop an e-mail listserver on coastal habitat management in Europe and the Mediterranean.

IRISH SAND DUNE SURVEY:
A BASIS FOR FUTURE MANAGEMENT

K. GAYNOR and G.J. DOYLE

Department of Botany, University College Dublin, Ireland

ABSTRACT

Most Irish dune systems display a progression through communities of the strand-line, foredunes, yellow dunes and fixed grey dunes. Depending on the nature of the sandy substrate, fixed grey dunes may give way to dune grasslands, dune heaths, dune slacks and/or dune scrub. Seventy-two Irish coastal dune sites were visited during the 1994–97 field seasons and assessed for a) extent, b) habitat diversity, c) intactness, d) apparent management practices and e) vegetation, flora and species diversity. Results indicate regional differences in the formation and structure of the dune systems, together with geographical differentiation in plant species composition of the dune communities. Such variations may be accounted for by a combination of geographic, edaphic, climatic and anthropogenic factors.

Introduction

Ireland's economic and social activities have long been concentrated in the coastal zone (Brady et al., 1972). Since Mesolithic times sand dunes have provided habitable areas, as evidenced by the presence of shell middens and dwelling sites on a number of western systems. In the past, marram *Ammophila arenaria* has been cut for thatching and used as bedding for livestock. Irish dunes have been grazed for centuries, so that dune grasslands support close-cropped vegetation, with grazers that provide natural control of invasive dune scrub. Excessive grazing of dune systems alters the species composition and may destroy the vegetation cover (Hewett, 1982). Recreational pressure is an increasing threat to the structure and stability of a number of sites, particularly along the Irish east coast (Carter, 1980; Mawhinney, 1971).

Despite such extensive anthropogenic influences, Irish sand dune systems have been relatively unaffected by intensive agriculture and many have been spared from the impact of severe recreational exploitation. As a result, Ireland has a number of coastal dune sites that remain relatively intact (Gaynor, 1998). Areas of fixed grey dune and machair, both designated as priority habitat types under the Habitats Directive (1992), are prominent features of the Irish coastline (Curtis, 1991). Effective management for the sustainable use of these complex systems demands an understanding of their overall ecology.

The work presented here is part of a doctoral study that involved a country-wide survey of Irish sand dune systems. The principal aim of this study is to develop a comprehensive classification of Irish sand dune systems. Vegetation and soil interactions are also under investigation. These data will provide a scientific basis for the assessment of the conservation status of Irish sand dune systems. Such information should underpin effective management strategies that will guarantee the preservation of the best examples of these fragile and ecologically important habitats.

Methods

During the period 1994–97, 72 coastal sites throughout the Republic of Ireland were surveyed (Fig. 1). Vegetation data have been analysed using VESPAN II (Malloch, 1988), allowing the identification of distinct relevé groups. Classification of these groups follows the Braun-Blanquet approach to phytosociology, and uses the syntaxonomic scheme for Ireland presented by White and Doyle (1982). Soil samples, collected from quadrats at 50 of the sites, were assessed in the laboratory for their nutrient status. Past management at each site was investigated through reference to historical accounts in literature and detailed records made available by the National Parks and Wildlife Service in Dublin.

Results

Preliminary analysis of the data highlights regional variation in Irish sand dune systems. The vegetation supported by east coast dunes has a species composition that is somewhat different to that found on west coast systems. A noticeable calcicole element is a feature of many western sites, while dune annuals are more typical on the east coast sites. Some species, such as wild asparagus *Asparagus officinalis* ssp. *prostratus* and sharp rush *Juncus acutus,* are confined to east coast systems. Machair vegetation is confined to the north west (Akeroyd and Curtis, 1980; Bassett and Curtis, 1985). These differences may be explained by a combination of geographic, edaphic, climatic and anthropogenic factors.

Geographic factors
A wide range of dune types are found along the highly indented west coast, including bay dunes, sand spits, tombolos, offshore islands and machair. East coast systems comprise bay dunes and sand spits. This variation is a direct result of the nature of the underlying geology, indentation of the coastline and the offshore sediment supply.

21. Inchydoney
22. Barley Cove
23. Rosbeigh
24. Inch
25. Castlegregory
26. Banna Strand
27. Ballyheige
28. White Strand
29. Lahinch
30. Fanore
31. Inisheer
32. Inishmaan
33. Eararna
34. Dog's Bay
35. Aillebrack
36. Mannin Bay
37. Omey Island
38. Aughrusbeg
39. Dooaghtry
40. Keel Lough
41. Doogort
42. Kinrovar
43. Aghleam
44. Cross Lough
45. Termoncarragh
46. Garter Hill
47. Lackan
48. Ross
49. Bartragh Island
50. Inishcrone
51. Strandhill
52. Streedagh Point
53. Trawlua
54. Bunduff
55. Fintragh

56. Glen Bay
57. Maghera
58. Sheskinmore
59. Clooney
60. Lettermacaward
61. Keadue
62. Carnboy
63. Lunniagh
64. Magheraroarty
65. Ballyness
66. Rinclevan
67. Dunfanaghy
68. Marble Hill
69. Tranarossan
70. Melmore
71. Lough Nagreany
72. Magheradrumman

1. Baltray
2. Laytown
3. Rush
4. Portrane
5. Malahide Island
6. Portmarnock
7. Bull Island
8. Killiney
9. Kilcoole
10. Brittas Bay
11. Mizen Head
12. Courtown
13. Cahore Point South
14. Tinnabearna
15. Kilmuckridge
16. Ballynamona
17. Curracloe
18. Rosslare
19. Ballyteige Burrow
20. Tramore

Figure 1. *Geographical location of the Irish coastal sites visited during this survey (1994–97).*

Edaphic factors

Irish east coast systems tend to be siliceous in nature, with sediment derived from local rock or glacial deposits, while west coast systems generally have higher shell fragment content. Values of up to 35 per cent carbonates were recorded from the west coast soils. A maximum value of 5 per cent was recorded from the east coast samples. Consequently west coast dune soils tend to be more alkaline than their eastern counterparts. The pH values recorded at west coast sites ranged from pH 6.0 on an acidified dune heath at Maghera to pH 9.45 on a machair plain at Lunniagh, both in County Donegal. Values in the range pH 6.0–7.0 were recorded from the majority of eastern sites.

Climatic factors

Owing to the Atlantic influence the west coast of Ireland is subject to higher annual rainfall and lower temperatures than the east coast. The east coast is less exposed and experiences higher maximum and lower minimum temperatures and lower rainfall amounts. Wind direction is predominantly onshore along the western seaboard and offshore along the eastern coast.

Anthropogenic factors

Recreation is the main environmental pressure on the east coast due to large-scale urban development, with an associated increase in demand for holiday homes and leisure facilities, such as golf courses. Heavy pedestrian traffic leads to the mobilization of dunes through an increase in the amount of bare sand. Manipulation of dune systems for links golf courses involves mowing and application of fertilizers, which combine to alter the vegetation dramatically. Recreational pressure on western sites has been restricted by distance from the major conurbations, and in many places by limited accessibility due to the remoteness of sites and the nature of the road infrastructure. Such pressures are replaced in importance by agricultural impacts in many western areas.

Conclusions

A very generalized boundary can be recognized in Ireland between 'the physically better-endowed lands of the south and east' and 'the physically harsher country to the north and west' (Collins and Cummins, 1996). This pattern is reflected in the plant communities found on Irish sand dune systems. It is hoped that further investigation of the vegetation and soil data will help to determine the extent to which the factors outlined above contribute to the regional variation observed.

The current work provides a detailed ecological baseline study of the vegetation of Irish dunes. Such information will a) identify the best sites for conservation purposes, b) allow monitoring of changes in site conditions in vulnerable dune systems, c) provide insights into the effects of past management strategies (grazing intensities, impact of different grazers, etc.) at sites that are physically similar, d) identify the physical and biological indicators of over-grazing or recreational damage, and e) provide an understanding of the overall ecology of Irish dune systems. The data provided by this study will allow site managers to ensure effective conservation of a geographically and ecologically representative selection of dune systems. Without such detailed baseline work, management decisions might be taken that could lead to the destruction of key representative dune sites.

Acknowledgments

We thank the National Parks and Wildlife Service for financial assistance of the fieldwork and acknowledge the particular support provided by Dr Tom Curtis and Dr Colmán Ó Críodáin.

References

Akeroyd, J.A. and Curtis, T.G.F. (1980), 'Some observations on the occurrence of machair in western Ireland', *Bulletin of the Irish Biogeographical Society*, **4**, 1–12.

Bassett, J.A. and Curtis, T.G.F. (1985), 'The nature and occurrence of sand-dune machair in western Ireland', *Proceedings of the Royal Irish Academy*, **85B**, 1–20.

Brady, H., Shipman, P., Martin, A. and Hyde, N. (1972), *National Coastline Study, Vols. 1–3*, An Foras Forbartha, Dublin.

Carter, R.W.G. (1980), 'Human activities and coastal processes: the example of recreation in Northern Ireland', *Actes de Colloques, Publications Sciences et Techniques*, **9**, 17–27.

Collins, J.F. and Cummins, T. (1996), *Agriclimatic Atlas of Ireland*, Agmet, Dublin.

Curtis, T.G.F. (1991), 'A site inventory of the sandy coasts of Ireland', in M.B. Quigley (ed.), *A Guide to the Sand Dunes of Ireland*, EUCC, Dublin.

Gaynor, K. (1998), 'Sand dunes – habitats under threat', *The Badger*, **66**, 7–9.

Hewett, D.G. (1982), 'Grazing studies on sand dunes', *Annual Report of the Institute of Terrestrial Ecology 1981*, 78–79.

Malloch, A.J.C. (1988), *VESPAN II. A computer package to handle and analyse multivariate species data and handle and display species distribution data*, University of Lancaster, Lancaster.

Mawhinney, K.A. (1971), *Brittas Bay Study – Part 2. Conservation and Recreational Use of the Beach and Dunes*, An Foras Forbartha, Dublin.

White, J. and Doyle, G.J. (1982), 'The vegetation of Ireland: a catalogue raisonné', *Journal of Life Science, Royal Dublin Society*, **3**, 289–368.

FLORA AND VEGETATION OF THE ATLANTIC DUNES OF THE NORTHWEST COAST OF PORTUGAL

F.B. CALDAS and J.J. HONRADO

Department of Botany, University of Porto, Portugal

The vascular plant species occurring in the area were named after Tutin et al. (1968–1980; 1993), except for *Rhynchosinapis johnstonii* (Heywood, 1964), *Coincya monensis* ssp. *cheiranthos* (Leadlay, 1996) and *Ulex europaeus* ssp. *latebracteatus* for. *humilis* (Coutinho, 1939).

ABSTRACT

The northwest coast of Portugal includes the southernmost Atlantic dune system in Europe. As part of the site selection process for the Natura 2000 network, we studied the flora and the vegetation of this critical coastal dune area and assessed the conservation value of the dune ecosystem. The most significant botanical features of the area are presented in this paper.

This dune system contains several species which are endemic to different degrees: *Rhynchosinapis johnstonii* is a Portuguese endemic, *Jasione lusitanica* (Annex II of the Habitats Directive) is endemic to the northwestern Iberian Peninsula, and *Corema album* is a plant of the western Iberian Peninsula and a very good indicator of the Mediterranean influence on the vegetation.

The plant communities occurring in the area are the following, identified according to the Braun-Blanquet phytosociological approach: *Salsola kali-Cakile maritima* community (drift vegetation); *Euphorbio paraliae-Agropyretum junceiformis* (embryonic shifting dunes); *Otantho maritimi-Ammophiletum australis* (white dunes); *Scrophulario frutescentis-Vulpietum alopecuris* (grey dunes); *Cisto salvifolii-Ulicetum humilis* (decalcified fixed dunes). The *Scrophulario-Vulpietum* and *Cisto-Ulicetum* communities are endemic to the northwest coast of the Iberian Peninsula.

Introduction

The Portuguese dunes are known for the richness of their flora, which is mostly due to the diversity of climatic influences derived from their special biogeographical situation (Braun-Blanquet et al., 1972).

The 120km shoreline of northwest Portugal, at the southern limit of the Atlantic biogeographical sub-region (Eurosiberian region) (Rivas-Martínez, 1987), is an area of great botanical significance. The climate is considered to be submediterranean with a strong oceanic influence (Penas, 1997). Here, the dominant Eurosiberian element of the flora receives significant input from the Mediterranean world (see Braun-Blanquet et al., 1957), resulting in the presence of several unique plant communities. An example is the climax forest formation in the area (*Rusco aculeati-Quercetum roboris*) which is a mixed oak forest with several Mediterranean species, such as *Ruscus aculeatus* and the cork oak *Quercus suber*. The presence of several species endemic to the northwestern Iberian Peninsula (the southwestern Atlantic element) is another factor in producing the unusual features of the flora and the vegetation of this region (Braun-Blanquet et al., 1957).

Along this part of the Atlantic coastline, some plant species and communities are particular to the Galaico-Portuguese biogeographical sector (which extends to the coasts of Galicia; Rivas-Martínez, 1987), while others, with broader distributions, have their southern or northern limits within this sector.

Methods

As part of the site selection process for the Natura 2000 network, we studied the flora and the vegetation of this critical coastal dune area and assessed its botanical value on the basis of the Habitats Directive (Annex I for habitat types, Annexes II, IV and V for plant species).

For the vegetation we used the Braun-Blanquet phytosociological approach for the study of plant communities (Braun-Blanquet, 1932; Westhoff and Maarel, 1973; Géhu and Rivas-Martínez, 1980). The syntaxa were named according to Costa et al. (1994), Espírito-Santo et al. (1995) and Rivas-Martínez et al. (1997).

Results and Discussion

Flora

Although the flora in this dune system is not very rich in endemic species it does include several notable taxa:

1. *Jasione lusitanica* (*Campanulaceae*) is endemic to the Galaico-Portuguese Atlantic dunes; because of its narrow distribution, it is listed in Annex II of the Habitats Directive.
2. *Rhynchosinapis johnstonii* (*Cruciferae*) is a Portuguese endemic which is now considered to be a variety of *Coincya monensis* ssp. *cheiranthos* (Leadlay, 1996). Future studies will probably show that it deserves a higher status, because, as well as its distinct habitat, it is the only tetraploid taxon in the group.
3. *Corema album* (*Empetraceae*) is endemic to the western Iberian Peninsula and a

very good indicator of Mediterranean influence. *Cistus psilosepalus* (*Cistaceae*), *Crocus serotinus* ssp. *clusii* (*Iridaceae*), *Genista triacanthos* (*Leguminosae*) and *Ulex europaeus* ssp. *latebracteatus* for. *humilis* (*Leguminosae*) are also Iberian endemics.

4. *Linaria caesia* ssp. *decumbens* (*Scrophulariaceae*), *Rumex bucephalophorus* ssp. *hispanicus* (*Polygonaceae*), *Anthyllis vulneraria* ssp. *iberica* (*Leguminosae*) and *Helichrysum italicum* ssp. *serotinum* (*Compositae*) are particular to the southwestern European coasts.

Vegetation

The following associations were identified from this dune system (Caldas, Honrado and Paiva, in press; Caldas, Honrado and Pedrosa, 1998):

1. **Drift vegetation** (Habitat type 1210): species-poor halonitrophilous communities dominated by *Salsola kali* and *Cakile maritima*, assignable to the class *Cakiletea maritimae*; *Honckenya peploides* and *Polygonum maritimum* are also present.

2. **Embryonic shifting dunes** (Habitat type 2110): association *Euphorbio paraliae-Agropyretum junceiformis*; widely distributed Atlantic community, dominated by the rhizomatous grass *Elymus farctus* ssp. *boreali-atlanticus*, but also with several other taxa typical of *Ammophiletea* (*Euphorbia paralias*, *Eryngium maritimum*, *Calystegia soldanella* and *Polygonum maritimum*). The first primary dunes are colonized by *Otanthus maritimus*, defining the sub-association *Otanthetosum maritimi*, which is then considered to be a transitory stage in the formation of the characteristic white dune communities.

3. **White dunes** (Habitat type 2120): association *Otantho maritimi-Ammophiletum australis*, an *Ammophila*-dominated community at its optimum on the Atlantic coasts of the Iberian Peninsula. These formations include many taxa characteristic of the *Ammophiletea* communities, such as *Medicago marina*, *Pancratium maritimum*, *Otanthus maritimus*, *Eryngium maritimum*, *Artemisia campestris* ssp. *maritima*, *Euphorbia paralias* and *Aetheorhiza bulbosa* ssp. *bulbosa*. There are two variants in the area: the typical sub-association, *Ammophiletosum arundinaceae*, colonizes the tops of the dunes, and on inland-facing slopes the perennials *Artemisia campestris* ssp. *maritima* and *Crucianella maritima* become more abundant and the community becomes species-richer, forming a particular sub-association, *Artemisietosum crithmifoliae*.

4. **Thermo-Atlantic grey dunes** (Habitat type 2133 – priority habitat): association *Scrophulario frutescentis-Vulpietum alopecuris*; a species-rich community developing on more stabilized sand, mostly composed of perennial herbs and small shrubs and the optimal community for both *Jasione lusitanica* and *Rhynchosinapis johnstonii*. It is endemic to the northwest coast of the Iberian Peninsula. This community is dominated by *Artemisia campestris* ssp. *maritima*, but 18 other species characteristic of the *Ammophiletea* communities are consistently present (*Crucianella maritima*, *Ammophila arenaria* ssp. *arundinacea*, *Vulpia alopecuros*, *Malcolmia littorea*, *Medicago marina*, *Leontodon taraxacoides* ssp. *taraxacoides*, *Anthyllis vulneraria* ssp. *iberica*,

Table 1. Syntaxonomic scheme for the Atlantic dunes on the northwest coast of Portugal

CAKILETEA MARITIMAE Tüxen and Preising *in* Tüxen, 1950
    Salsola kali – Calile maritima community

AMMOPHILETEA Br.-Bl. and Tüxen *ex* Westhoff, Dijk and Passchier, 1946
  Ammophiletalia Br.-Bl. and Tüxen, 1933
    Agropyro-Minuation peploidis Tüxen *in* Br.-Bl. and Tüxen, 1952
      Agropyro-Minuartienion peploidis
        Euphorbio paraliae-Agropyretum junceiformis Tüxen *in* Br.-Bl. and Tüxen,
        1952 *corr.* Darimont, Duvigneaud and Lambinon, 1962
          agropyretosum junceiformis
          otanthetosum maritimi Rivas-Martínez, Lousã, Díaz, Fernandez-González
          and J.C. Costa, 1990
    Ammophilion australis Br.-Bl. 1921 *em.* J.M. Géhu, Rivas-Martínez and R.Tx. *in*
    Rivas-Martínez, Costa, Castroviejo and Valdés, 1980 *corr.* Fdez.-Prieto and Díaz,
    1991
      Ammophilenion australis
        Otantho maritimi-Ammophiletum australis Géhu and Tüxen, 1975 *corr.*
        Fernández Prieto and T.E. Díaz, 1991
          ammophiletosum arundinaceae
          artemisietosum crithmifoliae Rivas-Martínez, Costa, Castroviejo and
          Valdés, 1980
  Crucianellietalia maritimae Sissingh, 1974
    Crucianellion maritimae Rivas Goday and Rivas-Martínez, 1963
      Scrophulario frutescentis-Vulpietum alopecuris Br.-Bl., Rozeira and P. Silva,
      1972

CALLUNO–ULICETEA Br.-Bl. and Tüxen, 1943 *em.* Rivas-Martínez, 1979
  Ulicetalia minoris Quantin, 1935
    Dactylo-Ulicion maritimi Géhu, 1975
      Cisto salvifolii-Ulicetum humilis Br.-Bl., P. Silva and Rozeira, 1964

Pancratium maritimum, *Seseli tortuosum* and others). Further inland, on more stabilized sand, the increasing abundance of the perennial *Cistus salvifolius* indicates successional development to the Atlantic heathland that follows.
5. **Dry coastal heaths with *Erica vagans*** (Habitat type 4040 – priority habitat): *Cisto salvifolii-Ulicetum humilis*, a shrubby community dominated by gorse (*Ulex europaeus* ssp. *latebracteatus* for. *humilis*) and several heath (*Erica*) species, endemic to this part

of the Portuguese coast and a small area in Galicia. It is distinct from all other Atlantic coastal heathlands due to the presence of the small Mediterranean shrub *Cistus salvifolius*, along with several other thermophilous plants. It includes 32 species, with an average of 21 per sample. This *Calluno-Ulicetea* community represents the final stage of succession in this region near the coast, as no forest establishment is possible because of the influence of the sea (Braun-Blanquet et al., 1964).

6. **Pine plantations on stabilized dunes** (Habitat type 2270 – priority habitat): long-established thermophilous pine plantations (almost exclusively of *Pinus pinaster*) on stabilized dunes colonized by Atlantic heathlands.

Due to the botanical significance and the general threatened condition of this part of the Portuguese coast, we have proposed that it be integrated in the national list of sites for the Natura 2000 network under the name 'Litoral Norte' (North Coast). Other forms of conservation, such as the governmental National Programmes for the Coastline, will surely be very important in preserving this crucial area of the Portuguese territory.

A syntaxonomic scheme for the Atlantic dunes on the northwest coast of Portugal is given in Table 1.

Acknowledgments

This study was partially funded by the LIFE project LIFE 94/P/A221/P/01043LIS. The authors would like to thank the APPLE (Esposende Protected Coastal Landscape) for assisting in the field work in the Esposende dune system.

References

Braun-Blanquet, J. (1932), *Plant Sociology*, Hafner, London.

Braun-Blanquet, J., Braun-Blanquet, G., Rozeira, A. and Pinto da Silva, A.R. (1972), 'Résultats de trois excursions géobotaniques à travers le Portugal septentrional et moyen. IV. Esquisse sur la végétation dunale', *Agronomia Lusitanica*, **33**, 217–34.

Braun-Blanquet, J., Pinto da Silva, A.R. and Rozeira, A. (1957), 'Résultats de deux excursions géobotaniques à travers le Portugal septentrional et moyen. II. Chenaies à feuilles caduques (*Quercion occidentale*) et chenaies à feuilles persistantes (*Quercion fagineae*) au Portugal', *Agronomia Lusitanica*, **18,** 167–235.

Braun-Blanquet, J., Pinto da Silva, A.R. and Rozeira, A. (1964), 'Résultats de trois excursions géobotaniques à travers le Portugal septentrional et moyen. III. Landes à cistes et ericacées (*Cisto-Lavanduletea* et *Calluno-Ulicetea*)', *Agronomia Lusitanica*, **23(4)**, 229–313.

Caldas, F.B., Honrado, J.J.. and Paiva, A.P. (in press), 'Vegetação da area de paisagem protegida do litoral de Esposende', *Quercetea*.

Caldas, F.B., Honrado, J.J. and Pedrosa, F. (1998), '"Litoral Norte", proposta de Sítio para a Lista Nacional (Rede "Natura 2000"): comunidades vegetais num sistema de dunas atlânticas do Portugal eurossiberiano', *Actas do Seminário Dunas da Zona Costeira de Portugal*, Associação EUROCOAST-Portugal.

Castroviejo, S. et al. (1986–1997), *Flora Iberica – Plantas Vasculares de la Península Ibérica e Islas*

Baleares. Vols. I–V, Real Jardín Botánico, CSIC, Madrid.

Costa, J.C., Espírito-Santo, M.D. and Lousã, M. (1994), 'The vegetation of dunes of southwest Portugal', *Silva Lusitana*, **2(1)**, 51–68.

Coutinho, A.X.P. (1939), *Flora de Portugal (Plantas Vasculares)*, Bertrand (Irmãos), Lda., Lisbon.

Directiva 92/43/CEE do Conselho, *Jornal Oficial das Comunidades Europeias* No. L: 206/7–206/49.

Espírito-Santo, M.D., Costa, J.C. and Lousã, M.F. (1995), *Sinopsis da vegetação de Portugal Continental*, Instituto Superior de Agronomia, Lisbon.

Franco, J.A. (1971), *Nova Flora de Portugal (Continente e Açores). Vol. I*, published by the author, Lisbon.

Géhu, J. and Rivas-Martínez, S. (1980), 'Notions fondamentales de phytosociologie', in H. Dierschke (ed.), *Syntaxonomie*, J. Cramer, Vaduz.

Heywood, V.H. (1964), '*Rhynchosinapis* Hayek', in T.G. Tutin, N.A. Burges, A.O. Chater, J.R. Edmondson, V.H. Heywood, D.M. Moore, D.H. Valentine, S.M. Walters and D.A. Webb (eds.), *Flora Europaea. Vol. I*, Cambridge University Press, Cambridge, 411–12.

Leadlay, E.A. (1996), '*Coincya* Rouy', in S. Castroviejo, C. Aedo, C. Gomez-Campo, M. Lainz, P. Monserrat, F. Munoz Garmendia, G. Nieto Feliner, E. Rico, S. Talavera and L. Villar, *Flora Ibérica – Plantas Vasculares de la Península Ibérica e Islas Baleares. Vol. IV (Cruciferae-Monotrapaceae)*, Real Jardín Botánico, CSIC, Madrid.

Penas, A. (1997), *Mapa bioclimático de Portugal Continental*, I Encontro de Fitossociologia ALFA, Bragança.

Rivas-Martínez, S. (1987), *Memória del Mapa de las Séries de Vegetación de España 1:400000*, ICONA, Madrid.

Rivas-Martínez, S., Fernández-González, F. and Loidi, J. (1997), *Syntaxonomical Check-List of the Iberian Peninsula and Balearic and Canary Islands (Spain and Continental Portugal)*, Phytosociological Research Center (CIF), Madrid.

Sampaio, G. (1990), *Flora Portuguesa*, 4th edition, INIC, Lisbon.

Tutin, T.G., Burges, N.A., Chater, A.O., Edmondson, J.R., Heywood, V.H., Moore, D.M., Valentine, D.H., Walters, S.M. and Webb, D.A. (eds.), (1964), *Flora Europaea. Vol. I*, Cambridge University Press, Cambridge.

Tutin, T.G., Burges, N.A., Chater, A.O., Edmondson, J.R., Heywood, V.H., Moore, D.M., Valentine, D.H., Walters, S.M. and Webb, D.A. (eds.) (1968–1980), *Flora Europaea. Vols. II–V*, Cambridge University Press, Cambridge.

Tutin, T.G., Burges, N.A., Chater, A.O., Edmondson, J.R., Heywood, V.H., Moore, D.M., Valentine, D.H., Walters, S.M. and Webb, D.A. (eds.) (1993), *Flora Europaea. Vol. I* (2nd edition), Cambridge University Press, Cambridge.

Westhoff, V. and Maarel, E. van der (1973), 'The Braun-Blanquet Approach', in R.H. Whittaker, *Handbook of Vegetation Science. Part V*, Junk, The Hague.

Section 6

MONITORING:
METHODS AND APPLICATIONS

THE EVOLUTION OF NEWBOROUGH WARREN DUNE SYSTEM WITH PARTICULAR REFERENCE TO THE PAST FOUR DECADES

P.M. RHIND, T.H. BLACKSTOCK, H.S. HARDY, R.E. JONES and W. SANDISON
Countryside Council for Wales, Bangor, Wales

ABSTRACT

Changes in sand dune vegetation cover at Newborough Warren (Anglesey) between the early 1950s and 1991 are quantified. There has been a trend from open and dynamic community composition to greater stability and closed vegetation. This development is closely linked to afforestation of part of the system and a reduction in rabbit and livestock grazing. The findings are considered in a historical context, and the need to take long-term flux into account for conservation management planning is emphasized.

Nomenclature follows: Stace (1997) for vascular plants, Hill, et al. (1991–1994) for bryophytes, and Purvis et al. (1992) for lichens.

Introduction

Newborough Warren is the largest of several sand dune systems situated on the south-west coast of Anglesey (Robinson, 1980). Sand deposits cover some 1300ha lying between the Cefni estuary and the western end of the Menai Strait (see Fig. 1). It is one of the major calcareous dune systems on the west coast of Britain and, although the western half of the site has been modified by the establishment of a conifer plantation, it is recognized as a site of outstanding conservation value. The unplanted open dunes were notified as a Site of Special Scientific Interest (SSSI) in 1955, and in the same year a National Nature Reserve (NNR) was declared at Newborough Warren. In 1957, the 720ha planted with conifers was also notified as a SSSI. The system is valued both for its sand dune habitat, with a wide range of rare taxa, and its geomorphological history and setting. More recently, the site has been given international recognition through its proposed designation as a European Special Area of Conservation (SAC) under the Habitats Directive.

Although there is a long historical record of the site based on observations and

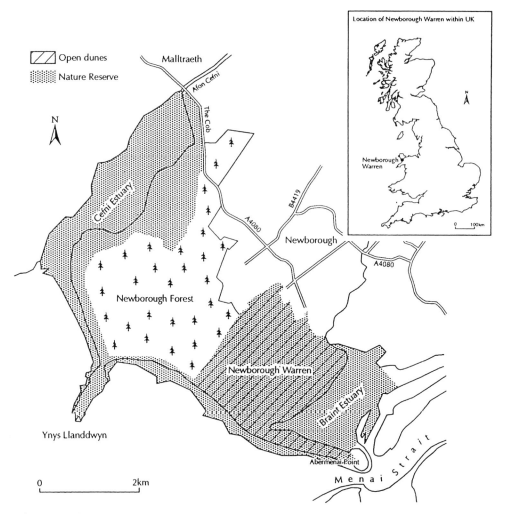

Figure 1. *The location of Newborough Warren in relation to the Menai Strait and to Britain as a whole.*

anecdotal records, scientific investigation at Newborough Warren began with the ecological work of Derek Ranwell in the early 1950s. The pioneer studies by Ranwell (1955; 1958; 1959; 1960a; 1960b) coincided with the preparatory stages of the conifer plantation and with the demise of the rabbit population following an outbreak of myxomatosis, both of which have led to profound modification of the ecological components of the dune system. His studies provide a baseline for future comparative reassessment of the vegetation.

Later research has helped to characterize more recent habitat change including sediment accumulation and saltmarsh development in the Cefni estuary (Packham and Liddle, 1970), impact of the afforestation on soils (Hill and Wallace, 1989) and the effects of livestock grazing on fixed dune grassland (Hewett, 1985).

The most recent survey in 1991 of the unplanted, open, eastern part of the dunes (Ashall et al., 1992) was part of a national inventory of sand dune vegetation (Dargie, 1995). We have used this data set to analyse the extent of ecological change since the 1950s. The main aim of this paper therefore is to summarize the observed changes and to consider implications for nature conservation planning at this key coastal NNR.

Geomorphological Context

The dune system is bisected by a rock ridge and flanked on one margin by the Menai Strait and on the other by the Cefni estuary (see Fig. 1). The Menai Strait has a residual current that flows to the southwest and acts as a barrier to the northward transport of sediment. The flows entering and leaving the Strait exert a major control over the local sediment transport regime, and much of the sediment is composed of sand (Robinson, 1980). The location of the offshore ebb and flood channels has varied through time and has had a significant influence on the evolution of the coastline. Though the maximum fetch is large, the incoming wave energy is limited by the offshore sandbanks and there is a limited spring tidal range of approximately 4.3m.

The Cefni estuary was heavily modified in the eighteenth century by the construction of a barrage known as 'The Cob'. The resulting reduction in size of the estuary caused rapid accretion and a narrowing of its mouth, which in turn is likely to have contributed to the growth of Newborough Warren dune system and the associated saltmarsh habitat.

The general alignment of the coast is at approximately right angles to the prevailing incoming wave energy, except where this is modified by the strong tidal streams at the mouth of the Menai Strait and the Cefni estuary. Landsberg (1956) examined the alignment of the dunes and showed that the orientation of parabolic dunes was very strongly correlated with the calculated wind resultant.

Many dune slacks are flooded in winter (Ranwell, 1959). More recent observations have shown that summer water table levels are significantly depressed around the forest plantation by the high evaporation rates of the pines (Cottingham, 1994). Calcium carbonate concentrations in the plantation soils range from 4 per cent in the youngest dunes to less than 1 per cent on older substrates (Hill and Wallace, 1989).

Methods

Vegetation data recorded in the early 1950s by Ranwell (1955; 1958; 1959; 1960a) were compared with the 1991 data of Ashall et al. (1992). Plant communities (and sub-communities) for the latter work are based on the National Vegetation Classification (NVC) (Rodwell, in press); sampling was undertaken using 2 × 2m quadrats and the cover of each vascular plant, bryophyte and lichen species per sample was assessed using the Domin scale.

Ranwell adopted a different approach to sampling the dune vegetation. He selected sampling plots of 20 × 20m in homogeneous stands of different vegetation types ('plant associes') which were more or less equivalent to NVC communities. Within each plot at least twenty 25 × 25cm quadrats were regularly positioned, and the frequency and percentage cover of each plant species present were recorded; summary data for different associes were tabulated in Ranwell, 1955; 1960a. For comparative purposes, each of the major associes has been allocated to its most appropriate NVC community, using TABLEFIT (Hill, 1992) to guide the conversion. Phytosociological codes follow the NVC. The detailed floristic records compiled by Ranwell in his PhD thesis (Ranwell, 1955) were the primary source of the data used in our analysis; it should be noted that these differ in certain detailed respects from the more condensed information in Ranwell, 1960a.

Plant communities were mapped by Ashall et al. (1992) at a 1:10,000 scale, while Ranwell (1958) published a small-scale sketch map of the major dune habitats based on a 1945 aerial photograph. These were adapted to provide a broad quantified estimate of habitat change between the early 1950s and 1991 using a Geographical Information System. Cover data were obtained for the open unplanted section of the dune system by grouping communities mapped in 1991 into categories corresponding as closely as possible with Ranwell's plant associes.

Changes in Vegetation Cover at Newborough Warren, 1950s–1991

Comparison of vegetation cover shows that the system has changed almost beyond recognition since the early 1950s. Vegetation maps based on what now broadly equates to the open dune habitat (excluding the now forested area) are shown in Figures 2 and 3. Nearly 75 per cent of the total dune area in the 1950s consisted of mobile dunes and embryonic dune slacks with open vegetation, while by 1991 (even if the forest is excluded) only about 6 per cent of the site could be classed as mobile and open, and embryonic dune slacks were more or less non-existent. In addition, there have been small losses in the north and northeastern parts of the dune system due to agricultural improvement; in the 1950s these included stands of dune heath, dry slack and fixed dune grassland. By 1991 the vegetation of Newborough Warren had become dominated by semi-fixed dune grassland and mature dune slack vegetation. For the most part, the system has therefore undergone successional development towards greater stability. Table 1 provides an overall summary of the areas of the different vegetation types found in the early 1950s and 1991. In the following sections changes that have taken place among the various habitat components are outlined, with reference to their constituent plant communities.

Strandline communities
Examples of these very variable and ephemeral vegetation types were found in the

Figure 2. *Habitat map of the unplanted dune system at Newborough Warren in the 1950s.*

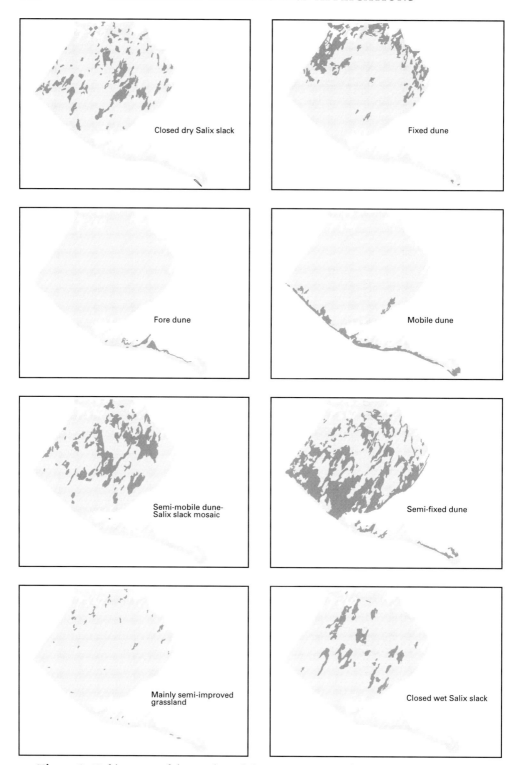

Figure 3. *Habitat map of the unplanted dune system at Newborough Warren in 1991.*

Table 1. The cover of vegetation types found at Newborough Warren in the 1950s and in 1991.
The areas were calculated from the habitat maps (Figs. 2 and 3) (see Table 9 for NVC communities)

Sub-habitats	Area (ha) 1950s	1991	Components of Ranwell's classification	NVC communities
Foredune vegetation	0	5		SD4 SD5
Mobile dune vegetation	?	24	Mobile dune Ammophiletum	SD6
Mobile dune – *Salix* slack mosaic	376	0	Mobile *Ammophila* and *Salix* dune and slack complex	SD6 SD13 SD14 SD16
Semi-fixed dune grassland	?	165	Semi-fixed Ammophiletum	SD7
Semi-fixed dune – *Salix* slack mosaic	0	77		SD7 SD16
Fixed dune grassland	42	59	*Festuca-Agrostis* turf	SD8 SD9
Open wet *Salix* slack	29	0	Open wet *Agrostis* and *Salix* associes	SD13 SD14 SD16d
Closed wet *Salix* slack	5	34	Closed wet associes (see Table 8)	SD13 SD16d
Closed dry *Salix* slack	27	52	Closed *Salix repens* associes	SD16
Dune heath	3	0	*Calluna-Salix* associes	H11
Totals	482	416		

1950s as scattered elements on the foreshore in front of the less eroded parts of the foredunes, and would probably have represented a *Honkenya peploides-Cakile maritima* strandline community (SD2) (Table 2). A scattered strandline community also occurs today, and in some years can be quite extensive, but it is often difficult to allocate it to any one of the recognized NVC strandline communities, since it often has affinities with both the SD2 and the SD3 (*Tripleurospermum maritimum – Galium aparine* strandline) communities (Table 2).

Table 2. Species recorded within the strandline vegetation at Newborough Warren (see Table 9 for NVC communities).
Abbreviations are as follows: f = frequent, o = occasional, r = rare, p = present in stand

	1990s	1950s
NVC community	**SD2/3**	**SD2/3**
Ammophila arenaria		o
Atriplex laciniata	p	r
Atriplex littoralis	p	
Atriplex patula		o
Atriplex prostrata	p	
Cakile maritima	p	r
Crambe maritima	p	
Elytrigia juncea	p	r
Eryngium maritimum	p	
Euphorbia paralias		o
Euphorbia portlandica	p	
Honkenya peploides	p	o
Plantago maritima	p	
Salsola kali	p	f
Sonchus sp.	p	
Silene uniflora	p	
Tripleurospermum maritimum		r

Foredune communities

In 1951 *Elytrigia juncea* was noted for its absence and *Leymus arenarius* does not appear to have been recorded (Ranwell, 1955). Foredune communities (SD4 and SD5) dominated by these two species therefore appear to have been absent at that time. Later, *Elytrigia juncea* was found to occur sporadically along the foreshore (Ranwell, 1960a), and had become well established on the northwest coast where foredune development had been encouraged by fencing associated with forestry operations. Generally, however, the foredunes at that time were found to be extremely mobile, being mainly composed of bare sand, associated with stands of almost pure marram *Ammophila arenaria*. In 1991, both *Leymus arenarius* and *Elytrigia juncea* had become well established along parts of the foredunes. The floristic composition of these communities is shown in Table 3. The absence of these two species in the early 1950s suggests that during times of high mobility neither were instrumental in the dune-building process.

Mobile dune communities

Much of Newborough Warren was dominated by bare sand in the 1950s. The windward faces of most dunes were denuded, but there were extensive stands of *Ammophila arenaria* on dune crests. In some of the slightly more sheltered areas, such as hollows

Table 3. Summary floristic table of 1991 quadrat data from foredune vegetation at Newborough Warren.
Roman numerals I–V: 1–20% frequency (that is, up to one sample in 5) = I, 21–40% = II, 41–60% = III, 61–80% = IV and 81–100% = V.
Figures in brackets show range of Domin values (see Table 9 for NVC community types)

NVC Community	SD4	SD5b
Number of quadrats	10	2
Herb height (range in cm)	8–20	20–60
Bare soil/sand/litter (range in %)	30–75	40–80
Ammophila arenaria	I (2-3)	II (4)
Armeria maritima	I (4)	
Atriplex laciniata	I (4)	
Beta vulgaris maritima	I (5)	
Cirsium arvense		II (1)
Elytrigia juncea	V (2–7)	V (3–5)
Eryngium maritimum	I (1)	
Euphorbia paralias	I (2–4)	
Festuca rubra	I (6–7)	
Glaux maritima	I (1)	
Halimione portulacoides	I (1)	
Leymus arenarius		V (3–8)
Plantago coronopus	I (5)	
Plantago lanceolata	I (2)	
Sagina nodosa	I (3)	
Senecio jacobaea	I (2)	
Silene vulgaris maritima	II (1–3)	
Sonchus arvensis	I (2)	
Spergularia media	I (3)	
Taraxacum seedling/sp.	I (1)	

and at the base of lee slopes, a community analogous to the *Ammophila arenaria* mobile dune NVC community (SD6) was present (Table 4). In other areas, usually with high sand mobility, a community described as an open *Salix repens* associes occurred, dominated by *S. repens* and often with no other associated species (Plate 1). These very

Table 4. *Summary floristic table of 1991 and 1950s quadrat data from mobile dune vegetation at Newborough Warren. Roman numerals I–V: 1–20% frequency (that is, up to one sample in 5) = I, 21–40% = II, 41–60% = III, 61–80% = IV and 81–100% = V. Figures in brackets show range of Domin values (see Table 9 for NVC community types). Abbreviations are as follows: a = abundant, d = dominant, f = frequent, o = occasional, r = rare, va = very abundant, p = present in stand*

	1991 Survey						1950 Survey
							Species recorded within *Ammophila* dunes
NVC Community	SD6e	SD6b	SD6d	SD6f	SD6g	SD6a	SD6 ?
Number of quadrats	13	2	2	2	1	2	5
Herb height (range in cm)	10–45	24–45	28–60	6–25	6	20–60	
Bare soil/sand/litter (range in %)	0–70	50	2–40	0–10	90	50–60	
Agrostis capillaris							va
Agrostis stolonifera	I (2)						f
Ammophila arenaria	V (2–9)	V (5–7)	V (5–8)	III (5)	2	V (5–8)	a
Anagallis arvensis				III (2)			o
Arctium lappa							r
Armeria maritima			III (1)				
Artemisia vulgaris							p
Atriplex laciniata						I (1)	
Atriplex littoralis	I (1)						
Barbula fallax				III (5)			
Brachythecium rutabulum				III (2)			
Cakile maritima	I (1)					I (p)	
Carex arenaria	I (2)			III (4)	4		f
Cirsium arvense	I (5)					I (1)	va
Cirsium vulgare	I (1)						o
Cladonia rangiformis				III (5)			
Cochlearia officinalis	I (2)						
Crepis capillaris				III (1)			
Cynoglossum officinale	I (p)						f
Desmazeria marina				III (3)			
Elytrigia juncea	III (3–5)	III (5)	III (2)			V (3–6)	
Erodium cicutarium							o
Euphorbia paralias	IV (p–4)	III (1)				I (2)	va
Euphorbis portlandica						I (2)	
Eurhynchium praelongum				III (2)			
Festuca rubra	V (1–7)		III (8)	V (8)			va
Galium verum							va
Holcus lanatus	I (4)						
Hypochoeris radicata	I (2–3)					I (1)	
Leontodon saxatilis	I (3)						o

Table 4 cont.	1991 Survey						1950 Survey Species recorded within *Ammophila* dunes
NVC Community	**SD6e**	**SD6b**	**SD6d**	**SD6f**	**SD6g**	**SD6a**	**SD6 ?**
Leymus arenarius		III (5)					
Tripleurospermum maritimum	I (4)						
Ononis repens				III (1)			va
Phleum arenarium							r
Plantago lanceolata	I (1)						
Poa pratensis	I (2)		III (3)	V (3–5)			
Pseudoscleropodium purum	I (5)						
Ranunculus bulbosus							o
Rhytidiadelphus squarrosus				III (1)			
Rubus caesius							o
Rubus fruticosus agg.	I (3)						
Rumex crispus	I (4)						
Salix repens							p
Salsola kali						I (1)	
Sedum acre							o
Senecio jacobaea	II (p–1)						f
Senecio vulgaris	I (1)				1 (1)		
Silene uniflora	I (1–3)			III (4)			
Sonchus arvensis	I (2–4)		III (1)				
Sonchus asper	I (1–3)						
Spergularia marina				III (1)			
Taraxacum seedling	II (p–3)					I (2)	r
Tortula ruralis ssp *ruraliformis*	I (1–2)						
Trifolium repens	I (1)						
Viola tricolor	I (4)						
Vulpia fasciculata	I (3)						

open *S. repens* zones never reached the height of the surrounding dunes. In somewhat more stable regions, occasional plants of other species occurred scattered among the *S. repens* shoots, including *Epipactis leptochila* var. *dunensis*. By the 1990s, there does not appear to have been any equivalent open *Salix repens* vegetation at Newborough Warren, and there is no recognized equivalent within the NVC. Furthermore, most of the modern mobile dune vegetation, which is mainly composed of the *Ammophila arenaria* mobile dune community (SD6), is restricted to the coastal fringe (see Fig. 3), whereas in the 1950s mobile dune vegetation extended in some areas for over 1km inland (see Fig. 2). It is interesting to note that one of the first colonizers of mobile *Ammophila* dunes in the 1950s was *Artemisia vulgaris*, but this species was not recorded from mobile dunes in the 1991 survey.

Semi-fixed dune communities

In the early 1950s fixed or semi-fixed vegetation at Newborough Warren was apparently only poorly developed (Ranwell, 1960a), but it was possible to distinguish an

Plate 1. *Mobile dune* Salix repens *community, Newborough Warren, 1950s.*

Ammophila arenaria semi-fixed community. This vegetation (see Table 5) mostly appears to have been analogous to the *Ammophila arenaria – Festuca rubra* semi-fixed NVC community (SD7), which is now the most extensive vegetation type at Newborough Warren, occupying about 36 per cent of the site. Certain species recorded within the semi-fixed dune grassland in the 1950s, such as the bryophyte *Racomitrium canescens* (sensu lato) and the lichen *Cetraria islandica,* do not occur today. Ranwell also recognized a *Salix repens* semi-fixed community, but the species composition and the percentage cover of certain species (see Table 8) suggest that it would now be considered a dry slack community, and it was identified as having affinities with the *Ononis repens* sub-community of *Salix repens – Holcus lanatus* dune slack (SD16a) vegetation using TABLEFIT.

Fixed dune grassland
The nearest equivalent to a fixed dune grassland in the 1950s was restricted to a zone bordering the landward limits of the central and eastern sections of the unplanted dunes. In the central section this was described as a *Festuca-Agrostis* turf, which was found to be rich in mosses, especially *Rhytidiadelphus triquetrus* (up to 35 per cent cover); the lichen *Cladonia furcata* was also a major component of the vegetation (contributing about 15 per cent of the cover). The abundance of mosses and lichens was partly attributed to the fact that these areas were heavily grazed by rabbits, which resulted in a very short sward. In the eastern section of the dune system adjacent to

Table 5. *Summary floristic table of 1991 and 1950s quadrat data from semi-fixed dune grassland vegetation at Newborough Warren. Roman numerals I–V: 1–20% frequency (that is, up to one sample in 5) = I, 21–40% = II, 41–60% = III, 61–80% = IV and 81–100% = V. Figures in brackets show range of Domin values (see Table 9 for NVC community types). Abbreviations are as follows: a = abundant, d = dominant, f = frequent, o = occasional, r = rare, va = very abundant, p = present in stand*

	1991 Survey			1950 Survey Species recorded within the semi-fixed *Ammophila* dunes
NVC community	SD7b	SD7c	SD7d	SD7 ?
Number of quadrats	5	5	5	
Herb height (range in cm)	6–45	25–40	4–30	
Bare soil/sand/litter (range in %)	0–1	0–15	1–10	
Achillea millefolium	II (1)			
Agrostis capillaris	I (1)			a
Aira praecox			I (2)	
Ammophila arenaria	V (1–8)	V (5–7)	V (3–7)	d
Anacamptis pyramidalis	I (3)			
Anagallis arvensis				o
Angelica sylvestris	I (2)	I (2)		
Anthoxanthum odoratum	I (3)			
Anthyllis vulneraria	I (3)		I (2)	
Aphanes arvensis				o
Arenaria serpyllifolia	I (1)			
Arrhenatherum elatius		II (2–3)		
Barbula cylindrica				o
Barbula unguiculata				o
Bellis perennis				o
Brachythecium albicans	I (2)	II (2)	IV (1–3)	f
Brachythecium glareosum				f
Bromus hordeaceus				o
Cardamine hirsuta				f
Carex arenaria	III (3–4)	I (2)	II (3)	a
Carlina vulgaris	I (1)		II (1)	
Centaurium erythraea	II (p–1)		I (1)	
Cerastium diffusum				f
Cerastium fontanum	I (1)	II (2)		
Cerastium semidecandrum			I (2)	
Ceratodon purpureus				f
Cetraria islandica				o
Cirsium arvense		III (1–2)		
Cirsium vulgare				o

| Table 5 cont. | 1991 Survey | | | 1950 Survey |
| | | | | Species recorded within the semi-fixed *Ammophila* dunes |
NVC community	SD7b	SD7c	SD7d	SD7 ?
Cladonia fimbriata	I (2)			
Cladonia furcata				a
Cladonia rangiformis	II (p–3)			
Crepis capillaris	I (1)	III (3)		
Cynoglossum officinale		I (1)		f
Dactylis glomerata	I (2)			
Dicranum scoparium	III (1–2)	II (1–2)		
Elytrigia juncea		II (3)		
Epilobium angustifolium	I (1)	I (p)	I (1)	
Epilobium montanum	I (1)	I (1)		
Epipactis leptochila var dunensis			I (2)	
Erodium cicutarium				a
Erodium maritimum				f
Euphorbia paralias		I (2)		
Evernia prunastri	I (3)			
Festuca rubra	V (2–7)	V (5–6)	V (3–6)	va
Galium verum	III (p–3)	IV(3)	II (3)	a
Geranium molle				f
Glechoma hederacea				o
Heracleum sphondylium	I (2)			
Hieracium 'indeterminate'	I (3)			
Pilosella officinarum	II (2–5)	I (1)	IV(1–4)	
Holcus lanatus	II (2)	III (2)	I (1)	
Homalothecium lutescens				o
Hylocomium splendens		I (2)		
Hypnum cupressiforme	V (2–5)		I (3)	a
Hypochoeris radicata	III (p–2)	I (2)	III (2)	
Jasione montana	I (2)		II (1–3)	
Leontodon hispidus			I (2)	
Leontodon saxatilis	II (2)	I (2)	I (3)	f
Leucanthemum vulgare	I (2)			
Lophocolea bidentata	I (1)			
Lotus corniculatus	I (3)	II (2)	II (3–5)	f
Luzula campestris	II (2–3)		II (1)	
Myosotis ramosissima				o
Myosotis discolor				o
Ononis repens	III (2–3)	V (3–4)	V (2–3)	o
Peltigera canina	I (2)			a

Table 5 cont.	**1991 Survey**			**1950 Survey** Species recorded within the semi-fixed *Ammophila* dunes
NVC community	**SD7b**	**SD7c**	**SD7d**	**SD7 ?**
Peltigera rufescens	I (1)			
Phleum arenarium			I (1)	
Plantago lanceolata	II (1–2)			
Pleurozium schreberi		I (2)		
Poa annua				o
Poa pratensis	III (3–5)	I (3)	I (2)	f
Polygala vulgaris	I (1)			
Polypodium vulgare		I (p)	II (1–2)	
Potentilla anserina		I (1)		
Pseudoscleropodium purum	I (2)	I (2)		
Racomitrium canescens				r
Ranunculus bulbosus	I (1)	I (3)		o
Rhytidiadelphus squarrosus	II (p–4)			f
Rhytidiadelphus triquetrus	II (2–3)	II (2–3)		f
Rubus caesius		I (7)	II (3–4)	
Rubus fruticosus agg.	I (2)	II (2–4)		
Rumex acetosella		I (2)		
Sagina procumbens				o
Salix repens agg.		I (4)	I (5)	
Sedum acre	II (2–4)		III (1)	a
Senecio jacobaea	I (1)	III (p–1)	I (p)	a
Sonchus arvensis	II (1)			
Taraxacum seedling/sp.	II (1)	I (2)		o
Thymus polytrichus	III (1–4)	I (1)	IV (p–4)	a
Tortella flavovirens	I (1)			
Tortula ruralis ssp. *ruraliformis*	I (1)		V (1–4)	o
Trifolium campestre	I (1)			
Tripleurospermum maritima	I (1)			
Veronica chamaedrys	I (1)	I (1)		
Viola canina	I (2)		I (1)	
Viola tricolor ssp. *curtisii*	II (1–2)	III (2)	II (p–1)	f

the River Braint, a *Salix repens–Festuca rubra–Carex arenaria* association was identified where the intensity of rabbit grazing was less pronounced. The species composition of this vegetation type (see Table 6) suggests that it had much in common with a *Festuca rubra–Galium verum* fixed dune grassland NVC community (SD8), which was still mostly restricted to the landward limits of the Warren at the time of the later survey. In general, there appears to have been a somewhat patchy expansion of fixed dune grassland between the 1950s and 1991, with an overall increase of about 17ha (see Table 1), although some of this had become rather rank by 1991, and was described as an *Ammophila arenaria–Arrhenatherum elatius* fixed dune grassland (SD9).

Dune slacks

Species recorded in dune slack vegetation in the 1950s and in 1991 are shown in Tables 7 and 8. The instability of Newborough Warren during the 1950s meant that damp sand was being continually exposed by erosion and covered by deposition, and unstable dunes and slacks extended inland for up to 0.8km or more. On newly exposed wet sand, *Agrostis stolonifera* was usually the first colonist, but in the event of any sand accretion it was usually succeeded by *Salix repens*. Hope-Simpson and Yemm (1979) also found *Agrostis stolonifera* to be one of the first colonists of newly created dune slacks at Braunton Burrows. The majority of dune slacks at Newborough Warren in the 1950s (see Plate 2a) appear to have had open, successionally young communities, with up to 80 per cent bare ground in some cases (see Table 8).

Table 6. Summary floristic table of 1991 and 1950s quadrat data from fixed dune vegetation at Newborough Warren. Roman numerals I–V: 1-20% frequency (that is, up to one sample in 5) = I, 21–40% = II, 41–60% = III, 61–80% = IV and 81–100% = V. Figures in brackets show range of Domin values (see Table 9 for NVC community types).
1950s survey: Domin values for species exceeding 5% cover; p = present in stand

| NVC community | 1991 Survey | | | | 1950s Survey |
	SD8a	SD8b	SD8e	SD9a	NVC equivalent (TABLEFIT) SD8b
Number of quadrats	5	5	1	8	
Herb height (range in cm)	3–15	3–5	7	25–55	
Bare soil/sand/litter (range in cm)	0–5	0–25	0	0	
Achillea millefolium	I (3)		3	II (3)	
Agrostis capillaris	I (3)	II (3–5)			p
Agrostis stolonifera	I (5)			I (2)	4
Aira praecox		II (1–3)			
Ammophila arenaria	III (1–3)	III (1–3)		V (1–7)	
Anagallis arvensis					p

| Table 6 cont. | 1991 Survey | | | | 1950s Survey NVC equivalent |
NVC community	SD8a	SD8b	SD8e	SD9a	SD8b
Anthoxanthum odoratum	II (1–2)	III (3)	4	II (1–4)	p
Arenaria serpyllifolia	I (3)	I (1)			
Arrhenatherum elatius			1	V (2–8)	
Bellis perennis	I (1)				p
Brachythecium albicans	I (2)				
Brachythecium glareosum					p
Brachythecium rutabulum				I (3)	
Bryum spp.					p
Calliergon cuspidatum				I (1)	
Carex arenaria	V (3–6)	IV (2–4)	2	II (2–5)	4
Carex flacca		I (2)	3		p
Carlina vulgaris				I (1)	
Centaurea nigra				II (1–4)	
Centaurium erythraea	II (1)				
Centaurium littorale	I (1)				p
Cephalozia bicuspidata				I (2)	
Cerastium diffusum	I (2)	I (4)			
Cerastium fontanum	I (3)	II (1–3)		I (1)	p
Cerastium semidecandrum	I (3)				
Ceratodon purpureus	I (1)				
Cirsium arvense				IV (1–2)	
Cirsium palustre					p
Cirsium vulgare	I (1)				p
Cladonia digitata	I (1)				
Cladonia fimbriata		I (2)			
Cladonium furcata					5
Cladonia rangiformis			2		p
Climacium dendroides		II (1)			4
Crepis capillaris	II (1–2)	I (1)	1	I (3)	
Dactylis glomerata				I (1)	
Danthonia decumbens					p
Daucus carota				I (2)	
Desmazeria marina		I (3)			
Dicranum scoparium	I (2)	III (2)			p
Drepanocladus revolvens		I (2)			
Epilobium montanum				I (2)	
Equisetum arvense				I (3)	
Equisetum palustre	I (3)				
Equisetum variegatum	I (1)				

Table 6 cont.	1991 Survey				1950s Survey NVC equivalent
NVC community	SD8a	SD8b	SD8e	SD9a	SD8b
Erica tetralix					p
Erodium cicutarium					p
Euphorbia portlandica	I (1)				
Euphrasia officinalis	III (1–3)		2		p
Eurhynchium praelongum				I (2)	
Evernia prunastri		I (3)			
Festuca rubra	V (4–6)	V (5–6)	2	IV (3–6)	5
Frullania tamarisci					p
Galium verum	V (2–5)	V (3–5)	3	II (1–3)	p
Gentianella amarella	I (1)		2		
Geranium molle	I (2)	III (1–2)			p
Glechoma hederacea				I (2)	
Heracleum sphondylium				I (2)	
Hieracium 'indeterminate'	I (2)				
Pilosella officinarum	I (3)	IV (1–4)	3		
Holcus lanatus	III (2–3)	III (2–3)	3	III (2–4)	p
Homalothecium lutescens	I (1)	I (1)			
Hydrocotyle vulgaris	I (1)				p
Hylocomium splendens			1		p
Hypnum cupressiforme	III (3–7)	II (3–5)	2		4
Hypochoeris radicata	II (2–3)	III (1)	3	II (1–3)	p
Hypogymnia physodes		I (2)			
Lathyrus pratensis				I (1)	
Leontodon autumnalis		I (1)			
Leontodon saxatilis	II (1–2)	II (1–2)			p
Linum catharticum					p
Lotus corniculatus	II (4)	IV (3–4)	3	I (1)	p
Luzula campestris	II (2–3)	V (1–4)	3	I (2)	p
Ononis repens	III (3–5)	II (1)		II (1–2)	
Peltigera canina	I (4)				p
Peltigera rufescens	II (1–2)				
Phleum arenarium		I (3)			
Plagiothecium sp				I (1)	
Plantago coronopus		I (3)			p
Plantago lanceolata	I (1)			I (2)	
Pleurozium schreberi	I (3)	I (2)		II (2)	
Poa pratensis	II (3–6)	IV (2–4)		II (2–4)	p
Poa subcaerulea	I (4)		5	II (4)	
Polygala vulgaris					p

| *Table 6 cont.* | **1991 Survey** | | | | **1950s Survey** NVC equivalent |
NVC community	SD8a	SD8b	SD8e	SD9a	SD8b
Polypodium vulgare				I (2)	
Potentilla anserina	I (3)				p
Potentilla erecta					p
Potentilla reptans	I (3)			I (2)	
Prunella vulgaris					p
Pseudoscleropodium purum	III (1–2)	IV (1–3)	1	IV (2–5)	p
Ranunculus acris	I (3)			I (1)	
Ranunculus bulbosus		I (1)			p
Rhinanthus minor	I (1)				
Rhytidiadelphus squarrosus		III (1–2)	1	II (2–3)	p
Rhytidiadelphus triquetrus	I (2)	III (3–4)	2		5
Rubus caesius	I (1)			II (5–8)	
Rumex acetosa	I (1)			I (2)	
Rumex acetosella					p
Sagina procumbens					p
Salix repens agg.		III (1–2)	1	II (3–4)	p
Sedum acre		I (2)			p
Senecio jacobaea	III (1–2)			I (3)	p
Sonchus arvensis		I (1)		III (1–3)	
Sonchus asper					
Taraxacum seedling	II (1)	I (1)	1	II (1)	p
Thymus polytrichus	III (3–4)	IV (2–4)			4
Tortula ruralis ssp. *ruraliformis*		I (2)			p
Tragopogon pratensis					
Trifolium campestre	I (2)				
Trifolium dubium		I (4)			
Trifolium repens		II (2–3)	2		p
Tussilago farfara				I (2)	
Veronica chamaedrys	I (2)	I (1)	3	III (1–3)	p
Veronica officinalis		II (2)			p
Veronica polita					p
Veronica serpyllifolia ssp. *serpyllifolia*	I (4)				
Vicia angustifolia					p
Viola canina	I (2)	II (2)	2		p
Viola tricolor ssp. *curtisii*		IV (1–4)		II (1–2)	p

Table 7. Summary floristic table of 1991 quadrat data from dune slack vegetation at Newborough Warren. Roman numerals I–V: 1–20% frequency (that is, up to one sample in 5) = I, 21–40% = II, 41–60% = III, 61–80% = IV and 81–100% = V. Figures in brackets show range of Domin values; unbracketed figures are Domin values for single quadrats (see Table 9 for NVC community types)

NVC community	SD13b	SD14a	SD14c	SD14d	SD15a	SD15b	SD16a	SD16b	SD16d	SD17a
Number of quadrats	2	1	2	2	1	2	2	2	2	1
Herb height (range in cm)	7–10	8	5	12–30	30	4–20	3–11	4–30	5–25	10
Bare soil/sand/litter (range in %)	2	0	0–2	0	0	0–2	0–30	0	0–2	0
Agrostis stolonifera	V (1–2)	2	III (2)	III (4)		III (1)	III (1)		V (2–4)	4
Aira caryophyllea	III (1)									
Amblystegium serpens	III (3)					III (2)				
Ammophila arenaria							III (2)			
Anagallis tenella			III (4)						III (3)	
Anthoxanthum odoratum	III (1)									
Anthyllis vulneraria	III (5)									
Bellis perennis						III (2)				
Brachythecium rutabulum				III (4)						
Bryum capillare			III (3)			III (2)				
Bryum pseudotriquetrum	III (3)		V (2–3)							
Calliergon cuspidatum		4		III (3)	8	III (3)			2	
Campylium stellatum		3	V (2–3)	III (3)						
Carex arenaria	III (3)			V (4–6)		V (2–3)	V (2)	III (4)	3	
Carex flacca	V (2–5)	4	V (4)	III (4)			III (3)	III (3)	III (3)	
Carex hirta									III (2)	
Carex nigra					4					3
Carex panicea				III (4)	2	III (5)				3
Carlina vulgaris									III (1)	
Centaurium littorale			III (1)	III (1)						
Cerastium fontanum triviale				III (3)						
Cirsium arvense					3			III (2)		
Cirsium palustre									III (2)	

Table 7 *cont.*

NVC community	SD13b	SD14a	SD14c	SD14d	SD15a	SD15b	SD16a	SD16b	SD16d	SD17a
Crataegus monogyna (s)			III (1)							
Crepis capillaris			III (1)							3
Dactylis glomerata									III (3)	
Dicranum scoparium							III (2)			
Drepanocladus sendtneri		2								
Drepanocladus sp						III (2)				
Epilobium angustifolium						III (5)				
Epipactis palustris	III (3)		III (1)			III (1)			III (2)	
Equisetum arvense									III (3)	
Equisetum palustre			III (3)							
Equisetum variegatum	V (3–5)	4	V (2–4)	V (1–4)	1	V (5)	III (3)	III (3)	V (3)	
Erigeron acer				III (2)			III (2)			
Euphrasia officinalis agg.	V (2–3)		V (1–4)			V (1–4)		III (2)	III (3)	
Festuca ovina				III (1)						
Festuca rubra	III (2)			V (3)			III (2)	III (4)	III (3)	3
Galium palustre										2
Gentianella amarella	III (2)		III (2)			III (1)				
Pilosella officinarum	III (2)		III (1)				III (3)			
Holcus lanatus	III (1)		III (1)	V (2)				III (2)	V (2–7)	6
Hydrocotyle vulgaris		4	V (4)		4	V (2–5)	III (4)	III (3)	V (3)	3
Hypnum cupressiforme	V (2)									
Hypochoeris radicata	V (1–2)		III (1)				III (2)	III (3)		
Hypogymnia physodes							III (2)			
Jasione montana							III (1)			
Leontodon autumnalis	III (1)	5	III (2)	III (2)		III (4)	III (2)			
Leontodon saxatilis			V (1–3)			III (1)			III (3)	
Linum catharticum	III (3)		III (2)			III (1)	III (2)			
Listera ovata									III (1)	
Lotus corniculatus	V (2–3)	4	V (4)			V (3)	III (1)	V (3–4)	V (2)	3

Table 7 cont.

NVC community	SD13b	SD14a	SD14c	SD14d	SD15a	SD15b	SD16a	SD16b	SD16d	SD17a
Mentha aquatica		3	V (2)	III (1)		III (4)		III (2)	III (2)	3
Ononis repens	III (4)						V (2)	III (2)	III (2)	
Parnassia palustris	V (1)		V (2;			III (1)		III (2)	III (2)	
Plantago coronopus			III (1)							
Pleurozium schreberi							III (1)			
Poa pratensis				V (4–5)	1			III (4)		
Poa subcaerulea				III (2)						3
Polygala vulgaris	III (1)									
Potentilla anserina		2	III (2)	III (2)	3			III (4)	III (2)	5
Prunella vulgaris	III (3)					III (2)		III (1)	III (2)	3
Pulicaria dysenterica									III (4)	
Pyrola rotundifolia	V (2–4)		III (4)				III (3)		III (5)	
Ranunculus acris		2	III (1)	V (3)		III (2)		III (1)	III (3)	
Ranunculus flammula								III (1)		
Ranunculus repens	V (1–2)			III (5)				III (6)		3
Rhinanthus minor										
Rubus caesius										
Salix repens agg.	V (7–9)	5	V (6)	III (9)	9	V (5–6)	V (6–9)	V (4)	V (4–6)	4
Schoenus nigricans						III (5)		III (2)		
Selaginella selaginoides			III (5)							
Senecio jacobaea			III (1)					III (1)		
Sonchus arvensis				III (2)				III (3)		
Sonchus asper									III (2)	
Taraxacum seedling										1
Thymus polytrichus							III (3)			
Trifolium pratense				III (1)					III (2)	
Trifolium repens				III (4)						3
Veronica chamaedrys		4							III (2)	
Viola tricolor							III (1)			

Table 8. Species recorded in dune slack vegetation in the early 1950s. Numbers refer to Domin values of species occurring at percentage covers exceeding 5%. p = present in stand (see Table 9 for NVC communities)

	Open slacks			Semi-closed wet slack		Closed wet slacks		Closed wet-dry slacks		Closed dry slacks		
Associes number	1	2	3	4	5	6	7	8	9	10	11	12
NVC equivalent (TABLEFIT)	SD16d	SD13a/ SD16d	SD16d	?	SD16d	SD14a	SD14c	SD16d	SD16d/ SD13b	SD16a	SD8b	H11
Ranwell's classification	Open wet Agrostis stolonifera associes	Open wet Salix repens associes	Open wet Salix repens associes	Littorella –Samolus associes	Salix–Carex Hydrocotyle associes	Juncus maritimus Salix associes	Schoenus– Salix associes	Calluna– Salix associes	Salix– Pyrola associes	Salix repens associes	Festuca– Agrostis associes	Calluna– Salix associes
Bare ground (%)	80	63	80	7	1	1	1	1	3	1	1	1
Soil pH	8.4	8.4	8	7.7	7.5	6.7	7.4	5.6		6.6	6.6	5
Agrostis capillaris	5					p	p	p		p	4	p
Agrostis stolonifera		p	p	4	5	p	5	5	p	p	p	
Anagallis tenella		p		p	4		p	p	p			
Bryum spp.	p	p		p	p		p	p	4		p	
Calliergon cuspidatum				p	4		p					
Calluna vulgaris								5				5
Campylium stellatum	p	p		p	5	5	7	4				
Carex arenaria	p	p	p					p	p	p	p	4
Carex flacca		p	p	p	4	p	4	p	p	p	p	p
Carex nigra		p		p	4	p	p	p	p	p		p
Carex pulicaris							p	5				p
Drepanocladus lycopodioides				6								
Drepanocladus revolvens		p		p	4	p	p		p			
Equisetum variegatum	p	p		p		p	p	p	4			p

Table 8 cont.

Associes number	Open slacks			Semi-closed wet slack	Closed wet slacks			Closed wet-dry slacks		Closed dry slacks		
	1	2	3	4	5	6	7	8	9	10	11	12
Eriophorum angustifolium									4			
Festuca rubra		p	p		p	p	4	p	p	p	5	4
Hydrocotyle vulgaris	p	p	p	4	4	5	p	p	p	p		p
Juncus maritimus					5	5						
Littorella uniflora	p			7								
Lotus corniculatus	p	p	p		p	p	p	p	5	p	p	p
Plantago coronopus	4	p	p		p		p	p	p			p
Potentilla anserina	p	p		p	p	4	p	p		p		
Pyrola rotundifolia								p	5			
Salix repens agg.	p	5	5	p	5	5	4	5	6	9		5
Schoenus nigricans						p	5					
Thymus polytrichus		p					p			p	4	
Agrostis canina												p
Aneura pinguis	p	p										
Barbula spp.	p	p										
Bellis perennis	p	p	p		p		p	p	p			p
Campylium chrysophyllum	p	p										
Cardamine pratensis					p	p		p				
Carex serotina				p	p		p					
Centaurium littorale	p	p			p		p		p		p	p
Cerastium diffusum	p	p	p						p			
Cerastium fontanum			p			p				p	p	p
Chara fragilis	p											
Cladonia rangiformis											5	
Climacium dendroides								p		p	p	p
Dicranum scoparium										p	p	p
Drepanocladus sendtneri					p				p		p	p

Table 8 cont.

Associes number	Open slacks			Semi-closed wet slack		Closed wet slacks		Closed wet-dry slacks		Closed dry slacks		
	1	2	3	4	5	6	7	8	9	10	11	12
Eleocharis palustris	p											
Erica tetralix								p				p
Festuca ovina							p	p				p
Fissidens adianthoides							p	p				
Galium palustre				p	p	p	p					
Galium verum						p				p	p	p
Gentianella amarella							p					
Glaux maritima		p		p	p							
Hocus lanatus		p				p	p	p	p	p		p
Homalothecium lutescens		p	p									
Hylocomium splendens								p			p	8
Hypericum maculatum						p						
Hypnum cupressiforme									p		4	
Juncus articulatus		p		p	p	p	p	p				
Juncus bufonius		p										
Leontodon saxatilis		p	p		p		p	p	p	p	p	p
Linum catharticum					p		p		p			
Lotus pendunculatus						p						
Luzula campestris		p	p	p			p	p	p	p	p	
Lychnis flos-cuculi							p					
Mentha aquatica	p	p		p		p	p	p	p	p		
Parnassia palustris					p		p					
Pellia endiviifolia		p										
Peltigera canina								p	p	p	p	
Petalophyllum ralfsii	p	p		p						p	p	p
Pinguicula vulgaris							p					
Plantago lanceolata						p	p					

Table 8 cont.

Associes number	Open slacks			Semi-closed wet slack		Closed wet slacks		Closed wet-dry slacks		Closed dry slacks		
	1	2	3	4	5	6	7	8	9	10	11	12
Plantago maritima							p					
Poa pratensis	p	p	p		p	p	p	p	p	p	p	p
Polygala vulgaris									p	p		p
Potentilla erecta						p	p	p		p	p	p
Preissia quadrata				p								
Prunella vulgaris		p	p		p	p	p	p	p	p	p	p
Pseudoscleropodium purum		p				p	p	p		p	p	p
Radiola linoides		p										
Ranunculus acris		p					p					p
Ranunculus flammula		p		p	p	p	p	p				
Rhytidiadelphus squarrosus						p	p			p	5	p
Rhytidiadelphus triquetrus						p	p	p		p	5	5
Sagina nodosa	p	p	p		p							
Sagina procumbens							p		p			p
Samolus valerandii				p	p							
Sedum acre		p	p								p	p
Selaginella selaginoides					p		p	p	p			
Senecio jacobaea	p	p	p						p	p		
Taraxacum sp.		p	p		p		p			p	p	p
Tortula ruraliformis										p	p	p
Trifolium repens		p			p	p	p	p	p	p	p	p
Veronica chamaedrys						p				p	p	p
Veronica officinalis								p		p	p	p
Viola canina										p	p	p
Viola riviniana		p			p	p						

Table 9. NVC communities referred to in the text and tables. Source Rodwell (in press)

H11	*Calluna vulgaris–Carex arenaria* heath
SD2	*Honkenya peploides–Cakile maritima* strandline
SD3	*Matricaria maritima–Galium aparine* strandline
SD4	*Elytrigia juncea* foredune
SD5	*Leymus arenarius* mobile dune
SD5b	*Elytrigia juncea* sub-community
SD6	*Ammophila arenaria* mobile dune
SD6a	*Elytrigia juncea* sub-community
SD6b	*Elytrigia juncea–Leymus arenarius* sub-community
SD6d	*Ammophila arenaria* sub-community
SD6e	*Festuca rubra* sub-community
SD6f	*Poa pratensis* sub-community
SD6g	*Carex arenaria* sub-community
SD7	*Ammophila arenaria–Festuca rubra* semi-fixed dune
SD7b	*Hypnum cupressiforme* sub-community
SD7c	*Ononis repens* sub-community
SD7d	*Tortula ruraliformis* ssp. *ruraliformis* sub-community
SD8	*Festuca rubra–Galium verum* fixed dune grassland
SD8a	Typical sub-community
SD8b	*Luzula campestris* sub-community
SD8e	*Prunella vulgaris* sub-community
SD9	*Ammophila arenaria–Arrhenatherum elatius* dune grassland
SD9a	Typical sub-community
SD13	*Salix repens–Bryum pseudotriquetrum* dune slack
SD13a	*Poa annua–Hydrocotyle vulgaris* sub-community
SD13b	*Holcus lanatus–Festuca rubra* sub-community
SD14	*Salix repens–Campylium stellatum* dune slack
SD14a	*Carex serotina–Drepanocladus sendtneri* sub-community
SD14c	*Bryum pseudotriquetrum–Aneura pinguis* sub-community
SD14d	*Festuca rubra* sub-community
SD15	*Salix repens–Calliergon cuspidatum* dune slack
SD15a	*Carex nigra* sub-community
SD15b	*Equisetum variegatum* sub-community
SD16	*Salix repens–Holcus lanatus* dune slack
SD16a	*Ononis repens* sub-community
SD16b	*Rubus caesius* sub-community
SD16d	*Agrostis stolonifera* sub-community
SD17	*Potentilla anserina–Carex nigra* dune slack
SD17a	*Festuca rubra–Ranunculus repens* sub-community

Ranwell (1959) described slacks as either wet, in which the free water table never fell below 1m of ground level in any season, or dry, in which the free water table in summer was 1–2m below ground level. He also recognized transitions between these two, and actually described 12 different categories of dune slack vegetation (see Table 8). In terms of the NVC much of this vegetation appears to have affinities with the *Agrostis stolonifera* sub-community of the *Salix repens–Holcus lanatus* dune slack (SD16), and the *Poa annua–Hydrocotyle vulgaris* sub-community of the *Salix repens–Bryum pseudotriquetrum* embryonic dune slack (SD13). Two of Ranwell's other categories seem to be closer to the *Salix repens–Campylium stellatum* dune slack (SD14), and two of his dry slack categories have more in common with fixed dune grassland (SD8) and dune heath (H11). The latter *Calluna vulgaris–Salix repens* heath had developed in a relatively rabbit-free zone in the north east of the dune system; it has since been reclaimed for agriculture, and there is no surviving dune heath.

Dune slacks at Newborough in the 1950s appear to have been rather less rich in plant species. In the 1991 survey, approximately 145 species were recorded in dune slacks, whereas Ranwell lists only 90 species, but this is not surprising considering the open and transitory nature of many of the dune slacks at that time. Although their species composition shows them to have had some affinities with SD13 and SD16, these comparisons are possibly somewhat misleading, as modern-day equivalents of some of the assemblages present in the 1950s are lacking. The nearest examples of embryonic dune slacks are at Aberffraw, a dune system approximately 5km north-west of Newborough. Here, *Juncus articulatus* is locally abundant, but unlike the

Plate 2a. *Northern end of coastal slack A, Newborough Warren, 1950s.*

Plate 2b. *Northern end of coastal slack A, Newborough Warren, 1990s.*

Newborough slack communities dominated by *J. articulatus* in the 1950s, *Agrostis stolonifera* is absent. Plates 2a and b show a clear example of the degree to which the vegetation has changed in a dune slack at Newborough over the past four decades.

Dune slack mosaics with other vegetation types

Ranwell described and mapped extensive areas of Newborough Warren as a mobile dune–*Salix* slack complex (see Fig. 2). In 1991 there were still large areas composed of a mosaic of hygrophytic and xerophytic vegetation (Fig. 3), but instead of mobile dune vegetation (SD6), a major part of the dune system was occupied by a combination of semi-fixed dune grassland (SD7) and mature, dry *Salix repens* slack (SD16). There has been a striking transformation from instability and mobility to stability and immobility.

Scrub encroachment

According to Ranwell (1955), trees did not occur naturally on the dunes, although he did later recall that prior to 1954 there were scattered and very sparse, heavily rabbit-chewed, hawthorn *Crataegus monogyna* stumps present on the Warren (Hodgkin, 1984). Shortly after the onset of myxomatosis in 1954, however, virtually all of the rabbits at Newborough Warren were eliminated (Ranwell, 1959; 1960b). The resulting reduction in grazing pressure gave rise to conditions much more suitable for scrub invasion, and by 1981 scrub had spread thinly over much of the Warren (Hodgkin, 1984). Hawthorn and to a certain extent birch *Betula* spp. were the most frequent species, but there were a further 17 species of trees and shrubs. The oldest birch dated from the year after the outbreak of myxomatosis, and there had been a more or less even rate of recruitment thereafter. The density of hawthorn declined seaward with distance from a plot of scrub on the northern landward edge, but it had spread to within 330m of the sea, and had colonized yellow dune, grey dune, dry slack and fixed dune grassland. No attempt to control hawthorn has been undertaken, but stands of sea buckthorn *Hippophae rhamnoides* and some birch have been eradicated.

Discussion

The historical context

Available recorded information on the early development of the dune system at Newborough has been collated and appraised by Ranwell (1955; 1959). Some of the key events are outlined here.

Throughout their history, Newborough Warren and many other dune systems in Britain and Europe (Boot and van Dorp, 1986; De Raeve, 1989; Pye, this volume) appear to have undergone cycles of mobility followed by intervening periods of relative stasis.

Although there is some evidence of prehistoric phases of dune activity at Newborough (Ranwell, 1959), historical references can be traced back to 1331 when a succession of southwesterly gales resulted in the loss of approximately 75ha of manorial land due to sand inundation. The area is thought to have previously been farmland on stable sand deposits.

During the fourteenth and fifteenth centuries, erosion and deposition events appear to have reached a further peak, and in the sixteenth century marram was planted in an effort to stabilize the mobile dunes. By the middle of the seventeenth century some degree of stability had been reached, and this appears to have extended into the latter part of the nineteenth century. Then in the years 1869, 1890 and 1896 the sand spit at Newborough Warren (Aber Menai Point) was breached by storm tides, and by 1890 the dunes were described as a mass of moving sand. In 1908 a visiting botanist (Jones, 1917) described the dunes to the west of the rock ridge, now planted with conifers, as floristically poor and bare except for marram; a few years later, the dunes were described as increasing in extent, and wind action was destroying *Salix* dunes in many places (Wortham, 1913). Greenly (1919) gives the following graphic account of an open, mobile dune system at Newborough Warren.

Between the sandy tracts of the two bays of Malldraeth and Llandwyn runs the ridge of spilitic lavas. It is about three miles wide in length and a quarter of a mile wide, a range of steep ice-worn bosses, 50 feet or more in height, standing out of boulder clay. Upon this the sands of both bays, creeping obliquely inland, have gradually encroached so that now the boulder clay is rarely to be seen and great drifts of sand shifting with every change of wind have gathered round the rocks, burying them sometimes nearly to their summits. The ridge is a piece of scenery, bringing to mind some views that have been published of the rocky deserts of the Sudan and of central Asia; and that by a paradox of Nature in a district with a rainfall of 37 inches in a year. The barrenness of some parts of the ridge is indeed remarkable. One may stand in some of the hollows between the bosses of Bryn Llwyd and not see so much as a blade of bent grass, not even a lichen on the clean swept rocks, nothing but knobs of green lava and drifts of yellow sand.

Military manoeuvres during both World Wars caused further destabilization of this already mobile dune system. It was within this extremely dynamic regime that Ranwell (1958) estimated that a dune developing along the shoreline would tend to travel landward at an estimated rate of 6.7m per year, although the rate of dune migration is likely to have varied considerably during that period (Robinson, 1980). Dune movement resulted from the fact that windward slopes were generally bare of vegetation, and acted as continuous erosion surfaces, while leeward slopes, which were partly vegetated, acted as continuous sand accretion surfaces. Superimposed upon this landward migration of dunes was the rapid production of parabolic dunes and dune slacks resulting from localized demolition of foredunes (blowouts) by high winds. The interior was also extremely unstable with many secondary blowouts. However, in the 1950s there was little sign of embryonic dune building and Ranwell (1955) suggested that the balance at that time appeared to have been in favour of stabilization – a trend which has continued to the present day.

Ecological change between the 1950s and the 1990s
There is clear evidence from changes in vegetation cover at Newborough Warren that the recent period since the early 1950s has been characterized by a transition to greater stability. This recent phase of dune stasis has been associated with the development of a conifer plantation since 1947, a decline in the numbers of rabbits following an outbreak of myxomatosis in 1954, a lack of farm livestock, the cessation of marram harvesting, and an increase in the deposition of atmospheric nitrogen.

The conifer plantation in the western half of the system is largely composed of Corsican pine *Pinus nigra* ssp. *laricio*, with smaller amounts of other conifers including lodgepole pine *Pinus contorta*, sitka spruce *Picea sitchensis* and some broad-leaved species (Mayhead, 1989). Various damp slacks were left unplanted. A range of methods were employed to stabilize the dunes within the area set aside for forestry. These methods included fencing and marram planting in 1951 to encourage dune development along the southwest dune foreshore to protect the immature plantations.

Following an investigation of the vegetation and soils in the maturing pine plantation in 1986, Hill and Wallace (1989) concluded that the major effects of afforestation had been to stabilize the dune morphology and to lower the water table, as well as to cause considerable changes in the floristic composition due to shading, drought, nutrient competition and needle litter accumulation. They also observed that the unplanted slacks had dried out. An interesting and unexpected feature of the forest is that populations of several rare plants, such as the dune helleborine *Epipactis leptochila* var. *dunensis*, are larger than in the unplanted dunes (Blackstock, 1985; Hill and Wallace, 1989).

Depression of the water table in the dune forest, with an effect extending into the open dune system, has also been observed by Cottingham (1994) and Bristow and Bailey (1998). Although the water table has not been consistently recorded, observations suggest that there has been a more widespread drying-out in the unplanted section. Ranwell (1959) reported extensive flooding during a wet winter in 1950–51, and Onyekwelu (1972) noted that some areas may remain under water for 3–4 months in high-rainfall winter periods. More recently, standing water has been much less frequently observed.

This trend towards stabilization was undoubtedly further enhanced in the 1950s by a massive reduction in the rabbit population as a result of myxomatosis. In the first three years after the disease reduced the rabbit population there was a considerable increase in the growth of most of the indigenous grasses and sedges at the expense of low-growing dicotyledonous herbs (Ranwell, 1960), and by 1965 several of the smaller plant species were being overrun by the vigorous growth of grasses, and the spread of vegetation was causing a reduction in the area of bare sand (Jones, 1965). Whether the present stabilization phase could have been offset by a flourishing rabbit population is open to debate.

The deposition of nutrients from atmospheric sources, particularly nitrogen, can also have a stabilizing influence on sand dunes by stimulating the growth of grasses and shrubs (Willis, 1963; van Dijk, 1992; van Boxel et al., 1997). This is particularly relevant to Newborough Warren since it is closely adjacent to a chicken farm which expels relatively large amounts of ammonia into the local atmosphere. The impact on the dune vegetation is currently under investigation by a consortium involving the Institute of Terrestrial Ecology (ITE), the University of Wales, Bangor and the Countryside Council for Wales.

Although the relative contributions of afforestation, grazing and geomorphological processes are difficult to segregate, and are to some extent interrelated, the combined impact has lead to closure of the vegetation cover, reductions in the exposure and movement of bare sand, and an overall lowering of the water table.

Conservation implications

The main short-term objectives for the open dune system managed as NNR are to maintain a wide range of early to late successional phases of dune and slack habitats, together with their characteristic herbaceous plant communities, and to maintain a range of rare and scarce sand dune taxa. Major management activities include recent reintroduction of livestock (cattle, horses and sheep) in several fenced compartments covering 60 per cent of the slacks and dune grassland in the NNR, together with scrub control.

Grazing experiments in the fixed dune grassland at Newborough Warren have shown that livestock grazing can promote increased species diversity (Hewett, 1985; Gibson, 1988). The extent to which the reintroduction of livestock will enhance the conservation of early successional communities, however, is not known. In due course, it may become necessary to undertake direct reactivation, as is being attempted in other stabilized dune environments (see van Boxel et al., 1997). Localized manipulation of vegetation cover to conserve particular rarities may also need to be considered, as for *Liparis loeselii* at Kenfig Burrows NNR in South Wales (Jones, 1998).

In the medium term, future post-harvest operations in the dune forest will require careful consideration in the overall context of environmental conservation at Newborough Warren. Notwithstanding its commercial value, the negative influence of afforestation on biological composition (including the spread of pines), water table behaviour and dune dynamics needs to be assessed alongside any apparent conservation gains. In view of the proximity of the open dune system as a source of colonizing propagules, there may be a potential for re-establishing open dune communities (Sturgess and Atkinson, 1993). The impacts of re-exposing these western dunes on adjoining agricultural land and settlement would also need to be considered.

In the long term, conservation ambitions may need to be tempered by unforeseen but potent climatic and other environmental impacts. The history of Newborough Warren and other young coastal dune ecosystems clearly demonstrates that the best-laid plans are likely to be prone to disruptive and unpredictable development. At Newborough Warren there is compelling evidence that the vegetation cover has been through major phases of change and that at some time in the future, it may undergo a complete reversal of its current trend. Are we ever therefore justified in attempting to either retard or reverse this process of change? Such action has certainly been a general characteristic of past conservation management. When dunes were more mobile there were attempts to stabilize them; now the emphasis is often more on destabilization. Furthermore, when dealing with an evolving system should any given point on the spectrum of change be regarded as optimal in terms of conservation value? For example, was Newborough as Ranwell knew it in the 1950s any better or worse than it is today? If there is a lesson to be learnt here, it is that sand dune conservation has to embrace the fact that flux is an integral part of the system. More emphasis should therefore be devoted towards assessing the direction of the current trends, and on extending the timescales over which future management is considered.

Acknowledgments

The authors would like to acknowledge with thanks the help of Keith Jones, Rob Jones, Dermot O'Leary and John Ratcliffe in preparing this account.

References

Ashall, J., Duckworth, J., Holder, C. and Smart, S. (1992), *Sand Dune Survey of Great Britain. Site Report No. 122. Newborough Warren and Forest, Anglesey, Ynys Mon, Wales, 1991,* Joint Nature Conservation Committee, Peterborough.

Blackstock, T.H. (1985), 'Nature conservation within a conifer plantation on a coastal dune system, Newborough Warren, Anglesey', in *Sand Dunes and their Management: Focus on Nature Conservation, No. 13,* Nature Conservancy Council, Peterborough, 145–49.

Boot, R.G.A. and Dorp, D. van. (1986), *De plantengroei van de Duinen van Oostvoorne in 1980 en veranderingen sinds 1934,* Stichting Het Zuidhollands Landscape, Schiedamsevest 44C, 3011 BA Rotterdam.

Boxel, J.H. van, Jungerius, P.D., Kieffer, N. and Hampele, N. (1997), 'Ecological effects of reactivation of artificially stabilized blowouts in coastal dunes', *Journal of Coastal Conservation,* **3**, 57–62.

Bristow, W. and Bailey, S. (1998), *Ground Penetrating Radar Survey of the Dunes at Newborough Warren,* Report to the Countryside Council for Wales by the University College of London.

Cottingham, A.M. (1994), 'Water table and afforestation of the Newborough dune system', BSc dissertation, University of Wales, Bangor.

Dargie, T.C.D. (1995), *Sand Dune Vegetation Survey of Great Britain, Part 3: Wales,* Joint Nature Conservation Committee, Peterborough.

De Raeve, F. (1989), 'Sand dune vegetation and management dynamics', in F. van der Meulen, P.D. Jungerius and J.H. Visser (eds.), *Perspectives in Coastal Dune Management: A Dynamic Approach,* SPB Academic Publishing, The Hague, 99–109.

Dijk, H.W. van (1992), 'Grazing domestic livestock in Dutch coastal dunes: experiments, experiences and perspectives', in R.W.G. Carter, T.G.F. Curtis and M.J. Sheehy-Skeffington (eds.), *Coastal Dunes: Geomorphology, Ecology and Management for Conservation,* Balkema, Rotterdam, 235–50.

Gibson, D.G. (1988), 'The relationship of sheep grazing and soil heterogeneity to plant spatial patterns in dune grasssland', *Journal of Ecology,* **76**, 233–52.

Greenly, E. (1919), *The Geology of Anglesey,* Memoirs of the Geological Survey, HMSO, London.

Hewett, D.G. (1985), 'Grazing and mowing as management tools on dunes', *Vegetatio,* **62**, 441–47.

Hill, M.O. (1992), *TABLEFIT – a program to identify types of vegetation by measuring goodness-of-fit to association tables,* Institute of Terrestrial Ecology, Abbots Ripton.

Hill, M.O., Preston, C.D. and Smith, A.J.E. (eds.) (1991–1994), *Atlas of Bryophytes in Britain and Ireland,* 3 volumes, Harley Books, Colchester.

Hill, M.O. and Wallace, H.L. (1989), 'Vegetation and environment in afforested sand dunes at Newborough, Anglesey', *Forestry,* **62**, 249–67.

Hodgkin, S.E. (1984), 'Scrub encroachment and its effects on soil fertility on Newborough Warren, Anglesey, Wales', *Biological Conservation,* **29**, 99–119.

Hope-Simpson, J.F. and Yemm, E.W. (1979), 'Braunton Burrows: developing vegetation in dune slacks, 1948–77', in R.L. Jefferies and A.J. Davey (eds.), *Ecological Processes in the Coastal Environment,* Blackwell Scientific Publications, London, 113–27.

Jones, D.A. (1917), 'The mosses and hepatics of the south-west of Anglesey', *Lancashire and Cheshire Naturalist,* **10**, 141–51.

Jones, P.H. (1965), 'Notes on some recent changes in the vegetation at Newborough Warren,

Anglesey', *Nature in Wales*, **9**,165–69.

Jones, P.S. (1998), 'Aspects of the population biology of *Liparis loeselii* (L.) Rich. var. *ovata* Ridd. ex Godfery (Orchidaceae) in dune slacks of South Wales, UK', *Botanical Journal of the Linnean Society*, **126**, 123–39.

Landsberg, S.Y. (1956), 'The orientation of dunes in relation to wind', *Geographical Journal*, **122**, 176–89.

Mayhead, G.J. (1989), 'Newborough Forest Site of Special Scientific Interest: a management plan for commercial working', *Quarterly Journal of Forestry*, **84**, 247–256.

Onyekwelu, S.S. (1972), 'The vegetation of dune slacks at Newborough Warren', *Journal of Ecology*, **60**, 887–98.

Packham, J.R. and Liddle, M.J. (1970), 'The Cefni saltmarsh, Anglesey, and its recent Developments', *Field Studies*, **3**, 331–56.

Purvis, O.W., Coppins, B.J., Hawksworth, D.L., James, P.W. and Moore, D.M. (eds.) (1992), *The Lichen Flora of Great Britain and Ireland*, British Natural History Museum.

Ranwell, D.S. (1955), 'Slack vegetation, dune development and cyclical change at Newborough Warren, Anglesey', PhD thesis, University of Wales, Bangor.

Ranwell, D.S. (1958), 'Movement of vegetated sand dunes at Newborough Warren, Anglesey', *Journal of Ecology*, **46**, 83–100.

Ranwell, D.S. (1959), 'Newborough Warren, Anglesey I. The dune system and dune slack habitat', *Journal of Ecology*, **47**, 571–601.

Ranwell, D.S. (1960a), 'Newborough Warren, Anglesey II. Plant associes and succession cycles of the sand dune and dune slack vegetation', *Journal of Ecology*, **48**, 117–41.

Ranwell, D.S. (1960b), 'Newborough Warren, Anglesey III. Changes in the vegetation on parts of the dune system after the loss of rabbits by myxomatosis', *Journal of Ecology*, **48**, 385–95.

Robinson, A.H.W. (1980), 'The sandy coast of south-west Anglesey', *Transactions of the Anglesey Antiquarian and Field Club (1980)*, 37–66.

Rodwell, J.S. (in press), *British Plant Communities. Volume 5: Maritime and Weed Communities*, Cambridge University Press, Cambridge.

Stace, C.A. (1997), *New Flora of the British Isles*, Cambridge University Press, Cambridge.

Sturgess, P. and Atkinson, D. (1993), 'The clear-felling of sand dune plantations: soil and vegetational processes in habitat restoration', *Biological Conservation*, **66**, 171–83.

Willis, A.J. (1963), 'Braunton Burrows: the effects on the vegetation of the addition of mineral nutrients to the dune soils', *Journal of Ecology*, **51**, 353–74.

Wortham, W.H. (1913), *Some Features of the Sand Dunes in the S.W. Corner of Anglesey*, Report of the British Association for the Advancement of Science, 1913.

COST-BENEFIT ANALYSIS OF VARIOUS ORTHOPHOTO SCALES IN THE EVALUATION OF GRAZING MANAGEMENT IN DUTCH COASTAL SAND DUNES

HARRIE G.J.M. VAN DER HAGEN
Dune Water Company of South Holland, The Netherlands
and MARK VAN TIL
Amsterdam Water Supply, The Netherlands
in co-operation with WIM J. DROESEN
Grontmij Geogroep, The Netherlands

ABSTRACT

A number of Dutch coastal sand dune managers evaluate their nature manage-ment by using false colour aerial photographs. Vegetation development is investigated by sequential photo analysis (see van Dorp et al., 1985). From about 1975, at 5-year intervals, photographs have been taken at a scale of 1:5000 and 1:2500. Aerial photographs are used for systematic vegetation mapping at a detailed scale, based on major vegetation structures or phytosociological syntaxa.

Visual photo interpretation yields valuable information on the vegetation development (Geelen, 1991) but has some major disadvantages. Traditional sequential photo analysis provides vegetation maps with discrete patterns, whereas many vegetation transitions in coastal dunes are continuous. Moreover, the subjectivity of the method increases when the spatial transitions in the vege-tation become less sharp (van Til and Loedeman, 1991).

From the beginning of the 1990s the Dune Water Company of South Holland and the Amsterdam Water Supply initiated research to overcome these problems by using digital false colour orthophotos. Optimal tuning of scanning density, unifying mapping units (Assendorp and van der Meulen, 1994) and fuzzy clas-sification of every raster cell of 25 × 25cm field size have recently been developed. A semi-automatic classification method of the vegetation structure is achieved by using high-performance computers. Processes such as blowout development or grass and scrub encroachment can be monitored (Droesen et al., 1995).

The spatio-temporal techniques have been put into practice on an area of approximately 400ha in Meijendel, a coastal dune area near The Hague. The consequences of five years of grazing by Nordic fjorden horses and Galloway cows have been evaluated.

Because of the high financial investment in these new techniques on a 1:5000 and 1:2500 photo scale, research has recently been executed to reduce costs by optimizing scanning density and photo scale. Two areas of approximately 250ha in Meijendel/Berkheide and the Amsterdam Watersupply Dunes were selected to compare photo scales of 1:5000, 1:10,000 and 1:15,000. The results are presented in terms of cost-benefit analysis.

Introduction

In recent decades nature conservation has become a more important issue in The Netherlands (Ecological Head Structure) and in Europe (EECONET and Habitats Directive). These issues are not only a concern of governments but also of local managers. Drinking water companies along the Dutch coast manage coastal dunes of about 3500–6500ha per company, accounting for almost one third of the total coastal strip of The Netherlands (see Figure 1).

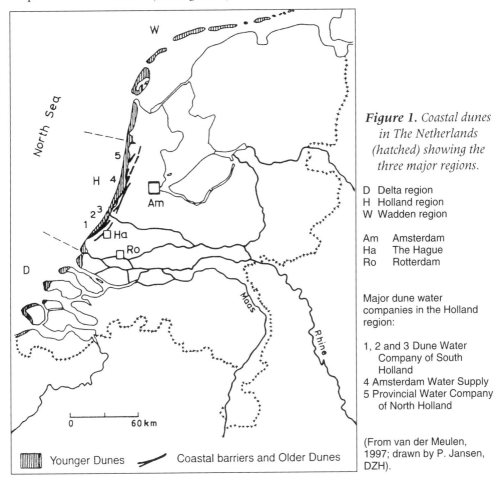

Figure 1. *Coastal dunes in The Netherlands (hatched) showing the three major regions.*

D Delta region
H Holland region
W Wadden region

Am Amsterdam
Ha The Hague
Ro Rotterdam

Major dune water companies in the Holland region:

1, 2 and 3 Dune Water Company of South Holland
4 Amsterdam Water Supply
5 Provincial Water Company of North Holland

(From van der Meulen, 1997; drawn by P. Jansen, DZH).

The coastal sand dunes in North and South Holland have suffered a decline in ecosystem variability due to human interference. Groundwater levels have dropped due to impoldering of the hinterland and groundwater extraction, as a result of which dune slacks have become desiccated. Furthermore, air pollution causes acidification and eutrophication of the dune ecosystems. Rejuvenation of dune landscapes had stopped due to reclamation of blowouts. In consequence pioneer plant communities have almost disappeared and grass and scrub communities now dominate Dutch dune landscapes.

The need for monitoring

Managers compensate for the decline of ecologically important dry and wet dune grasslands by (actively and passively) stimulating blowouts and by introducing cattle to counteract grass and scrub encroachment. These aims are defined in management plans. On several levels, from landscape to species, the dune manager wishes to know whether these actions are successful. Monitoring should provide the answers. In general, the manager's interest is particularly focused on the development of the vegetation or vegetation complexes (the temporal aspect), though the spatial arrangement is also an important factor in successful implementation of management. The finer the mosaic, the better an ecosystem works.

Landscape and species monitoring

From landscape to species level, one uses different (and in best practice interlocking) methods based on different scales. Examples include landscape monitoring on a 1:25,000–1:250,000 scale or above by SPOT or LANDSAT image processing, vegetation monitoring (in Dutch coastal dunes traditionally on a 1:2500–1:5000 scale) and species monitoring in permanent plots. Common practice in vegetation and landscape research is the use of aerial photographs in combination with permanent plots. Permanent plots or permanent plots in transects provide information on processes in time. They give results of a very limited space, but hopefully represent the development of a larger area. It is difficult, however, to determine the extent to which the sample truly represents the wider area. Aerial photographs add spatial information to the results drawn from the permanent plots. Temporal sequences of aerial photographs can be used to produce sequences of vegetation maps which can then be analysed and compared chronologically (see van Dorp et al., 1985).

From about 1975 aerial photos have been taken at five-year intervals at a scale of 1:5000 or 1:2500. Managers of Dutch coastal sand dunes evaluate their nature management by interpreting these aerial photographs manually to produce systematic vegetation mapping on a detailed scale. The legend for mapping is based on main vegetation structures or based on phytosociological syntaxa sampled by field work. The basic mapping unit is 5 × 5mm or 2 × 10mm on the photo image. The translation to field sizes is given in Table 1.

Although visual photo interpretation yields valuable information on vegetation and its development (Geelen, 1991) it has some major disadvantages. The interpretation

Table 1. Translation of photo area to field area

Basic mapping unit (photo area)	1:2500 scale (field area)	1:5000 scale (field area)
5 × 5mm	12.5 × 12.5m	25 × 25m
or	or	or
2 × 10 mm	5 × 25m	10 × 50m

of aerial photographs into discrete vegetation patches does not provide sufficient information to determine relevant processes such as wind erosion or grass and scrub encroachment, because natural landscapes often show both discrete and continuous variation in space and time. The spatial variation in vegetation results in discrete and continuous variation in both aerial photographs and satellite images. Nonetheless, these images have tended to be crisply classified into discrete patches regardless of whether the vegetation exists as a well-defined mosaic or as a series of continua. Consequently, gradients cannot be represented through crisp classification and many classification errors can be attributed to artificial boundaries. Obviously it is very subjective to draw sharp boundaries when in reality gradients exist. Moreover, the subjectivity of the method increases when the spatial transitions in the vegetation become less sharp (van Til and Loedeman, 1991). Furthermore, there are technical difficulties to cope with (e.g. geometric correction of the manual interpretation per photo to a composite vegetation map) which increase with a higher scale of image and an exact measure of the inaccuracy is impossible to give.

In a temporal analysis this disadvantage is emphatically manifest as many transitions appear to be artificial. Therefore it is not surprising that researchers (e.g. Geelen, 1991; Bisseling and van Ekeren, 1983) noted that crisp classification has to be treated with caution in the interpretation of images of natural landscapes. Crisp classification should only be applied to mapping discrete features in the terrain.

Aims of the research

In response to the issues outlined above the Amsterdam Water Supply and the Dune Water Company of South Holland initiated research into the spatial modelling and monitoring of natural landscapes. Furthermore, optimal tuning of scanning density, unifying mapping units (Assendorp and van der Meulen, 1994) and fuzzy classification (of raster cells of, for example, 25 × 25cm field value) was necessary. The research as a whole resulted in a generic method for the monitoring of gradient situations in landscapes by applying digital image interpretation techniques.

Central in the new approach is the use of fuzzy (or continuous) classification by which both discrete and continuous patterns can be effectively quantified. It is now possible to monitor processes such as blowout development or grass and scrub encroachment quantitatively (Droesen et al., 1995).

The aims of this paper are:

1. to elucidate the effectiveness of the generic method of orthophoto analyses in evaluating grazing management in Meijendel; and
2. to present a cost-effectiveness study on the use of orthophoto images with different scales.

Evaluation of Cattle Grazing at Meijendel

The spatio-temporal techniques have been put into practice on an area of approximately 400ha in Meijendel, a coastal dune area near The Hague. The area is divided into Helmduinen (123ha) and Kijfhoek/Bierlap (273ha). The stocking rate of Nordic fjorden horses and Galloway cows is about 1 animal per 12ha. The consequences of five years of grazing have been evaluated. The aim of this management was the reduction of *Calamagrostis epigejos*-dominated grass vegetation, the return of species-rich dune grasslands and pioneer communities, the development of blowouts and a finer vegetation mosaic. One series of false colour photos was taken just before introduction of the stock in 1990. These photos were compared with the series of 1995 (Van der Hagen, 1996; Grontmij Geogroep, 1996).

Material and methods

False colour aerial photographs (scale 1:2500) were scanned with a Zeiss Photogrammetric Scanner (PS1) with a resolution of 30 microns. The scan resolution results in pixels of 7.5cm field size. Two sets of 45 photographs for 1990 and 1995 were scanned to cover the project area of approximately 400ha. Subsequently, orthophotos with a 25cm resolution were produced based on an aerotriangulation and digital elevation model. Finally the orthophotos were radiometrically corrected and mosaiced to a single image. This orthophoto image formed the basis for the spatio-temporal evaluation of five years of grazing in Meijendel.

The interpretation of the digital orthophoto has been performed according to the hierarchy shown in Figure 2 (Droesen, 1999). The interpretation can be done automatically, apart from the distinction between *Hippophae rhamnoides* and the other main shrub species (*Salix repens* and *Ligustrum vulgare*). These scrub types are mapped by hand. The classification has an accuracy of more than 95 per cent (Grontmij Geogroep, 1996).

Grasslands determine a great deal of the natural value of the dune ecosystem (see also the Habitats Directive; Appendix I, no. 2130). The specification of grasslands into subtypes is needed to enable an adequate evaluation of the landscape development. Typically grassland types occur as continuous gradients throughout the dunes. Therefore the grassland types are mapped by applying fuzzy classification techniques. Table 2 shows the different grassland types distinguished.

Level

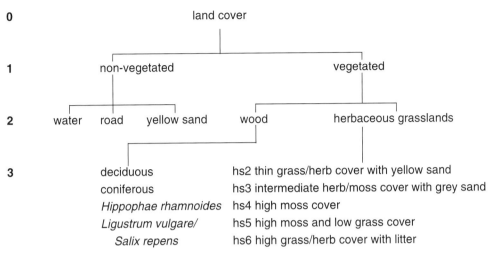

Figure 2. Land-cover hierarchy (after Assendorp and van der Meulen, 1994).

Results of the classification

In the Helmduinen area bare sand (blowouts and paths) and *Hippophae* scrub both increased in cover by just under 3ha (Fig. 3a) while the grasslands in general decreased. Trampling by the grazing animals opens the vegetation sward and this leads to pioneer communities and bare sand. The *Hippophae* scrub, which was mostly more or less dense scrub, was opened up by the cattle and it changed to a mixture of bare sand and *Hippophae*. The seemingly small decrease of just under 0.5ha of the area of *Salix* and *Ligustrum* scrub is relatively large considering the small average size of these shrubs. The decrease is due to actual trampling.

About 50 per cent of the Helmduinen area consists of grasslands. The increase of bare sand and blowouts is reflected in the increase of grassland type hs2, a thin grass/herb cover with yellow sand (Fig. 3b). A small increase in the subtype which is grazed by rabbits (hs5) is promising. Another indication of the opening up of the vegetation is the reduction of hs6, a coarse grassland type with accumulation of litter. The main reason for these changes is, as intended, the grazing by cattle. The reduction of encroachment by *Calamagrostis epigejos* following the introduction of grazing has therefore been fairly successful at this site.

A measure of an increase or decrease of the mosaic is determined by calculating the total border length between all vegetation types. Table 3 shows the absolute and relative increase of this border length. It indicates that the patches in the area became smaller and that the mosaic has become more fine-grained. Field research (De Bonte et al., 1999) in an assumed representative transect of 7ha in the Kijfhoek/Bierlap area shows that the patchiness increased by 14.1 per cent. One must conclude that with regard to this aspect, the transect was not representative of the grazed area as a whole.

Table 2. Grassland types (after Assendorp and van der Meulen, 1994)

Code/type	Description	Indicator species	Process
hs2 Thin cover with yellow sand	Yellow sand, i.e. sand with negligible amount of organic matter, has by far the largest contribution in this cover type. Herbs are annual as well as biennial. Grass types are mainly solitary and clonal which react more or less positively to wind activity. Tussock-forming grass types can be present.	*Festuca arenaria* *Ammophila arenaria* *Sedum acre*	Fresh blown sand
hs3* Intermediate herb/moss cover with grey sand	Largest contribution to the overall cover is by mosses which react more or less positively to, or can sustain, some geomorphological activity. Bare grey sand, i.e. sand with organic matter content, has a substantial contribution to the overall cover. Herbs are annual and biennial with locally some perennials. Some woody plants can occur at the sub-pixel level. Grasses are solitary and tussock forming.	*Tortula ruralis* var. *ruraliformis* *Tortella flavovirens* *Ceratodon purpureus* *Carex arenaria* *Corynephorus canescens* *Sedum acre* *Phleum arenarium* *Saxifraga tridactylites*	Lightly covered with blown sand
hs4 High moss cover	Total cover of the soil with mosses and lichens, very locally with annual and biennial herbs. Grasses are nearly absent.	*Dicranum scoparium* *Campylopus introflexus* *Hypnum cupressiforme* *Cladonia* spp. *Rumex acetosella*	Water repellent, (partly) decalcified
hs5* High moss and low grass cover	Soil is totally covered with mosses combined with low herbaceous vegetation. Herbs and grasses are mainly small though larger woody plants can occur at the sub-pixel level.	*Avenula pubescens* *Festuca ovina* ssp. *hirtula* *Rubus caesius* *Galium verum* *Galium mollugo* *Hypnum cupressiforme* *Dicranum scoparium*	Rabbit grazing
hs6 High grass/herb cover with litter	Total cover of mainly grasses with perennial/clonal herbs. Some woody plants at the sub-pixel level. Dead organic matter partly determines the nature of this type.	*Calamagrostis epigejos* *Ammophila arenaria* *Elytrigia* spp.	Grass encroachment

* Grassland type of the priority class of the Habitats Directive, Appendix I.

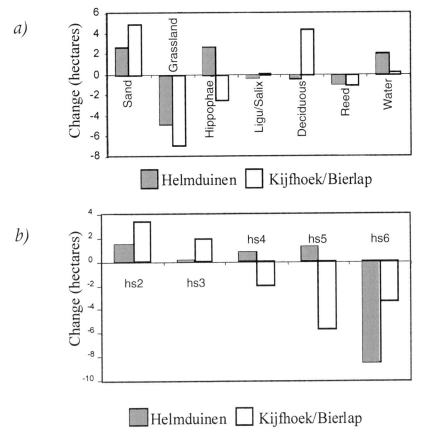

Figures 3a and 3b. *Absolute change (ha) after five years of grazing (1990–95), (a) of major land cover types as given in Figure 2 and (b) of the grassland sub-types as given in Table 2.*

Another application of false colour photos is habitat analysis of species such as the sand lizard *Lacerta agilis*, enabling investigation of the impact of nature management on animal species as well as the vegetation. The sand lizard is an Appendix IV species of the Habitats Directive. It was hypothesized that the potential habitats for the sand lizard would increase as a result of grazing management because of the increase of sandy patches around scrub. A lizard habitat can be defined as a combination of different vegetation structure elements. In our experiment we defined the sand lizard habitat as being *Hippophae* shrub (ideal place for thermoregulation) with some bare sand (potential egg-laying site) within a distance of 1m. The number of occurrences of this potential sand lizard habitat can be determined by searching on the orthophoto. In practice one has to visit the sites and possibly correct the definition of a sand lizard habitat. Table 4 gives the results for the two grazed areas, showing a substantial increase.

Table 3. Absolute and relative increase of border length between all vegetation types

	1990 (km)	1995 (km)	% increase
Helmduinen	156.9	176.1	12.2
Kijfhoek/Bierlap	397.5	422.1	6.2

Table 4. *Increase in number of potential sand lizard habitats.*

	1990	1995
Helmduinen	33	674
Kijfhoek/Bierlap	186	2666

Optimizing Scanning Density and Photo Scale

The fuzzy image interpretation techniques for the classification of vegetation into discrete patches and continuous gradients were tested as shown above with false colour photographs on a 1:2500 or 1:5000 scale. The latter scale is traditionally based on its suitability for manual interpretation. For digital analysis of photographs other criteria should be adopted:

1. the geometric and radiometric resolution; and
2. the reduction of cost through a reduction of the number of photos to cover an area.

By using smaller-scale photographs, e.g. 1:10,000 or 1:15,000, the costs for orthophoto production can be significantly reduced, thus the use of digital photography in a monitoring system becomes more cost-effective. A change to smaller scales might, however, affect the geometric and the radiometric characteristics of the photographs. An empirical test with several photo scales was designed to weigh the reduction in cost against the reduction in the radiometric quality of the photographs.

Material and methods

Two areas of about 250ha in Meijendel/Berkheide and the Amsterdam Water Supply Dunes were selected to compare the photo scales 1:5,000, 1:10,000 and 1:15,000. Minimum standards set for acceptable results were (1) the qualitatively correct mapping of the vegetation types distinguished by Assendorp and van der Meulen (1994), and (2) similarity of results from the radiometric correction and semi-automatic classification of the three orthophoto images. The hypothesis was that somewhere up to the scale of 1:10,000 an optimum could be achieved.

The same techniques were used as described in the evaluation of the grazing experiment, with the exception that the false colour photos were scanned with 15 microns (Zeiss PS1) to produce the three orthophoto images. The geometric resolution of the orthophotos was 25cm field size. Additionally, reference data of (fuzzy) grassland observations were obtained for the calibration and validation of the fuzzy and crisp image interpretation techniques. The plots were measured with a very accurate Global Positioning System. Image characteristics were directly related to field classification. The methods for crisp classification of the discrete vegetation types (Fig. 2) and fuzzy classification of the grasslands (Table 2) are already given above.

Results of the classification

The crisp classification of the discrete types of sand, woody vegetation and grasslands was very good, as could be expected from the results of the grazing experiment. Mapping of the different scrub types has to be done by a trained interpreter. Applying an on-screen scale of 1:2500 leads to the best results for these specific manual classification sessions. The overall conclusion is that all scales, even up to 1:15,000, were satisfactory to distinguish the discrete vegetation patterns (Grontmij Geogroep, 1998).

The fuzzy classification of the grasslands did not yield such good results. This was mainly caused by difficulties in timing the flight in relation to optimal weather conditions. The aerial photos were taken in the beginning of August 1997. Due to a rather long period of hot, dry weather the grassland types had a large component of dead organic material which hampered the separability. Timing of the flight can be optimized by using a field spectrometer for the determination of the spectral separability of the grassland classes. Despite the bad timing of the flight, all photo scales, even up to 1:15,000, were technically sufficient to pass the criteria for a fuzzy classification of the grasslands (Grontmij Geogroep, 1998).

Conclusions

Cattle grazing in Meijendel

The aims of the management (grazing by cattle) were reduction of grass encroachment of *Calamagrostis epigejos*, return of species-rich dune grasslands and pioneer situations, reintroduction of blowouts and increase of the vegetation mosaic structure. Orthophoto interpretation based on 1:2500 false colour photos of two time series (1990 and 1995) gives an exact change in the vegetation types in general and grassland types in particular. The shift in general is not spectacular, but is encouraging for the future. Several decades of deterioration of a dune ecosystem are not reversed in five years. The goals set for the grazing itself were only partly achieved within the five years of grazing. On the basis of the orthophoto interpretation, in combination with other monitoring results, recommendations were made to maintain the number of cattle in the Helmduinen area and reduce the stock in the Kijfhoek/Bierlap area. Hopefully the species rich-dune grasslands (hs3 and hs5) will extend their distribution. The next evaluation in 2000 should give some answers.

The technical goals set for handling large areas of a dune ecosystem for the evaluation of management techniques by orthophoto analysis were achieved with a very high percentage of performance and high accuracy (Grontmij Geogroep, 1996).

Cost-benefit analysis

The results from the comparison of the photo scales of 1:5000, 1:10,000 and 1:15,000 are impressive. All scales were suitable to the minimum qualifications set by Assendorp and van der Meulen (1994) on the correct mapping of the vegetation units distinguished. This accounts as well for the crisp as for the fuzzy image interpretation. Furthermore, there were no complications with the geometric and radiometric resolution.

Based on a photo scale of 1:15,000 (in comparison with 1:2500), a cost reduction of up to 16 times can be achieved for the production of an orthophoto! The dune areas managed by the Amsterdam Water Supply and the Dune Water Company of South Holland are each about 3500ha in size. This means that on a scale of 1:2500 about 350 false colour photos have to be scanned for the production of a composite orthophoto. Only about 20 photos are sufficient to cover the area based on a scale of 1:15,000. The more or less quadratic reduction in the number of false colour photos means that the costs of the scanning and production of a geometrically and radiometrically correct orthophoto image are greatly reduced.

Concluding remarks

Manual production of a detailed vegetation map from a large area can take years of work. A suitable scale, traditionally used for Dutch dunes for this type of work, is 1:2500 or 1:5000. There are also technical difficulties to cope with. For instance, only the central part of a (false colour) photo is geometrically more or less correct. The geometry of the manual interpretation of every photo therefore has to be corrected to produce a composite vegetation map. This difficulty increases with a higher scale.

The research resulted in a generic method for monitoring gradient situations in natural landscapes, applying digital image interpretation techniques. The orthophoto is geometrically and radiometrically correct beforehand. Central in the new approach is the use of fuzzy (or continuous) classification techniques by which both discrete and continuous patterns can be effectively quantified. Introducing this new technique means a considerable reduction in time (and costs). This is true for both spatial and temporal aspects of image interpretation.

Another and major advantage of the computerized monitoring system is that the resolution has improved considerably. The basic mapping unit from 5×5mm or 2×10mm by manual photo image processing, which is 25×25m (1:5000 scale) field size, is upgraded to 25×25cm field size based on the high quality of the false colour photos. Using orthophoto images of large areas, gigabytes of information are produced. Until recently, the management of gigabytes of information was a problem. Since the immense progress in computer hardware, these problems seem to be solved.

Instead of relying on an assumed representative transect to describe landscape development, large areas can now be analysed by the developed generic method. If necessary or considered useful the interpretation can be done per landscape zone (Dutch coastal dunes are in general about 2–5km wide) or an interpretation can easily be executed on parts of the orthophoto. Furthermore, subjectivity on the interpretation of a photo is excluded. Every time the same area is processed with the same algorithms, the vegetation map is the same.

The generic method of digital orthophoto interpretation with a high resolution forms, for the dry parts of the dune ecosystem, an important basis for landscape management in general and vegetation ecology in particular. The digital image can be applied in terrain inventories, the planning and evaluation of management techniques, calculating peak standing crop per vegetation type and so forth. Furthermore, because of the accurate temporal interpretation, it is an ideal tool for monitoring. A minor disadvantage of the method is the fact that it is 'only' based on the vegetation types distinguished. On the other hand important vegetation types of dry grasslands (Habitats Directive) can be accurately identified.

The research provided an intermediate mapping product and forms an ideal, cost-saving intermediate between the monitoring of the dune landscape as a whole and exact species composition of a specific patch, but with the difference that one is able to monitor large areas in great detail and with high levels of accuracy. The actual information on local species presence still has to be gathered by additional methods, because the exact species composition cannot be derived from aerial photographs. Collecting field data still needs to be done and time series of permanent plots form very useful additions.

References

Assendorp, D. and van der Meulen, F. (1994), *Habitat Classification of Dry Coastal Ecosystems: The Integration of Image and Field Sample*, Proceedings of EARSeL Workshop on Remote Sensing and IS for Coastal Zone Management, Rijkswaterstaat Survey Department 24–26 October 1994, Delft, The Netherlands.

Bisseling, C. and van Ekeren, A. (1983), *Vegetatiekartering en vegetatiestructuur in het duingebied 'Scheepje' en omgeving, Meijendel*, Dune Water Company of South Holland, The Hague.

De Bonte, A.J., Boosten, A., Van der Hagen, H.G.J.M. and Sykora, K.V. (1999), 'Vegetation development influenced by grazing in the coastal dunes near The Hague, The Netherlands', *Journal of Coastal Conservation*, **5(1)**, 59–68.

Dorp, D. van, Boot, R. and van der Maarel, E. (1985), 'Vegetation succession on the dunes near Oostvoorne, The Netherlands, since 1934, interpreted from air photographs and vegetation maps', *Vegetatio*, **58**, 123–36.

Droesen, W.J. (1999), 'Spatial modelling and monitoring of natural landscapes, with cases in the Amsterdam Waterworks Dunes', PhD thesis, Wageningen Agricultural University.

Droesen, W.J., van Til, M. and Assendorp, D. (1995), 'Spatio-temporal modelling of the vegetation structure in the Amsterdam water works dunes using digital false colour orthophotos', *EARSeL Advances in Remote Sensing*, **4(1)**, 106–14.

Geelen, L.H.W.T. (1991), 'Landscape succession in a Dutch dune area', *Proceedings of the Symposium on Global and Environmental Monitoring ISERS*, 363–69.

Grontmij Geogroep (1996), *Interpretatie en analyse van digitale kleur-infrarode orthofotos van de gebieden Helmduinen en Kijfhoek/Bierlap*, report, Dune Water Company of South Holland.

Grontmij Geogroep (1998), *Monitoring systeem AWD/Meijendel. Test van fotoschalen en validatie fuzzy graslandkartering*, report, Dune Water Company of South Holland.

Meulen, F. van der (1997), 'Dune water catchment in the Netherlands', in E. van der Maarel (ed.), *Dry Coastal Ecosystems, General Aspects*, Ecosystems of the World, 2c, Elsevier, Amsterdam, 533–56.

Til, M. van. and Loedeman, J.H. (1991), 'Nauwkeurigheidsaspecten van een sequentiele luchtfoto-analyse', *Deelrapport Oeco-hydrologisch onderzoek*, Gemeentewaterleidingen Amsterdam/Landbouwuniversiteit Wageningen, Amsterdam Water Supply.

Van der Hagen, H.G.J.M. (1996), *Paarden en koeien in Meijendel. Een evaluatie van vijf jaar begrazing in Kijfhoek/Bierlap en Helmduinen*, report, Dune Water Company of South Holland.

THE FLORA OF THE FLEMISH COASTAL DUNES (BELGIUM) IN A CHANGING LANDSCAPE

S. PROVOOST and W. VAN LANDUYT

Institute of Nature Conservation, Brussels, Belgium

ABSTRACT

As in many European countries, the Belgian coastal zone has been subject to quite drastic changes during the past centuries. These changes have taken a number of forms. Suburbanization associated with tourist development is considered the major cause of nature degradation; in fact almost half of the dune area, which originally covered a surface of approximately 7.6km², has disappeared at the expense of buildings, gardens and roads. Within the remaining dune sites, the natural landscape has changed as a result of different, often interacting, factors such as changes in agricultural land use and recreation. To investigate the influence of several 'natural' and anthropogenic processes on floristic distribution, digitized maps are compared using a GIS. Data can be extracted and put in a database by overlaying the maps with the km² grid used for floristic inventories.

Introduction

As in many European countries, the Flemish coastal zone has changed radically during the past century. Changes in landscape have, almost by definition, a negative connotation to conservationist ears, but this fear is not necessarily justified. Various factors such as rate of change and spatial scale have also to be considered. In this article, some general quantitative and qualitative aspects of recent developments in Flemish dune landscape and flora are discussed.

Methodology

The analysis of GIS (Geographical Information System) digital vegetation structure overlay maps is used to study vegetation change. Vegetation maps (scale 1:10000) are based on the work of De Vliegher (1989), who interpreted aerial photographs from 1948 and 1988 of the western part of the coastal dune area (Westkust, 3600ha). As only two dates are considered, the results sketch general features; no rates of change could be detected. Additional information on changes in dune landscape is based on the summary of Provoost (1996).

Changes in composition of the flora along the Flemish coast (dunes and saltmarsh) between two periods, 'before 1940' and 'after 1972', are analysed on a presence/absence basis. Rappé et al. (1996) compiled a substantial amount of the required information, mainly based on the former national flora inventory and atlas (Van Rompaey and Delvosalle, 1972; 1979), the *Prodrome de la flore belge* (De Wildeman and Durand, 1898–1907), the historic data of Massart (1908; 1912), and recent floristic research. Floral statistics are based on data in Cosyns et al.,1994. More detailed information on trends in species abundance is derived from flora inventories in 59 × 1km² grid cells, sufficiently investigated at different times. Unfortunately the national flora inventory began only in the 1930s, so trend analysis was carried out on data of the 1935–60 and 1980–98 periods. Some of these results have been discussed in Ameeuw, 1998.

Results

Landscape change

Structural change took place over at least 54 per cent of the landscape (see Table 1 and Fig. 1). Urbanization associated with tourist development caused substantial degradation of the natural environment in the Flemish coastal zone. Half of the dune area, which originally covered a surface of approximately 70km², disappeared under buildings, gardens and roads (see Fig. 2), a development which shows an exponential trend. Along the Westkust, the urban area doubled between 1948 and 1988 (see

Table 1. Landscape transition matrix for the Westkust 1948–88 (%)

	Mobile dune	Semi-fixed dune	Dune grassland	Scrub	Woodland	Pasture	Arable land	Urban area
Mobile dune	**16**	20	14	18	4	0	0	28
Semi-fixed dune	15	**15**	17	25	6	0	0	23
Dune grassland	3	10	**23**	29	12	1	0	22
Scrub	3	6	19	**31**	8	0	0	32
Woodland	0	1	4	2	**73**	1	1	19
Pasture	1	1	3	0	2	**38**	23	32
Arable land	0	0	3	0	2	10	**62**	23
Urban area	0	0	1	1	1	2	0	**94**

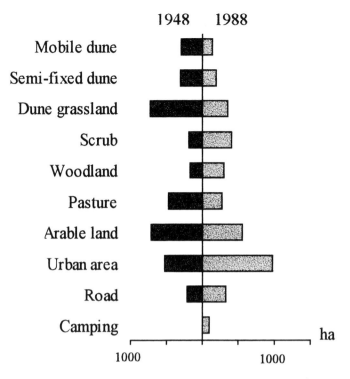

Figure 1. The Westkust dune landscape in 1948 and 1988.

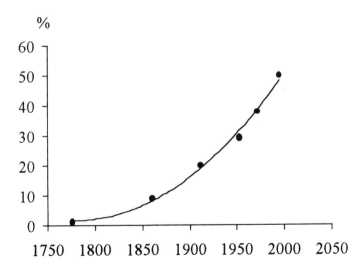

Figure 2. Evolution of the urban area along the Flemish coast.

Table 1 and Fig. 1). Houses (sometimes with botanically interesting gardens) comprise 55 per cent of its actual surface. Road corridors cover nearly a quarter of the urban area. Within the remaining dune areas the vegetation structure changed significantly. Scrub encroachment is probably the most striking phenomenon. A quarter of the open dune area was invaded by scrub, mostly *Hippophae rhamnoides* and to a lesser extent, *Ligustrum vulgare, Salix* spp., *Rosa* spp., *Crataegus monogyna* and others. Regression of scrub into species-poor grassland dominated by *Calamagrostis epigejos* and *Carex arenaria* occurs to some extent.

The increase in woodland area is mainly due to afforestation of dune grassland and farmland. Succession of scrub towards dense woodland takes place on a small scale and could not clearly be shown using the 1:10000 vegetation maps.

A large proportion of the semi-mobile dunes has changed to either urban area or scrub (see Table 1). A general reduction of aeolian dynamics seems to have occurred. Transition of mobile dune into semi-fixed dune and grassland appears to be in quantitative balance with the opposite process. In absolute figures, this stabilization and destabilization occurred over an area of 160 and 140ha respectively.

Changes in agricultural land use are considerable. Farming of slacks and grazing of dune grassland was still fairly common at the beginning of this century. During the inter-war period these farming activities diminished in favour of more lucrative tourism. Nowadays there is hardly any agriculture in the central part of the dunes.

Along the Westkust dune fringe, a quarter of the agricultural area was urbanized after World War II (see Table 1). The remaining farmland was subject to fertilization, drainage and removal of hedges and ponds. Only 40 per cent of the original pasture area was not ploughed or urbanized.

Several changes in land use cannot be detected with the GIS overlay method but are nevertheless important for flora and fauna. Increased recreational use of the dunes, for example, causes severe disturbance and soil damage at some sites.

Decline of the groundwater table is another process not detected by the methods used, but which has considerable ecological consequences. Groundwater extraction for drinking water supply, increased interception due to urbanization and agricultural drainage are considered to be the main causes. Nearly 10 per cent of the remaining dunes (287ha) are designated as 'water collection area' on planning maps. In some of these areas the water table has dropped by about 4m.

Coastal defence activities have a considerable impact on dune geomorphology. Only 15km of foredune (22 per cent of the coastline) is not dyked.

Floral shift

General features

During the past century, 917 taxa of vascular plants have been found along the Flemish coast. At present, 60 per cent of the Flemish flora is represented in this zone (698 taxa). This figure is fairly high considering that the area of coastal dune and saltmarsh only covers 70km², or 0.5 per cent of the Flemish region. According to the general species–area relationship for the north of Belgium calculated by Stieperaere (1980) (S = 119A$^{0.238}$, with S the number of species and A the area in km²), the coastal zone is twice as species-rich as the average.

The list derived from the 59 grid cells (covering 70 per cent of the dune area), contains 735 taxa or 80 per cent of the above-mentioned total number of plant taxa. The distribution of species abundance shows a characteristic community pattern (Magurran, 1983). Half of the flora consists of (very) rare taxa, present in 5 grid cells at the most. As a consequence, significant trends in abundance (chi-squared test, p<0.05) could only be calculated for a limited number of taxa.

Figure 3 illustrates the dynamics in floral composition. Within 100 years, the total number of taxa has increased by approximately 15 per cent. One hundred and twenty-four taxa were only found before 1940 ('local extinction'); 214 only after 1972 ('local appearance'). Considering the 93 species showing a significant trend ('increase' and 'decline') in the 59 1km² grid cells, we arrive at a flora-flux of at least 48 per cent. Species whose coastal populations are important for their survival in the Flemish region show a decline of 30 per cent.

Habitat preference
In accordance with the changes in vegetation structure (Fig. 1), a considerable increase of scrub and woodland species can be detected. Species of dry, ruderal places and grey dunes also show an upward trend. Substantial decrease in species richness can be noted for dune slacks, moist grasslands, water(sides) and agricultural habitats (see Fig. 4).

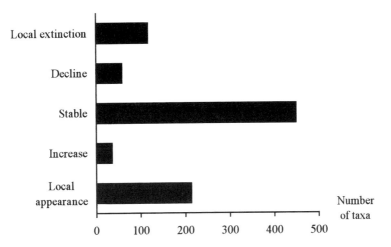

Figure 3. *Trends in the flora of the Flemish coast. Periods considered for local extinction and appearance are 'before 1940' and 'after 1972'. Decline and increase is based on 59 × km² grid cells, compared between 1935–60 and 1980–98.*

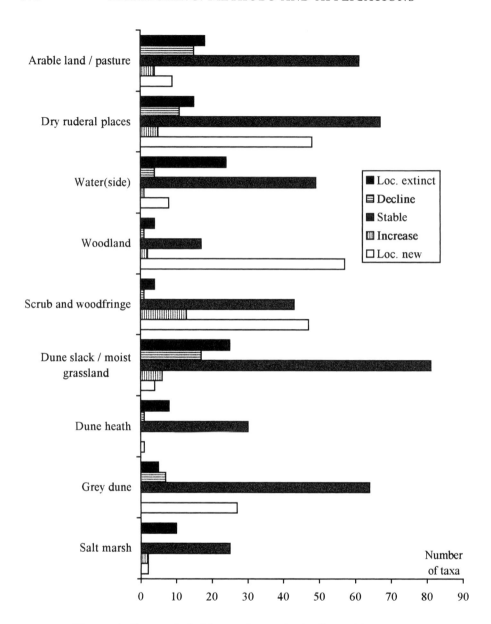

Figure 4. *Changes in habitat preference in the flora of the Flemish coast.*

The decline of groundwater-dependent taxa (according to Bakker et al., 1979) is striking. In this category, 65 taxa show a decline or have disappeared while only 20 show increase. The number of nitrophilous species rose amongst plants of woodland, scrub and ruderal places. A general decline of species indicating nitrogen-poor soil conditions can be observed except for woodland and scrub.

Of the 'new' flora, 46 per cent is not indigenous to Flanders. These exotic species mainly appeared in woodland (57 taxa), ruderal places (48 taxa) and scrub (47 taxa).

Discussion

The landscape and flora of the Flemish coastal area have changed radically. On the one hand, the declining agricultural pressure has resulted in more spontaneous successional developments in the relief-rich parts of the dunes. Conservationists have to accept this kind of change, both physical and biotic, as an inherent part of the coastal ecosystem. Management goals can be inspired by historical information but should not attempt to reflect 'past glory'. Some processes are irreversible; as De Raeve (1989) stated, you simply cannot stop time. On the other hand, however, human influence has increased in different ways. Therefore 'spontaneous' developments cannot be interpreted as 'natural' processes. People constantly, and often unconsciously, act upon wind and wave dynamics, hydrology, diaspore availability and many other features. Along the Flemish coast and in many other European countries, natural and cultural processes are so closely interwoven that nature conservation and management cannot but be resigned to certain anthropogenic limitations on the ecosystem's functioning. This conclusion should also be reflected in policy aims.

The above-mentioned figures suggest an enrichment of the coastal flora, mainly due to an increase of woodland and ruderal taxa. The rise in species number for grey dunes is surprising given the significant decline in grassland and semi-fixed dune area. The persistence of species-rich grey dune habitat in gardens could explain this phenomenon. Taking into account the qualitative aspects of change, we might conclude that flora specificity has declined. This is mainly caused by a decreasing number of dune slack and moist grassland species. Van der Laan (1985) came to similar conclusions for the dunes of Voorne (The Netherlands), quoting figures on rarity. This kind of evaluation of flora is avoided here. As Rabinowitz et al. (1986) pointed out, different forms of rarity can be distinguished, depending on geographic scale. The available information for the Flemish coast is not sufficiently detailed for such an approach. Additional literature research, as presented in De Raeve et al. (1983) for groundwater-dependent species, is essential to increase knowledge on the population size and dynamics of many rare plant species. The application of conclusions to site management requires caution. Nevertheless, this floristic overview provides a useful framework for a general vision on conservation (cf. Provoost and Hoffmann, 1996). Restoration and development of dune slack and moist grassland has priority, at least from a botanical perspective.

Ecosystem processes can be demonstrated using photographs and maps but will not be understood unless multivariate research is carried out. GIS and databases can help in coping with the enormous amount of data associated with cartographic material but do not solve the problems involved with the gathering of information. Vegetation or habitat mapping is particularly subject to personal bias. Application of photo-interpreting software (cf. Droesen et al., 1995) seems to be a promising solution and a useful technique to link vegetation as a habitat and (easier to determine) vegetation as an assemblage of plants with certain ecological demands. Differences in temporal and spatial scale, but also in basic characteristics of system components, also complicate the multivariate approach. Species behaviour, for instance, fundamentally complies with biological, genetic patterns whereas, for example, decalcification is a physical and chemical process. Integrating these worlds is the ecologist's challenge.

References

Ameeuw, G. (1998). 'Flora en flora-veranderingen in de 20ste eeuw aan de Vlaamse kustduinen in relatie tot enkele standplaatsfactoren en de bebouwingsgraad', dissertation, University of Ghent.

Bakker, T.W.M., Klijn, J.A. and van Zadelhoff, F.J. (1979), *Duinen en duinvalleien: Een landschapsecologische studie van het Nederlandse duingebied*, Pudoc, Wageningen.

Cosyns, E., Leten, M., Hermy, M. and Triest, L. (1994), *Een statistiek van de wilde flora van vlaanderen*, Free University of Brussels and Institute of Nature Conservation, Brussels.

De Vliegher, B. (1989), *Onderzoek naar de beheersrelevante milieufactoren in enkele sleutelgebieden, partim luchtfotoanalyse*, report of the research project 'The Nature Development Plan for the Belgian Coast', under the authority of the Ministry of the Flemish Community, University of Ghent.

De Raeve, R. (1989), 'Sand dune vegetation and management dynamics', in F. van der Meulen, P.D. Jungerius and J.H. Visser (eds.), *Perspectives in Coastal Dune Management*, SPB Publishing, The Hague, 99–109.

De Raeve, R., Leten, M. and Rappé, G. (1983), *Flora en vegetatie van de duinen tussen Oostduinkerke en Nieuwpoort*, Nationale Plantentuin van België, Meise.

De Wildeman, E. and Durand, T. (1898–1906), *Prodrome de la flore belge*, 3 vols., A. Castaigne, Brussels.

Droesen, W.L, van Til, M. and Assendorp, D. (1995), 'Spatio-temporal modelling of the vegetation structure in the Amsterdam water works dunes using digital false colour orthophotos', *EARSeL Advances in Remote Sensing*, **4(1)**, 106–14.

Magurran, A.E. (1983), *Ecological Diversity and its Measurement*, Croom Helm, London.

Massart, J. (1908), *Essai de géographie botanique des districts littoraux et alluviaux de la Belgique + Annexe*, Lamertin, Brussels.

Massart, J. (1912), 'La cinquantième herborisation générale de la Société royale de botanique de Belgique sur le littoral belge', *Bull. Soc. Roy. Bot. Belg.*, **5(1)**, 69–185.

Provoost, S. (1996), 'Bewonings- en landschapsgeschiedenis', in Provoost and Hoffmann, 1996, 140–66.

Provoost, S. and Hoffmann, M. (eds.) (1996), *Ecosysteemvisie voor de Vlaamse kust, 2. Natuurontwikkeling,* University of Ghent and Institute of Nature Conservation, under the authority of the Ministry of the Flemish Community, Nature Division, Brussels.

Rabinowitz, D., Cairns, S. and Dillon, T. (1986), 'Seven forms of rarity and their frequency in the flora of the British Isles', in E. Soulé (ed.), *Conservation Biology: The Science of Scarcity and Diversity,* Sinauer, Sunderland, 182–204.

Rappé, G., Leten, M., Provoost, S., Hoys, M. and Hoffmann, M. (1996), 'Biologie', in Provoost and Hoffmann, 1996, 167–372.

Stieperaere, H. (1980), 'The species–area relation of the Belgian flora of vascular plants and its use for evaluation', *Bull. Soc. Roy. Bot. Belg.,* **112**, 193–200.

Van der Laan, D. (1985), 'Changes in the flora and vegetation of the coastal dunes of Voorne (The Netherlands) in relation to environmental changes', *Vegetatio,* **61**, 87–95.

Van Rompaey, E. and Delvosalle, L. (1972), *Atlas van de Belgische en Luxemburgse flora, Pteridofyten en Spermatofyten,* Meise, Nationale Plantentuin van België.

Van Rompaey, E. and Delvosalle, L. (1979), *Atlas van de Belgische en Luxemburgse flora. Pteridofyten en Spermatofyten,* 2nd edition, Meise, Nationale Plantentuin van België.

WATER RESOURCES MANAGEMENT OF THE SEFTON COAST

P.A. BIRCHALL and A. ELLIOTT
Environment Agency, UK

ABSTRACT

Much of the dune area of the Sefton Coast, northwest England, has been designated as a candidate Special Area of Conservation (cSAC) through the Habitats Directive. The dune area includes extensive wetlands that are an important breeding ground for the rare natterjack toad *Bufo calamita*. The dune area also forms an important local water resource and has been exploited for many years for irrigation purposes by many of the links golf courses in this area, including the Royal Birkdale, Hillside, and Southport and Ainsdale. Changes in climatic conditions and a greater demand for water for irrigation purposes have significantly increased the pressures on the Sefton Coast aquifer.

The Environment Agency is one of the most powerful regulators in the world and has legal duties to protect and improve the environment and so contribute towards sustainable development. One of the Agency's aims is to manage water resources to achieve a proper balance between the needs of the environment and those of the abstractors and other water users. It is committed to improving wildlife habitats and conserving the natural environment.

As a 'competent authority' under the EC Habitats Directive the Environment Agency is required to review new or existing consents that may affect habitats or species listed under the Directive. The Agency also has a duty under the Directive to assess the impact of any 'project or plan' on the integrity of the site by carrying out an 'appropriate assessment'.

Work to determine and review applications for new and existing licences on the Sefton Coast is split into two main activities: 'Site Specific Assessments', which are aimed at improving our understanding of the hydro-ecological interactions, and 'Coastal Hydrological Assessments' to assess the collective impact of existing and potential increases in abstraction, and to provide a management tool to evaluate the Site Specific Assessments.

Throughout the determination and review periods the Agency is working with the golf courses and key partners at a national and local level. This consultation process is a vital element of developing a sustainable water resources strategy for the coast.

Introduction

Much of the dune area of the Sefton Coast, northwest England, has been designated as a candidate Special Area of Conservation (cSAC) through the Habitats Directive. The dune area includes extensive wetlands that are important breeding grounds for the natterjack toad *Bufo calamita*, an endangered species listed under Annex IV of the Habitats Directive. The Sefton Coast sand dunes also form an important local water resource, exploited for many years for irrigation purposes by the links golf courses in this area, including the Royal Birkdale, Hillside, and Southport and Ainsdale. The Royal Birkdale golf course recently played host to the Open Golf Championship in July 1998. Changes in climatic conditions and a greater demand for water have significantly increased the pressures on the Sefton Coast aquifer. This paper outlines the Environment Agency's role in meeting the balance between environmental protection and the needs of local economies.

The Environment Agency

The Environment Agency is one of the most powerful regulators in the world. Concerned with land, air and water, the Agency exists to provide high-quality environmental protection and improvement throughout England and Wales. It achieves this by emphasizing prevention, enhancement, education and vigorous enforcement whenever necessary. The Agency's responsibilities include water resources, pollution prevention and control, flood defence, fisheries, conservation, recreation, navigation and waste regulation. As guardian of the environment the Agency has legal duties to protect and improve the environment and so contribute towards sustainable development – that is, meeting the needs of today without compromising the ability of future generations to meet their needs.

Water Resources Role in the Environment Agency

One of the Agency's aims is to manage water resources to achieve a proper balance between the needs of the environment and those of the abstractors and other water users. The Environment Agency manages water resources to protect the environment and ensure secure supplies to agriculture, industry and the public. In doing this, it recognizes the importance of water to both national and local economies. The Agency has duties under Section 2 of the Environment Act 1995, and is charged with safeguarding and improving the environment, including responsibilities for rivers and groundwater. It is committed to improving wildlife habitats and conserving the natural environment in everything it does.

In balancing these needs the Agency has the following functions:

1. Control of water abstractions by issuing licences, setting limits on the amounts which can be taken, and prohibiting use in some cases;
2. Enforcement of licence restrictions when river and groundwater levels are below normal;
3. Monitoring and measuring of all water conditions including rainfall, weather, evaporation, river flows and underground water levels;
4. Analysis of information to find out where water is and how much is available;
5. Promotion of measures to manage the demand for water to ensure efficient water use and water conservation;
6. Investigation of the impacts of abstraction;
7. Raising awareness through the media to educate everyone of the need for careful water use.

Water Resources and Conservation Issues

The Agency has duties under Section 6(1) of the Environment Act 1995 to further conservation, wherever possible, when carrying out water management functions. In addition, under Section 6(2) of the Act, there are responsibilities for the Agency 'to take all such action as it may from time to time consider…to be necessary or expedient for the purpose of conserving, redistributing or otherwise augmenting water resources in England and Wales, and of securing the proper use of water resources'. The Agency's work on the sand dunes of the Sefton Coast is a prime example of addressing the balance between the needs of the local economy and the environment.

Water is a fundamental aspect of golf course management, and it is becoming increasingly important for golf clubs to use water efficiently and effectively. There are seven golf courses on the Sefton Coast that use groundwater to irrigate, including during the summer months when the environment is under stress. In recent years, particularly following the 1995 drought, water demand for spray irrigation on courses has dramatically increased. As a result the Agency is working with golf courses towards more effective use of water resources, while protecting the environment.

The Sefton Coast sand dune system covers c. 2100ha (Doody, 1991) and has the largest area of wet slacks in England. Some of the notable species associated with these slacks include the natterjack toad and the rare liverwort, petalwort *Petalophyllum ralfsii*. The conservation importance of the area is reflected in the variety and extent of its designations. This includes five Sites of Special Scientific Interest (SSSIs), a candidate Special Area of Conservation (cSAC), a Ramsar site (internationally important wetland), a Special Protection Area (SPA) and Local Nature Reserves (LNRs). However, even with this high level of statutory site protection, the dune coast is perceived to be at risk from groundwater abstraction.

What is required?

The Water Resources Act 1991 requires almost anyone who wants to take water from surface or underground sources to obtain a licence from the Environment Agency.

On the Sefton Coast a number of golf courses with existing licences have applied to increase the amount to be abstracted. In determining such variations the Agency requires that the whole licence is reviewed, not just that part which is up for variation.

One of the principal criteria considered when dealing with an application is the justifiable need of the golf course. Indeed, the Agency looks for evidence of demand management initiatives and water use efficiency, such as the optimization of operating strategies for irrigation.

The Environment Agency, as a 'competent authority' under the Habitats Directive, is required to review new or existing consents that may affect habitats or species listed under the Directive. Nationally agreed procedures have been drawn up between the Agency and English Nature for the review process, and are due for completion in March 2004. The Agency also has to assess the impact of any 'project or plan' on the integrity of the site by carrying out an 'appropriate assessment'. A 'project or plan' includes authorizations given by the Agency, in this case applications by golf courses for increases in groundwater abstraction.

Determining abstraction applications has become increasingly complex, requiring detailed environmental assessments by applicants. It is now necessary for Agency staff, along with the applicant and other interested organizations such as English Nature, to improve their own understanding of both the geological and ecological conditions on site and to understand the interaction between the two. It is important not to forget that while impacts may be modelled and predicted on a site-specific basis, the cumulative effects of the abstractions across the whole dune area must be understood.

What can be done?

Work to determine applications from golf courses on the Sefton Coast is split into two main activities: 'Site Specific Assessments' and 'Coastal Hydrological Assessments'. For example, during 1998 and 1999 the environmental impacts of a proposed increase in abstraction were assessed at the Royal Birkdale Golf Club under both constant rate test pumping and operational conditions. Firstly, both ecological and physical monitoring is conducted in order to provide baseline data against which any proposal can be assessed.

Ecological monitoring is conducted using National Vegetation Classification (NVC) methodology via a series of quadrats which are adjacent to the location of piezometers on the course. From this work an appropriate list of dune slack indicator plant species can be monitored that may indicate hydrological change.

Physical monitoring includes installation of piezometers, the collection of weekly groundwater level measurements, monitoring of climatic conditions, topographical surveys and fixed-point photography.

In addition to site-specific surveys, the Agency is increasing its own network of groundwater monitoring points along the coast, conducting a ground-penetrating radar survey to develop an understanding of the aquifer characteristics, and also developing a coastal groundwater model. The model will be required to consider other impacts on the proposed sites.

So often water abstraction is seen as the root cause of environmental problems, and yet sites are impacted on in many ways including land drainage and land management. It is important that these other factors are considered. The aim of this coastal work is to assess the collective impact of existing and potential increases in abstraction as well as providing a management tool with which to evaluate the site-specific assessments.

Throughout the determination period for the applications the Agency is working with the golf clubs themselves and key partners at a national and local level. These include English Nature, the Joint Countryside Advisory Service (ecological advisers to local authorities on north Merseyside) and the Sefton Coast Life Project (see Rooney, this volume). The consultation process is a vital element of developing a sustainable water resources strategy for the dune coast.

Key Messages

In managing water resources the Environment Agency takes a holistic approach to protecting the environment, and a long-term perspective on protecting and enhancing biodiversity. The Agency strives to achieve the right balance between the needs of the environment and other water users by:

- Changing the way we think of water;
- Assessing justifiable needs for water;
- Using water efficiently and wisely. Efficient and effective use of water is the key to environmentally friendly golf course management. This includes setting up accurate and reliable irrigation schedules, and implementing irrigation best practices;
- Assessing and reviewing the impacts of existing and future proposals;
- Understanding the cumulative effects of abstractions, not just those which are site specific;
- Recognizing that there are other influences on ecosystems including land management and land drainage;
- Conducting baseline monitoring, without which we cannot assess any environmental changes.

Groundwater levels play an important part in sustaining the ecological integrity of the Sefton Coast sand dune system. The Agency has a key role to assess the impacts of abstractions and, by working in close consultation with golf course managers, to ensure that conservation interests are properly taken into account. Water is not cheap and limitless but a valuable and finite commodity.

References

Water Resources Act (1991), HMSO, London.

Environment Act (1995), HMSO, London.

Doody, J.P. (1991), *Sand Dune Inventory of Europe*, Joint Nature Conservation Committee, Peterborough.

NON-INVASIVE INVESTIGATION OF WATER TABLE AND STRUCTURES IN COASTAL DUNES USING GROUND-PENETRATING RADAR (GPR): IMPLICATIONS FOR DUNE MANAGEMENT

CHARLIE S. BRISTOW
Birkbeck College, University of London, England
and SIMON D. BAILEY
University College London, England

ABSTRACT

Ground-penetrating radar (GPR) provides a unique insight into the internal structure of coastal dunes which cannot be achieved by any other non-destructive or geophysical technique. GPR equipment is portable and lightweight; it can be deployed by hand or in a wheelbarrow. The structure of coastal dunes from North Wales and North Norfolk has been investigated.

Results show good resolution of artificial and natural structures within coastal dunes. The water table is readily identifiable at almost all localities as a strong sub-horizontal reflection which can be mapped within dune fields. The water table is a strong reflector because there is a significant change in dielectric properties between wet and dry sand. The depth to the water table can be calculated from velocity measurements using common mid-point (CMP) gathers. In addition, artificial structures such as water pipes on a golf course are readily identified. Natural surfaces such as sets of cross-stratification and erosion surfaces are also observed.

By combining geomorphic and geophysical investigations into the structure and morphology of coastal dunes in Norfolk and West Wales, we have been able to determine more accurately their development and evolution. GPR profiles image cross-stratification and bounding surfaces in dunes, revealing their internal structure and relative chronology. The structures revealed by the GPR profiles indicate dune aggradation, landward migration, seaward progradation and trough cut and fill. In Norfolk, the lower parts of dune ridges tend to be dominated by seaward dips indicating accretion towards the beach. This seaward accretion is interpreted as the result of dune vegetation trapping sand blown onshore from the beach. The crest and rearslope of dune ridges contain sets of trough cross-stratification formed by offshore winds. Erosion surfaces identified on GPR profiles can be correlated with erosion scarps on dune ridges. Some of these can be correlated with low-angle reflectors dipping below the water table, which are interpreted as beach erosion surfaces resulting from erosion during storm events. In Wales, landward dips are more common where dunes have been driven inland by prevailing southwesterly winds. Greater understanding of coastal dune formation and the timing of dune movement is required to inform long-term dune management strategies.

A glossary of technical terms is provided at the end of the paper.

Introduction

Ground-penetrating radar (GPR) is a versatile tool for investigating the sub-surface. It is relatively easy to deploy and, importantly, is a non-invasive technique resulting in no disruption to the area surveyed. It produces cross-sections through the sub-surface of varying depth and resolution, and has a number of applications, particularly with regard to hydrogeologic and water table studies, the detection of buried objects, and coastal dune structure and evolution, as discussed in this paper.

Recent research has shown that GPR profiles can be used to study the depth to the water table, locate and resolve buried objects, and show the internal structure of dunes (Schenk et al., 1993; Bristow et al., 1996; Clemmensen et al., 1996; Harari, 1996). This paper presents selected results from recent GPR surveys of coastal dunes in Wales and Norfolk.

Methodology

The GPR surveys were conducted using a PulseEKKO 100 system with 100 and 200 MHz antennae. GPR emits short pulses of electromagnetic energy which propagate into the sub-surface from a transmitting antenna. The energy is reflected back when it encounters abrupt changes in dielectric properties within the ground. These typically represent sedimentary structures, the water table, or changes in composition. The reflected signal is amplified, recorded and processed, with the results displayed in real time. Propagation velocities, penetration depth and resolution of the radar vary according to the complex dielectric properties of the sediments and the frequency of antennae used (Davis and Annan, 1989; Jol, 1995). Survey design and deployment of the equipment in the field is relatively simple (Annan and Cosway, 1992).

Water Table

The change from dry or slightly damp to saturated sand results in a significant change in dielectric properties and usually produces a strong reflection on the GPR profiles (Harari, 1996). The water table reflection is usually strong, laterally extensive, and

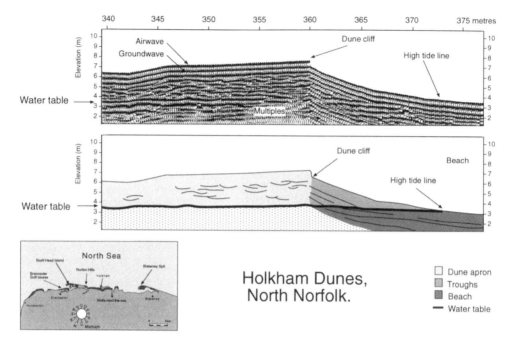

Figure 1. *GPR profile across coastal foredunes and down onto the beach. The water table forms a strong horizontal reflector at the level of the high tide mark at 3m OD. Conductive saline groundwaters increase attenuation, and GPR profiles extended onto the beach show attenuation at the beach water table (from Bristow et al., in press).*

sub-horizontal, and can be mapped within dune fields. Because the water table is a very strong reflector, other dipping reflectors may appear to be cross-cut by the horizontal water table reflector. However, sedimentary dips can be traced both above and below the water table. In some cases, dipping sedimentary reflectors are refracted at the water table and appear to increase in dip because the velocity within saturated sand is around 0.06m/ns, whereas the velocity in dry or damp sand is usually close to 0.11m/ns. Within dune systems the water table can control sedimentation (Fryberger et al., 1988), and there may be a change in reflection character due to changes in sedimentary structures. Dipping reflectors may downlap onto an interdune stabilization surface which could be interpreted as a water table or palaeo–water table stabilization surface. In some coastal dunes the water table is observed at, or close to, the elevation of the high tide line. In this case the water table may merge with beach top reflectors and mark the boundary between wave-lain beach deposits and wind-blown dune sand or beach platform and back beach washover deposits (Bristow et al., in press). The GPR may also be used in hydrogeologic studies for determining and monitoring groundwater levels, flows and forms, and to monitor leakages from broken pipes or waste disposal sites (Beres and Haeni, 1991; Jackson, 1996; Reynolds, 1997). GPR has been used to monitor the movement of contaminants through sand under controlled conditions (Brewster and Annan, 1994; Brewster et al., 1995), although it may be difficult to distinguish contaminant plumes from natural heterogeneity in field conditions.

The salinity of the groundwater is important because saline groundwaters have very high conductivity which leads to rapid attenuation and limited depth of penetration. Most of the dunes on the coasts of Wales and North Norfolk are underlain by a fresh water aquifer, allowing reasonable radar penetration and resolution beneath the water table. However, approaching the sea, or on a beach, the groundwater becomes saline, leading to rapid attenuation. A GPR profile at Holkham steps down from coastal dunes onto a beach and shows a strong horizontal reflector from the water table at the level of the high tide mark (see Fig. 1). Conductive saline groundwaters increase attenuation, and GPR profiles extended onto a beach show complete attenuation at the beach

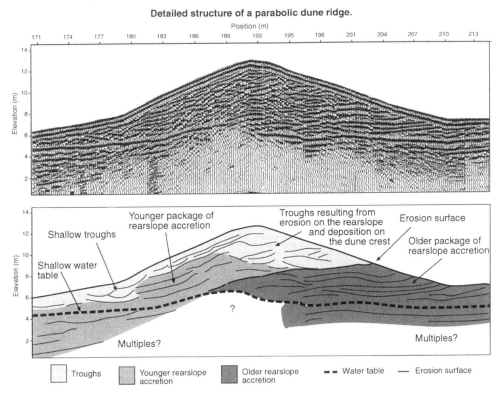

Figure 2. *Detailed radar transect (200 Mhz) across a parabolic dune ridge at Aberffraw, North Wales. Note the predominance of landward dipping reflectors representing rearslope accretion from sand blown over the dune crest by the prevailing southwesterly winds. The older package of reflectors is clearly separated from the younger package within the dune. Trough and fill structures is being created, particularly on the crest of the dune, due to erosion and deposition.*

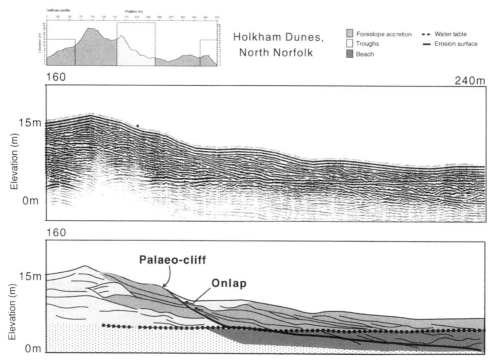

Figure 3. *GPR profile on the northern side of the main dune ridge at Holkham shows onlap onto a buried dune palaeo-cliff which is correlated with a marine erosion surface in the beach facies which occur below the water table.*

water table. The water table is clearly shown in Figures 2 and 3.

Coastal Dune Structure

GPR provides a non-invasive method for imaging the internal structure of dunes. Radar profiles of dunes commonly show the internal structure of dune ridges and foredunes, units of foreslope accretion, rearslope accretion, trough cut and fill, roll-over, erosion surfaces and major bounding surfaces, as well as the depth to the water table (Bristow et al., 1996; Bristow et al., in press; Clemmensen et al., 1996; Harari, 1996).

Figure 2 shows a GPR profile through a large parabolic dune at Aberffraw, Anglesey, North Wales. The structure of the dune consists predominantly of gently undulating, sub-planar, landward dipping reflectors (in two packages) and small sets of concave troughs. It is important to note that the landward dipping sets of accretion surfaces are present on both the rearslope and the foreslope (windward) side of the dune. An erosion surface divides the landward dipping reflectors into two packages – an older package of gently dipping, coherent reflectors (dark tone), and a younger package of less coherent, broken, landward dipping reflectors (light tone). The large area of trough cut and fill on the windward side of the dune and the crest suggests that sediment is being removed from the dune foreslope and inter-dune area on the seaward side, and is then accreting on the rearslope (landward) side of the dune. In this manner, some of the older landward dipping accretionary surfaces (dark tone) are preserved as the dune rolls over itself, migrating inland. This accounts for the landward, instead of seaward, dipping reflectors on the foreslope (windward side) of the dune. Here, the landward dips are a result of the strong prevailing onshore winds which drive the dunes inland.

In North Norfolk the accretionary structures of the dunes typically dip offshore, which is partly due to the prevailing offshore wind, and to dune vegetation trapping sand blown onshore from the beach (Bristow et al., in press). Figure 3 shows a profile through a dune ridge at Holkham, North Norfolk. Discontinuous, hummocky, low-angle reflectors dip towards the sea, and are interpreted as small (5–20m wide) sets of cross-stratification. Between 185m and 195m these low-angle, dipping reflectors are truncated at a radar sequence boundary. This surface is onlapped by overlying low-angle reflectors. The radar sequence boundary between the two sets of reflectors within the dune can be traced below the water table where it is correlated with a low-angle reflector which dips towards the north at between 10 and 20 degrees. This can be interpreted as a beach erosion surface. Its continuation from the dune and into the beach suggests a link between the formation of the erosion surface within the dune and the beach, probably a storm erosion surface which cut into the dune foreslope (Bristow et al., in press).

Artificial Structures

GPR may also be used for locating, surveying and determining the nature of buried objects. Figure 4 shows an example of a buried water pipe within dunes at Brancaster golf course, Norfolk. Due to the nature of the survey design, and the way in which the radar 'sees' structures within the sub-surface, the pipe appears as a strong, dipping reflector. The dip is artificial, and only represents the fact that the radar survey approaches the pipe at an oblique angle before crossing over the top of it. The apparent depth is related to the distance between the GPR profile and the pipe. Clearly the contrast between sand and the metal of the pipe creates a very strong reflection. GPR is therefore useful for pipe location and detection (Jackson, 1996; Bowers and Koerner, 1982). GPR can also be used to detect other buried objects and in archaeological studies (Vaughan, 1986; Bevan, 1991).

Discussion

The ability to image the water table, measure its depth and map it using GPR has useful

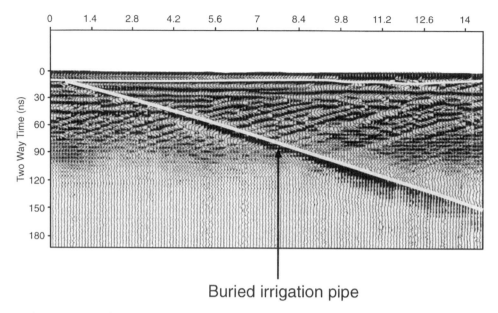

Buried irrigation pipe

Figure 4. Some buried objects produce a strong return when surveyed with the GPR. The strong reflector in the profile represents side-swipe generated by buried metal irrigation pipes within a links golf course at Brancaster, North Norfolk. The horizontal scale is in metres.

implications for the conservation and management of dunes. The water table is an important part of coastal dune systems; its level is particularly relevant to flora and fauna in dune slacks (Ranwell, 1972). Importantly, it also acts as a control on sand movement within a dune field (Fryberger et al., 1988) and can control sand deposition (Kocurek et al., 1990). Dry sand is prone to erosion by the wind, but damp or wet sand will have a higher cohesive value and resists movement. The advantage of using GPR to measure or monitor the water table is that GPR is non-invasive and leaves no mark on the dune system, unlike boreholes which are semi-permanent. Recent work at Newborough Warren, Anglesey, revealed evidence that the water table locally dips towards the forested section of the dunes, possibly indicating that the forest is drawing down the water table. The effect of this could be seen to extend over 100m into the dunes beyond the edge of the plantation forestry. More intensive surveys with the GPR could investigate whether the water table is perched or domed. A GPR survey would be quicker than sinking wells, and continuous profiling with GPR provides better spatial resolution than boreholes. Combining a GPR survey with borehole data provides an additional control on the results.

    GPR is a useful tool for examining coastal dune structure as the internal sedimentary structure of the dunes may be used to interpret their behaviour and evolution in the past. Sets of landward or seaward dipping reflectors may be interpreted as accretionary foresets representing progradation on the foreslope or the rearslope of the dune – periods of dune building and advancement. Similarly, troughs with cut and fill structures may be interpreted as periods when the dune sands are being eroded and reworked. The structures within the dune shown in Figure 2 indicate that this dune is rolling over itself as it migrates inland, with sand eroded from the foreslope (upwind) deposited on the rearslope (downwind). Periods of stabilization or erosion are represented by the bounding surface within the accretionary foresets. Bounding surfaces within the dune allow the construction of a relative chronology, based on cross-cutting relationships and stratigraphic position. It is possible to develop this further by determining an absolute chronology using Optically Stimulated Luminescence (OSL) dating. This technique may be used to date sand from specific packages identified from the GPR surveys, providing depositional ages for the various parts of the dune. This research will eventually allow a better understanding of coastal dune structure and evolution, with an aim of identifying important controls on dune development, such as climate change, storminess or sea-level change. A good example of the effect of storms on coastal dunes is shown in Figure 3. The large erosion surface interpreted as a beach erosion surface has cut back into the foot of the dune, creating a dune cliff. This has later been onlapped by a dune foreslope apron after beach accretion.

The ability to identify and interpret dune structures in the sub-surface without damaging the dunes will provide important information on dune formation with relevance to dune management and conservation. The timing of dune movement and the triggers for dune activity are still poorly understood. Previously, dune management policy has concentrated on stabilizing dunes (Ranwell and Boar, 1986). However, some sand movement is required to maintain dune habitats, and an improved understanding of dune activity in the past is required to inform future dune management.

Conclusions

GPR may be used to identify the depth to the water table, and changes in its characteristics. This has important implications for monitoring of the water table, particularly within the confines of dune management and conservation. GPR can also detect buried objects such as pipes, containers, or archaeological remains, as well as aiding the monitoring of sub-surface spills or leaks.

GPR can image internal dune structure. As such it is an important tool when reconstructing dune evolution and behaviour, particularly with regard to conservation and management. The technique is non-invasive with no damage to the habitat during data collection.

Acknowledgments

The authors would like to thank the Countryside Council for Wales for financial support of the Welsh fieldwork, and Brancaster Golf Club for access to their site. Thanks also to Neil Chroston for help with data collection in Norfolk, and Mike Bailey for field assistance in Wales. Simon Bailey is in receipt of NERC Grant GT04/97/189/ES. The GPR equipment used in this study was on loan from the NERC Geophysical Equipment Pool.

References

Annan, A.P. and Cosway, S.W. (1992), 'Ground penetrating radar survey design', PulseEKKO 100 System Technical Description, PEMD, **54**, 1–12.

Beres, M., Jr and Haeni, F.P. (1991), 'Application of ground-penetrating radar methods in hydrogeologic studies', *Ground Water*, **29**, 375–86.

Bevan, B.W. (1991), 'The search for graves', *Geophysics*, **56**, 1310–19.

Bowers, J.J., Jr and Koerner, R.M. (1982), 'Buried container detection using ground probing radar', *Journal of Hazardous Material*, **7**, 1–7.

Brewster, M.L. and Annan, A.P. (1994), 'Ground-penetrating radar monitoring of a controlled DNAPL release: 200MHz radar', *Geophysics*, **59**, 1211–21.

Brewster, M.L., Annan, A.P., Greenhouse, J.P., Keuper, B.H., Olhoeft, G.R., Redman, J.D. and Sanders, K.A. (1995), 'Observed migration of a controlled DNAPL release by geophysical methods', *Ground Water*, **33**, 977–87.

Bristow, C.S., Chroston, N.P. and Bailey, S.D. (in press), 'The structure and development of coastal dunes: insights from ground penetrating radar (GPR) surveys, Norfolk, England', *Sedimentology*.

Bristow, C.S., Pugh, J. and Goodall, T. (1996), 'Internal structure of aeolian dunes in Abu Dhabi determined using ground-penetrating radar', *Sedimentology*, **42**, 995–1003.

Clemmensen, L.B., Andreasen, F., Nielsen, S.T. and Sten, E. (1996), 'The late Holocene coastal dunefield at Vejers, Denmark: characteristics, sand budget and depositional dynamics', *Geomorphology*, **17**, 79–98.

Davis, J.L. and Annan, A.P. (1989), 'Ground penetrating radar for high-resolution mapping of soil and rock stratigraphy', *Geophysical Prospecting*, **37**, 531–51.

Fryberger, S.G., Schenk, C.J., and Krystinik, L.F. (1988), 'Stokes surfaces and the effects of near-surface groundwater-table on aeolian deposition', *Sedimentology*, **35**, 21–41.

Gawthorpe, R.L., Collier, R.E.L.I., Alexander, J., Bridge, J.S. and Leeder, M.R. (1993), 'Ground penetrating radar: application to sandbody geometry and heterogeneity studies', in C.P. North and D.J. Prosser (eds.), *Characterisation of Fluvial and Aeolian Reservoirs*, Geological Society Special Publication, **73**, 421–32.

Harari, Z. (1996), 'Ground-penetrating radar (GPR) for imaging stratigraphic features and groundwater in sand dunes', *Applied Geophysics*, **36**, 43–52.

Jackson, A.F. (1996), 'Applications and limitations of ground penetrating radar with specific reference to the water industry', unpublished MSc thesis, University of Leeds.

Jol, H.M. (1995), 'Ground penetrating radar frequencies and transmitter powers compared for penetration depth, resolution and reflection continuity', *Geophysical Prospecting*, **43**, 693–709.

Kocurek, G., Townsley, M., Yeh, E., Havholm, K., and Sweet, M.L. (1990), 'Dune and dune-field development on Padre Island, Texas, with implications for interdune deposition and water-table-controlled accumulation', *Journal of Sedimentary Petrology*, **62**, 622–35.

Mitchurn, R.M. Jr, Vail, P.R. and Thompson, S. III (1977), 'Seismic stratigraphy and global changes of sea level. Part 2: The depositional sequence as a basic unit for stratigraphic analysis', in C.E. Peyton (ed.), *Seismic Stratigraphy – Applications to Hydrocarbon Exploration*, American

Association of Petroleum Geologists Memoir, **26**, 53–62.

Ranwell, D.S. (1972), *Ecology of Saltmarshes and Sand Dunes*, Chapman and Hall, London.

Ranwell, D.S. and Boar, R. (1986), *Coastal Dune Management Guide*, Institute of Terrestrial Ecology, NERC, Monkswood, Huntingdon.

Reynolds, J.M. (1997), *An Introduction to Applied and Environmental Geophysics*, John Wiley and Sons, Chichester.

Schenk, C.J., Gautier, D.L., Olhoeft, G.R., and Lucius, J.E. (1993*)*, 'Internal structure of an aeolian dune using ground penetrating radar', in K. Pye and N. Lancaster (eds.), *Aeolian Sediments Ancient and Modern, Spec. Pub. Int. Ass. Sediment*, **16**, 61–69.

Vaughan, C.J. (1986), 'Ground-penetrating radar surveys used in archaeological investigations', *Geophysics*, **51**, 595–604.

Glossary

Bounding surfaces: a non-generic term which can be applied to surfaces which separate packages of sediment, and may be erosive or accretionary.

Common mid-point (cmp): measurements repeated with increased spacing around a central point, the common mid-point, used to determine the velocity of the wave propagating through the ground.

Downlap: successively younger strata (reflectors or bedding planes) terminate progressively downdip against an older surface. Corresponding terms are *toplap*: the tops of inclined strata terminate against an upper boundary, usually as a result of nondeposition; *onlap*: succesively younger strata (reflectors or bedding planes) terminate progressively updip against an older, dipping surface (based on Mitchurn et al., 1977).

Foreset: an inclined layer of sand produced by migration of a sand dune, commonly formed on the steep leeside of a dune when sand avalanches down a slipface. Corresponding terms are *toeset*: horizontal or gently dipping layers of sand deposited at the base of a dune. Toesets may be overlain by foresets as a dune advances or progrades; *topset*: gently dipping layers of sand deposited on top of a dune. The general term for inclined or gently dipping layers of sand within a dune is *cross-stratification*.

Foreslope: the side of a dune facing the sea; *rearslope*: corresponding term for the side of a dune facing inland (away from the sea).

Radar sequence boundary: a surface where reflectors terminate or are truncated; radar sequence boundaries define genetically related packages of strata termed radar sequences (Gawthorpe et al., 1993).

COMPREHENSIVE AND EFFECTIVE RECORDING OF EDAPHIC CHARACTERISTICS OF DUNE ECOSYSTEMS AS APPLIED IN THE MONITORING PROJECT OF THE FLEMISH COASTAL DUNES

C. AMPE and R. LANGOHR

Soil Science Unit, University of Ghent, Belgium

Introduction

A project was set up in 1996 to monitor management regimes (shrub removal, grazing and mowing) implemented in different nature reserves along the west Belgian coast. The project integrates botanical, zoological, hydrological and pedological research at several levels (Bonte et al., 1997). The soil study aims to produce a comprehensive and effective system of soil description, and in this paper we propose the methodology used for recording soil characteristics at the most detailed research level. The initial period of the monitoring project runs for 3 years.

Methodology

Soils are described on the basis of both field and laboratory characteristics.

Field description
Two levels of detail are used.

1. Soils in 80 permanent quadrats (3 × 4m) situated in (approximately) 50 × 50m plots representing 16 management regimes (including the controls) and spread over six sites, situated in the Westhoek Nature Reserve and the Flemish Nature Reserve of Houtsaegerduinen, were investigated in great detail. Mini-soil profiles, 50 × 40cm and approximately 50cm deep, were dug for which vertical and horizontal sections were studied.
2. At a number of other sites in the Westhoek Nature Reserve, Dune Marchand, Ter Yde, Oostvoorduinen and Schuddebeurze soil profiles were investigated in less detail by observations of profiles produced by using a hand auger.

First a more *general description of the soil profile* was recorded. This description partly followed the Food and Agriculture Organisation (FAO) guidelines (1990), but this standard technique for soil profile description provides insufficient data for an understanding of the ecosystem dynamics. Based on literature, mainly from The Netherlands, and personal experience we added a number of parameters to be described.

Dune soils are often shallow for root penetration and the *humus form* will play a very important role in increasing the available nutrients and moisture for plant growth. The terminology according to Green et al. (1993) was employed for the description of the humus form. The parameters considered were horizon designation and depth, moisture status, colour, fabric including structure, texture and character, roots, soil flora and fauna.

The *thickness of the biological active layer ('bi' horizon)*, the sum of the horizons where the majority of the roots are concentrated, is often limited in sandy dune soils. Former research along a topo-chronosequence in the Westhoek Nature Reserve has shown that in hollows the thickness of the biological active layer was often limited to about 20cm. This biological active layer was underlain by a more *compact ('d') horizon* with almost no more roots. Bulk density measurements resulted in the following values: 1.35 (n = 49), 1.50 (n = 121) and 1.56 (n = 436) g per cm^3 for the Abi, Bbi and Cd horizons respectively. On the micro-ridges the biological active layer became thicker, often due to bioturbation caused, for example, by rabbits, and rooting was more developed in the resultant air spaces (Ampe and Langohr, 1993). The thickness of the biological layer was measured from the vertical section or determined with a penetration rod of surface area 1cm^2.

The limited thickness of the biologically active layer has important consequences for water and nutrient supply. Firstly, the water available to plants in dune soils amounts to only a few per cent of the total amount of water present and is only higher in those surface horizons with somewhat greater organic matter (OM) content. Secondly, if the groundwater table is at a greater depth than the sum of the thicknesses of the biological active layer and the capillary rise, plants encounter great problems in water uptake. Due to a limited thickness of the biological active layer, the system is very sensitive to disturbances.

In this study particular attention was paid to *rooting characteristics*. Size (diameter), abundance, orientation and species are determined on the vertical as well as on the horizontal section. In some profiles live roots were observed growing in old humified root galleries. It can be deduced that on the one hand we have plants which grow with the accreting dune such as *Ammophila arenaria* and *Salix repens*, and on the other hand there are plant species which make use of the presence of these old humified root galleries in later successional stages. If these old root galleries are missing, further colonization of the dune will be very slow.

Specific rooting characteristics of different plant species have also been investigated as these may explain why some plant species are more competitive than others in their struggle for water and nutrient supply.

Extent of decalcification is determined with HCl (2N). Decalcification can vary within the profile due to:

- the presence of buried decalcified humic horizons;
- the active process of deposition whereby the soil is regularly showered with fresh calcareous sand;
- lateral supply with calcareous seepage water;
- turbation activity by animals (rabbits, moles, mice) or people (forest management, thinning).

The pattern of decalcification can also be very irregular over short distances in space, resulting in a mosaic of calcicole and calcifuge vegetation.

The soil is tested for *water repellency* under field conditions and also on air-dry samples in the laboratory. Soils which are water repellent show a certain resistance to water penetration, a common characteristic of dune soils (Dekker and Jungerius, 1990). The model of homogeneous wetting fronts cannot be applied in such soils. Preferential flows develop in tongues or fingers along roots, in microdepressions or less repellent areas, or follow buried horizons. Along these preferential flow routes water can reach the groundwater table much faster than would normally be expected. In the field, patches of moist soil surrounded by completely dry patches can be observed. Water repellency is therefore another factor causing dune soils to be drier than expected.

Oxidation-reduction features were described. Abundance, size, contrast, boundary, colour, the activity status and the related distribution of the mottling were recorded. Active mottling can be represented by the presence of a rusty rim especially along *Carex* roots. On the horizontal section, sharp concentric rust-coloured mottling surrounding a live root can be seen; on the vertical section this kind of mottling forms rusty vertical streaks along the root. U-shaped mottles can very clearly be seen on horizontal sections, their morphology consisting of a dark, sometimes black nucleus – a dead root channel – surrounded by the U-shaped rusty colouring. Indeed, due to lateral water flow the original concentric mottle becomes U-shaped with the closed end showing the direction from where the water flow comes. Apart from these very striking types of mottling, a rust-coloured soil matrix can be observed in some horizons.

Other special features can be observed on the vertical and horizontal sections. Plough layers or thin bands that are probably buried algal bands are examples of such features.

Sampling strategy

Siting of the *mini-profile* and sampling took place on one of the four sides of the permanent quadrat with a preference for the southern. In the mini-profile bulk samples were taken for each horizon. For some representative horizons undisturbed samples (5 replicates of cores of $100cm^3$) were taken on the horizontal section for bulk density determination.

For the *surface mineral horizon* between 0–5 and 0–10cm, a composite sample was taken consisting of 5 subsamples. Because of persisting dry conditions of the soil or the presence of thick roots (*Salix repens*) or rhizomes (*Iris pseudacorus*), neither the gouge auger nor the grass plot sampler (a small auger designed for grass-covered terrain, pressed into the soil by stepping on the container which subsequently collects the sample) proved to be practical. Therefore rings with a height of 5 and 10cm were used to sample the soil.

The *humus profile* was sampled at some selected sites. For this a stainless steel cylinder with a diameter of 25cm and a height of 20cm was hammered into the soil. The L, F, H and A horizons were taken off separately and placed in bags. For each permanent quadrat this sampling was repeated three times.

Laboratory analysis

The following analyses are provisionally being considered as appropriate.

Main horizons: colour, water repellency, bulk density, pH H_2O, pH KCl, $CaCO_3$, organic carbon.

Samples between 0–5 and 0–10cm depth based on mixed sampling: water repellency, pH H_2O, pH KCl, $CaCO_3$, organic carbon, N, Cation Exchange Capacity.

Samples of the humus profile: air dry weight, pH H_2O, pH KCl, organic carbon and N.

Acknowledgments

This research is co-financed by the LIFE nature project ICCI.

References

Ampe, C. and Langohr, R. (1993), 'Distribution and dynamics of shrub roots in recent coastal dune valley ecosystems of Belgium', *Geoderma*, **56**, 37–55.

Bonte, D., Ampe, C., Hoffmann, M., Langohr, R., Provoost, S. and Herrier, J.L. (1997), 'Monitoring research in the Flemish dunes: from a descriptive to an integrated approach', in C.H. Ovesen (ed.), *Coastal Dunes – Management, Protection and Research*, report from a European seminar, Skagen, Denmark, August 1997. National Forest and Nature Agency, Geological Survey of Denmark and Greenland.

Dekker, L.W. and Jungerius, P.D. (1990), 'Water repellency in the dunes with special reference to The Netherlands', *Catena Supplement*, **18**, 173–83.

Food and Agriculture Organisation (1990), *Guidelines for Profile Description*, 3rd edn, Food and Agriculture Organisation, Rome.

Green, R.N., Trowbridge, R.L. and Klinka, K. (1993), *Towards a Taxonomic Classification of Humus Forms, Forest Science*, Monograph 29, Society of American Foresters, Bethesda, MD.

MORPHOLOGICAL CHARACTERIZATION OF HUMUS FORMS IN COASTAL DUNE SYSTEMS: EXPERIENCE FROM THE FLEMISH COAST AND NORTHWEST FRANCE

C. AMPE and R. LANGOHR

Soil Science Unit, University of Ghent, Belgium

ABSTRACT

Humus profiles of dune soils on 46 sites along the Flemish coast and in northwest France were classified according to three classification systems (Delecour, 1980; Green et al., 1993 and Jabiol et al., 1995). These classification systems are not well adapted to young ecosystems on almost pure sand. Problems arose in the keys themselves and in the terminology and criteria which are used to classify the soils.

Introduction

Dune soils are characterized by a shallow available soil volume for rooting (Ampe and Langohr, 1993) and a low nutrient and water supply. The presence of organic horizons is therefore of great importance to ecosystem dynamics, and a workable, comprehensive classification of humus form is needed.

The humus profiles studied are in dune areas along the Flemish coast and in northwest France. The sites are imperfectly to well drained, situated on a calcareous substratum, and vary in vegetation cover from almost bare sand to forest, including sites under grasses and shrubs. On the Flemish coast, 12 profiles were investigated from the Domeinbos Klemskerke-Vlissegem-Wenduine (Baes, 1997) and 4 from the Hannecart forest. In the Réserve biologique domaniale de la côte d'Opale (Merlimont, northwest France) 30 sites were investigated (Ampe, 1998).

The aim of the study is to test 3 published classification systems for humus profiles (Delecour, 1980; Green et al., 1993 and Jabiol et al., 1995).

Methodology

The organic horizons of the 46 sites were described according to the guidelines of

Green et al. (1993). The following factors are used for description: moisture status, colour, fabric including structure, texture and character, roots, soil fauna and vegetation.

Thirty-eight sites were sampled by using a stainless steel cylinder (25cm diameter and 20cm long) which was hammered into the soil. The sample was separated horizon by horizon and the procedure was repeated two or three times at each site. The subsites were sampled to show the amount of variation in the humus profiles.

The humus type was determined for all 46 sites according to the three classification systems.

The Classification Systems

The main criteria for classifying humus types, in imperfectly to well drained conditions, defined at the highest level, i.e. mull – moder – mor, are summarized in Table 1. The approach in the three classification systems is substantially different.

The classification system of Delecour (1980) distinguishes between forest and herbaceous formations at the start of the key. For the forest systems the mull and moder/mor humus types are distinguished on the basis of the morphology of the holorganic layer, which should be composed of either a mainly thin (thickness is not specified) L horizon or of an L, F and H horizon respectively (the holorganic horizon is the organic layers of the humus profile: the L, F and H horizons for well to imperfectly drained soils and the O horizon for the poorly drained soils). The distinction between moder and mor is made on the basis of the thickness of the holorganic horizons (greater or less than 10cm). Mulls are further subdivided on the occurrence of mottling, reaction with HCl or pH of the A horizon. Moders are subdivided on the relative proportion of the F and H horizon, the type of substratum, effervescence with HCl and structure of the A horizon. Within the mors, humus forms are differentiated upon the relative proportions of the F and H horizons. For the herbaceous systems only the mull and mor types exist. Mulls are distinguished from the mors on the basis of the distinctness of the boundary between the holorganic horizons and the A horizon, and the structure of the A horizon, and although not mentioned explicitly, absence or presence of the H horizon. In the mulls, subdivisions are made according to thickness of the L and F horizons and the thickness and structure of the A horizon. The differentiating criteria in the mors are thickness of the L and F horizon, structure of the A horizon sequence within the holorganic horizons and the root morphology.

In the system of Green et al. (1993) the first criterion used is the combined thickness of the F+H horizons. Then within the orders of the mors and moders, the type of F horizon (mycogenous, zoogenous or an intergrade between myco- and zoogenous) determines the group. In the mull order the group is determined by the type of A horizon (rhizogenous or zoogenous).

The system of Jabiol et al. (1995), which includes some of the concepts of Delecour, has the type of A horizon, the boundary between the holorganic and mineral horizons

Table 1. Criteria (simplified – relevant for dune soils) for the main orders for well to imperfectly drained sites (horizon symbols as they are used in the respective classification systems)

	Delecour (1980)	Green et al. (1993)	Jabiol et al. (1995)
Mull	Forest: • absence or very thin Ol, Of, Oh Herbs: • gradual transition between ecto- and endorganic horizons • Ah de 'complexation'	F+H ≤ 2cm with Ah > 2cm	• A de 'biomacrostructuré', horizon with crumb structure • discontinuity between OL, OF, OH and A horizon
Moder	Forest: • Ol+Of+Oh horizons < 10cm thick • Ah de 'diffusion' • transition 'OAh' Herbs: • not used	F+H > 2cm or F+H ≤ 2cm if Ah < 2cm And F is zoogenous and/or an intergrade between zoo- and mycogenous	• A horizon of type 'juxtaposition' – pepper and salt • gradual transition from OL, OF, OH and A horizons • horizonation of OL-OF-A or OL-OF-OH-A
Mor	Forest: • Ol+Of+Oh horizons > 10cm thick • no 'OAh' horizon Herbs: • abrupt boundary between ecto- and endorganic horizons • no Ah de 'complexation'	F+H > 2cm or F+H ≤ 2cm if Ah < 2cm And F is mycogenous	• OL, OF and OH horizon are thick • abrupt boundary between OL, OF, OH and mineral horizon • in the mineral horizon: organic matter de 'diffusion'

and the horizon sequence as diagnostic criteria. Mulls are characterized by an A horizon with crumb structure and a clear boundary between the L, (F) and A horizons. They are further subdivided according to the thickness and type of L horizon and whether or not the F horizon is present. The moders have an A horizon with the pepper and salt morphology (A *de 'juxtaposition'* – the bleached sand particles occur next to the particles of organic matter), a more gradual limit between L, F, H and A horizons and a horizon sequence of L, F, (H) and A horizon. The moders are subdivided according to the thickness of the F and H horizon and the structure of the A. Mors have thick L, F and H horizons and an abrupt boundary between the holorganic and mineral horizons.

Results

Table 2 provides an example of 6 sites that have been classified according to the three systems. For a total of 46 sites, the three systems classify only 24 sites as the same type at the highest level, i.e. mull–moder–mor. If we compare the systems in pairs then the agreement between systems is greater: Delecour and Green 32, Delecour and Jabiol 30 and Green and Jabiol 31. Discrepancies occur between mull/moder and moder/mor.

Discussion and Conclusions

When applying these three classification systems it must be remembered that they were originally designed for forest systems. Delecour (1980) studied the beech forests of the Ardennes (Belgium). The system of Green et al. (1993) is based on existing classifications in use in North America, Great Britain and Europe. The background of this system is again forestry, but the classification is aimed at use in a wider range of ecosystems. Jabiol et al. (1995) state that their proposed system should be applied to forests.

Our testing clearly demonstrated that the classification systems are not well suited to young ecosystems on almost pure, sandy soils. Problems arise in the keys themselves and in the terminology and criteria for classification.

In the key of Delecour's system a subdivision is made according to vegetation type: forest or herbs. What is the position of the shrubs and combined systems of herbs and trees or herbs and shrubs? In our testing they were considered as forest formations. This leads to very similar humus forms being classified differently. A second problem is that a thickness of L+F+H of more than 10cm is required for a mor humus type. In dune soils this 10cm proves to be a very severe criterion. Furthermore, the different humus types are not described, which makes checking of the obtained result very difficult.

One of the main problems in Green et al. (1993) is the obligatory presence of Fm (mycogenous) for mor humus type. In the mull order only 2 types are distinguished, rhizo- and vermimull. Some humus forms show neither characteristic.

In the system of Jabiol et al. (1995) there is a fundamental problem. Mulls are char-

Table 2. Comparative table showing the results of the classification
(horizon symbols as they are used in the respective classification systems)

Profile	Delecour (1980)	Green et al. (1993)	Jabiol et al. (1995)
MEP3	Of: 4.5cm, Ah: 8cm mull *feuilleté*	well to imperf. drained, F+H > 2cm, Ah: 8cm, no Fm, Fz? mullmoder	OF: 4.5cm, OH: -, A, rootmat dysmull
MEP7	Ol: 1.5cm, Ah: 9cm mull *grenu mince*	well to imperf. drained, F+H < 2cm, Ah: 9cm rhizomull	OL: 1.5cm, OLn cont., OF: -, OH: -, A: single part mésomull
MEP10	Ol: 0.5cm, Ah: 7cm mull *carbonaté*	well to imperf. drained, F+H < 2cm, Ah: 7cm rhizomull	OL: 0.5cm, OLn cont., OF: -, OH: -, A: single part mésomull
MEP16	Ol: 2–2.5cm, Of: 2.5cm, Oh: 3cm, E+(H), Of = Oh, Ol+Of+Oh < 10cm dysmoder	well to imperf. drained, F+H > 2cm, Fm, Hr, Ah: 3cm, Hr > F resimor	OL: 2–2.5cm, OF: 2.5cm, OH: 3cm, E+(H): pepper and salt, weak granular mor
KL109a	Ol: 0.5–0.9cm, Of: 1.3–3cm, Ah: 8-18cm dystrophic mull	well to imperf. drained, F+H < 2cm, Ah: 8–18cm vermimull; well to imperf. drained, F+H > 2cm, Ah: 8–18cm, Fz mullmoder	OL: 0.5–0.9cm, OF: 1.3–3cm, H+E: pepper and salt, weak granular dysmull
VL206b	Ol: 1.8–3.2cm, Of: 4.5–7.8cm, H+F: 0–2.8cm, Ol+Of+Oh mostly > 10cm thick, AE fibrimor	well to imperf. drained, F+H > 2cm, Ah: 26-46cm, Fa, Hr mormoder	OL: 1.8–3.2cm, OF: 4.5–7.8cm, OH: 0–2.8cm, H+E: pepper and salt mor

acterized by an A with a crumb structure and an intimate mixture of organic and mineral material caused by the activity of earthworms and described as the A *de 'biomacrostructuré'*. In dune soils, as soon as some decalcification takes place, the morphology of 'pepper and salt' or the A *de 'juxtaposition'* is observed. This type of horizon is, however, a characteristic of the moder humus form. Another problem is that neither moders nor mors show any crumb or micro-crumb structure; the A horizons should be massive or single-grained. However, the A horizon of the *'juxtaposition'* type with a crumb structure has been observed.

There are terminology and criteria problems in the three classification systems:

- clear definition of the diagnostic criteria, especially to distinguish L and F, is missing in the three systems;
- definition of horizon A *de 'complexation'* and A *de 'diffusion'* (Delecour) is missing;
- differentiating criteria in the keys are too vague; thickness of horizons is not specified (e.g. Delecour: *mince* [thin], *très mince* [very thin]);
- use of two criteria at one point in the key which are not always mutually exclusive (e.g. Delecour: thickness of Ah and rooting density and thickness of Ol+Of+Oh with sharpness of boundary with the mineral horizons and type of A horizon and substratum);
- clear illustrations of the diagnostic horizons such as the Fz, Fa and Fm, and the Hh, Hz and Hr, are missing (Green et al., 1993).

In the three systems little or no attention is paid to the seasonal variability in the dynamics of the organisms in the humus profile. This poses problems for measuring the thickness of the holorganic horizons and the presence of mycelia and soil fauna.

Based on the experience explained above, the following recommendations can be made:
- in the definition of the L and F a clear distinction should be made between the degree of fragmentation and the degree of mixing;
- size of fragments of needles, leaves and grasses should be specified;
- specification of description and sampling period is necessary;
- development of more precise and illustrated diagnostic criteria (thickness, sharpness, traces of biological activity) is required.

Acknowledgments
This research is co-financed by the LIFE nature project ICCI.

The Merlimont data were collected as part of a project on the Réserve biologique domaniale de la côte d'Opale financed by the Office National des Forêts, France. Action CNM 1997–N°97–0100–02.

References

Ampe, C. and Langohr, R. (1993), 'Distribution and dynamics of shrub roots in recent coastal dune valley ecosystems of Belgium', *Geoderma*, **56**, 37–55.

Ampe, C. (1998), *Etude édaphologique de la Réserve biologique domaniale de la côte d'Opale. Office national des forêts*, Action CNM 1997–N°97–0100–02.

Baes, S. (1997), *Verband bodem-vegetatie langsheen topotransecten in het domeinbos Klemskerke-Wenduine (West-Vlaanderen, België)*, Licentiaatsverhandeling, RUG, 116.

Delecour, F. (1980), 'Essai de classification pratique des humus', *Pédologie*, **2**, 225–41.

Green, R.N., Trowbridge, R.L. and Klinka, K. (1993), *Towards a Taxonomic Classification of Humus Forms, Forest Science*, Monograph 29, Society of American Foresters, Bethesda, MD.

Jabiol, B., Brêthes, A., Ponge, J.-F., Toutain, F. and Brun, J.-J. (1995), *L'humus sous toutes ses formes*, Ecole Nationale du Génie Rural des Eaux et des Forêts, France.

MONITORING THE CHANGING ECOLOGY OF SAND DUNES AT GEAR SANDS SSSI AND CANDIDATE SAC IN CONJUNCTION WITH MANAGEMENT OF ACCESS, RECREATION AND WILDLIFE

STEVE CRUMMAY
Cornwall County Council, England
CATRIONA NEIL and ADRIAN SPALDING
Spalding Associates, England

ABSTRACT

Since the 1970s, Cornwall County Council Countryside Services has developed a high degree of expertise in sand dune management. Initially this stemmed from a requirement to take action to prevent habitat loss in the more mobile foredune areas of sites that were particularly vulnerable to visitor pressure. Many areas of dune were stabilized and access management measures put in place to prevent erosion in the future. This has been followed by an increasing amount of positive management activity on more established dune habitat. At Gear Sands, following discussion with the site owners, the County Council Countryside Service negotiated and established a Countryside Stewardship Agreement funded by the Ministry of Agriculture, Fisheries and Food (MAFF). A management agreement with the Rank Organisation was negotiated that established the Countryside Service as the site managers and administrators of this agreement. This aimed to establish a monitoring programme and subsequent management of the site and its habitats as an aid to species conservation and the provision of recreational access. The monitoring is implemented by Spalding Associates (Environmental) Ltd for Cornwall County Council. This complex project brings together ecological survey and monitoring work with a detailed programme involving the control of recreational access and habitat management by mowing and grazing regimes. The project is the result of co-operation between a range of partners, including local user groups, the parish and district councils, nature conservation organizations (including English Nature), but most notably Cornwall County Council and the Rank Organisation.

Site Description

Penhale Sands form part of the most extensive sand dune system in Cornwall, located on the North Cornwall coast and stretching for some 4km from Perranporth to Ligger Point. The west-facing coastal margin rises steeply to form a 50m sand cliff, behind which the dunes rise to a maximum of 93m on Gear Sands.

The total system area is 1070.4ha of which Gear Sands, the area covered by this project, comprises approximately 200ha. Gear Sands comprises a full range of dune habitat from mobile foredune through dune slack to fixed calcareous grassland and scrub. The majority of the project area is owned by Haven Leisure, part of the Rank Organisation, and contains an extensive holiday complex that houses up to 5000 visitors at any one time for a season of 12 weeks during the summer months.

There is a history of de facto open access and the area is widely used by the public with horse riding, beach access, general walking and notably large numbers of dog walkers making use of the site. Informal consultation revealed that some dog owners make a round trip of up to 40km in order to exercise their pets. Also, the historical interest of St Piran's Oratory within the site attracts a large number of visitors. To the north east the site is bounded by land owned by the Ministry of Defence over which there is no public access.

The whole of the Penhale Sands system is designated as a Site of Special Scientific Interest (SSSI) under the Wildlife and Countryside Act 1981 (as amended) and is a candidate Special Area of Conservation (cSAC) under the Habitats Directive.

There is significant archaeological interest at the site, which includes two Scheduled Ancient Monuments, both of which are owned by Perranzabuloe Parish Council.

Development of Management

Despite the importance of the site and the presence of many nationally rare species, prior to 1993 there had been no significant access or conservation management of this part of the Penhale dune system. Cornwall County Council Countryside Services had spent a number of years implementing dune stabilization schemes and managing access to control erosion at several dune sites throughout the county, and in 1993 established a new Countryside Service that included Penhale Sands.

Habitat and species diversity have been shown to be affected more seriously by pollution and development than recreational pressures at the national and regional levels (Sidaway, 1988; 1995). However, at a regional level it has been recognized that the impacts of recreation may require management action and much work has been done that demonstrates the effects of recreational pressure on dune vegetation and dune structure (Sothern et al., 1985; Liddle, 1973). At Gear Sands the combination of high recreational pressure and a lack of habitat management led the Countryside Service to set up initial survey work to investigate any habitat change or change in populations of rare species within the site. Evidence suggested that vegetation and habitat change could be threatening species typical of dune grassland habitat. For example, 13 out of 16 nationally or locally scarce key indicator invertebrate species recorded here are associated with short, rabbit-grazed turf.

Initial surveys included a full National Vegetation Classification (NVC) survey and comparison of aerial photographs covering the period from 1947 to 1994. Results of this work showed an increase in the amount of scrub habitat (Holyoak, 1995) and, in most recent times, a reduction in the area of blown-out dune despite a perception locally that the dunes were being destroyed by visitor pressure. Out of a total of 22 blowouts within the project area 17 had reduced in size between 1988 and 1995.

Following the establishment of the new Countryside Service, the Countryside Commission agreed to include sand dunes within the targeted habitats for Countryside Stewardship, a grant scheme that pays landowners fixed rates for particular management prescriptions and includes an objective for the maintenance and provision of public access. The Countryside Service approached the Rank Organisation as landowners and initiated applications for Stewardship and the setting up of management agreements allowing the Service to implement any scheme that was approved.

It is widely recognized that grazing is the optimum management method in maintaining species diversity within fixed dune grassland (Boorman, 1989; van Dijk, 1992; Houston, 1997). A second and possibly less effective option is mowing with removal of cuttings (Anderson and Romeril, 1992). This was recognized by the Countryside Commission, who established preferential payment rates for grazing of dune grassland.

A range of management actions have now been initiated in response to the initial monitoring programme. Following lengthy consultation, in excess of 3.5km of fencing has been erected, with over twelve access points, to control damaging illegal access by four-wheel-drive vehicles, motorbikes and horses. Fencing has allowed grazing to be established in trial areas at low stocking rates using six Shetland sheep per hectare. Other areas will be mown and control quadrats have been established in areas of managed and unmanaged vegetation so that the long-term effects of different management regimes can be researched.

In addition to habitat management, a full archaeological survey was commissioned and carried out by the Cornwall Archaeological Unit (Cornwall Archaeological Unit, 1997).

Public Consultation

After initial discussions with the local parish council the Countryside Service commissioned the survey work that established the need for management of the site. The management preference for grazing was introduced to the public through a series of exhibitions and meetings during 1995 and 1996. The preferred grazing regime involved temporary fenced areas that would allow targeted grazing and accurate monitoring of the results. This was rejected by the public, concerned about fencing within the site, and so free-ranging stock on the site was proposed. At this stage dog walkers concerned about stock worrying organized a leaflet campaign. As a result of this the management proposal reverted to having enclosures for grazing while still retaining a new perimeter fence.

The first grazing enclosure was erected in an area adjacent to the South West Coast Path (see Plate 1) that crosses the site. This allowed managers to gauge public reaction to fencing and stock within the dunes. The initiative has been well received and the perimeter fencing has resulted in a reduction in damaging activities such as off-road motorcycling. Signs have been erected that explain the rationale behind the management work and also background information on the Shetland sheep. The use of a rare breed has provided a useful interpretative subject and added to the public interest.

At all times the strength of the project's public appeal and success has been the partnership structure between public and private sectors, along with central government grant support through the Stewardship Agreement. The extent to which research, allowing understanding of basic processes operating within an area, has been in place prior to management plan formulation has been questioned (Williams and Randerson, 1989). However, in this case the Countryside Service has been seen to adapt to public opinion while still maintaining policies that will meet habitat management objectives set in response to research results.

Monitoring

The most important element of the initiative has been the establishment of a comprehensive monitoring regime that was initiated prior to any site management. It is a primary objective that management is directed by monitoring rather than providing subject matter to be monitored. With this in mind the following regime was developed by the Countryside Service and Spalding Associates.

An initial NVC survey is to be repeated at five-year intervals. In 1995, ten NVC communities were recorded, including the dominant communities of *Ammophila arenaria–Festuca rubra* semi-fixed dune (SD7) and *Festuca rubra–Galium verum* fixed dune grassland (SD8), with extensive areas of wetland comprising mainly *Potentilla anserina–Carex nigra* dune slack (SD17), mobile *Ammophila arenaria* dune (SD6) and *Festuca ovina–Carlina vulgaris* calcicolous grassland (CG1). Extensive areas of scrub are present in a range of communities, mainly bramble *Rubus fruticosus* agg., European gorse *Ulex europaeus*, blackthorn *Prunus spinosa* and communities similar to *Rubus fruticosus* agg.–*Holcus lanatus* underscrub (W24). Comparison with a previous survey in 1987 (Radley et al., 1987) showed that the mobile dune community (SD6) at the seaward end of the site was much reduced. Fixed-point photography is being used to record changes in physical features including increases in scrub cover and variations in the seasonal ponds.

Following the vegetation survey and a desk study to provide details of species known to be present, species were selected for monitoring which respond to changes in the

Plate 1. *The South West Coast Path alongside the first grazing enclosure fence on Penhale Sands.*

key physical and biotic features of the dunes: the exposed sand leading to temperature extremes, the mobility of habitat, the general lack of water and the lack of cover for small animals and birds. Four classes of species were selected (see Table 1). Future work may involve more detailed inventories of fauna, such as have been carried out on arthropods at Ynyslas dunes (Miles, 1987), to show the relationship between invertebrates, habitats and phases in dune formation.

Many species appear in more than one category. Butterfly transect methodologies allow annual and longer-term comparisons to be made for butterflies occupying different ecological niches. Transect data have been supplemented by population estimates for the nationally scarce silver-studded blue *Plebejus argus,* estimated as 2152 on the site in 1996 (Haes et al., 1996). The dragonfly transect was discontinued after one year as the results are not relevant to the management of the dune system. The nationally declining skylark *Alauda arvensis* is monitored every year in accordance with Common Bird Census techniques. The number of estimated territories at Gear has varied from 11 in 1996 and 16–18 in 1997 to 13–15 in 1998; territory size varies between 0.5–1ha, which is roughly comparable to territory size reported from dunes in Cumbria (Delius, 1965). Skylark territories here are mainly associated with the marram-dominated dunes and largely restricted to the less disturbed areas, a pattern replicated in the Hayle-Gwithian dune system (Spalding, 1998; Tunmore, 1998).

Table 1. Species selection criteria for monitoring the changing ecology of Gear Sands
1995–98

- Taxonomic groups with a well-established monitoring methodology: butterfly transects (Pollard et al., 1986); dragonfly transects (Moore, 1953); NVC surveys of successional changes; Common Bird Census.

- Indicator species which respond rapidly to changes in management regimes: butterflies, beetles, molluscs and birds. Some invertebrates may be used to detect subtle changes within a site, which plants and vertebrates may not respond to measurably for several years (Eversham, 1994). Many bird species are particularly responsive to changes in scrub cover and the amount of disturbance. Species chosen include brown argus *Aricia agestis*, silver-studded blue *Plebejus argus*, glow-worm *Lampyris noctiluca*, skylark *Alauda arvensis*, early gentian *Gentianella anglica* and petalwort *Petalophyllum ralfsii*.

- Nationally rare and locally important species as listed nationally in the *Biodiversity Steering Group Report, Volume 2* (1995) and locally in the *Red Data Book for Cornwall and the Isles of Scilly* (Spalding, 1997): shore dock *Rumex rupestris*, petalwort, early gentian, solitary bee *Andrena hattorfiana*.

- Easily identified species that can be easily monitored in the field: butterflies, skylark, glow-worm and the dune villa bee-fly *Villa modesta*.

Specialist species, such as thermophilous and psammophilous invertebrates, occupy warm sheltered hollows with abundant bare sand and can respond rapidly to changes in management regimes. The dune villa bee-fly *Villa modesta* was found to be widespread in these areas and was added to the monitoring programme in 1997. Little is known about its ecology (Stubbs, 1997) and the project has contributed to knowledge of this species. A total of 244 were recorded at Gear in 1997, many of which were nectaring on wild thyme *Thymus polytrichus* and common stork's-bill *Erodium cicutarium*. Work on glow-worms *Lampyris noctiluca* in 1998 shows a close association with fixed and semi-fixed dune grassland (77 per cent of glow-worms recorded were in this habitat) with a mean sward height of 14cm. Detailed population estimates have been made for the nationally rare solitary bee *Andrena hattorfiana* which in Cornwall is restricted to two sites, both dune systems. These bees form compact nesting aggregations in a variety of substrates but are entirely dependent on field scabious *Knautia arvensis* as a source of pollen for rearing their grubs. On Gear Sands, field scabious is restricted to the sunny sides of open banks and mounds at the edges of the site, where 16 *Andrena hattorfiana* individuals were seen in 1996 (Haes, 1996), 18 in 1997 and 13 in 1998.

The endemic, nationally scarce early gentian *Gentianella anglica* has been recorded at Gear Sands in previous years. The plants previously identified as Cornish gentian

Gentiannella anglica ssp. *cornubiensis* are now considered to be the hybrid *G.* x *davidiana* which is a cross between early gentian and autumn gentian *Gentianella amarella* (Rich et al., 1997), even though *G. amarella* does not occur here. There has been some concern that all gentians here are hybrids, but analysis of morphological measurements taken from permanent quadrats in 1998 indicates that at least three colonies are true *G. anglica*. The population means for corolla length, internode number and proportion of overall height contributed by the terminal internode conform to the ranges defined for separating *Gentianella anglica* from hybrids (Rich et al., 1997). Annual fluctuations in the distribution and abundance of sub-colonies make monitoring difficult. For example, the total number of plants recorded in 1996 was 692, but in 1997 was 11,590, in keeping with national fluctuations recorded that year. Sub-colonies are large: 1080 individuals were estimated in 900m^2 in 1995 but none were seen in this colony in 1996 (Neil, 1997). Fixed quadrats have been established in areas of managed and unmanaged vegetation, so that the long-term effects of different management regimes can be researched.

Nationally rare species present include petalwort *Petalophyllum ralfsii* and shore dock *Rumex rupestris*, both of which are named species in the designation of the Penhale Dunes as a cSAC. Five small colonies of petalwort totalling 139 thalli grow on damp sand in and near dune slacks (Holyoak, 1998). Early indications are that these colonies fluctuate considerably in size, but would be at risk from scrubbing over consequent to declines in rabbit grazing and reduction in groundwater levels following changes in seasonal hydrological regimes.

The nationally rare shore dock *Rumex rupestris* was discovered in 1997 for the first time at Gear during monitoring activities, and it occurs in significant numbers in two slacks at the centre of the site. This plant appears to have declined noticeably in Britain in recent years (Daniels et al., 1998). Monitoring at monthly intervals of the water table at the slack containing the main concentration of plants began in winter 1997/1998. Observed changes in water levels have been uneven, indicating that the site has a complex hydrology, perhaps due to previous mine workings now buried beneath sand. This hydrological pattern will be investigated in more detail in 1999. Monitoring of above-ground growth is based on counting rosettes and crowns more than 5cm apart as separate plants. However, it is impossible to be sure of the spread of individual shore dock plants since underground shoots (ramets) are sometimes produced tens of centimetres from the parent crown; at coastal sites where the plants are often widely spaced, a 50cm distance has been proposed, but is not suitable at Gear where the plants are closely crowded together. Counts for rosettes and crowns indicate an increase between 1997 and 1998 from 50 to 95, whereas the number of flowering stems decreased from 141 to 126. The largest, most vigorous plants growing on drier ground at the edge of the hollow are the hybrid with clustered dock *Rumex conglomeratus*. Care has to be taken to differentiate between these and pure shore dock during counts. The threat to the shore dock from introgression has not been finally assessed. Invertebrate predation on shore dock is also being investigated: the larvae of small copper *Lycaena phlaeas* and the weevils *Hypera rumicis* and *Apion miniatum* have been recorded in adjacent colonies.

Management Recommendations

A number of management recommendations have been made following the initial stages of the monitoring programme. The trend towards an increase in the amount of scrub habitat has been shown to be generally harmful to the nature conservation value of the site. Scrub encroachment adversely affects a range of species, especially the nationally rare and the seven nationally scarce bryophytes present, but also skylark, early gentian and a range of invertebrates including glow-worm, *Andrena hattorfiana* and *Villa modesta*. The association of shore dock with the surrounding plant community (vegetation height and NVC type) is currently under investigation. In many places, areas are kept open by trampling, rabbit grazing or exposure to strong, salt-laden winds. Monitoring rabbit populations will start in spring 1999, as these are generally the most influential grazing animal on British sand dunes (Boorman, 1989). The management regime is low key, with the aim of restoring a dynamic landscape and creating suitable conditions in which targeted wildlife can flourish but avoiding over-interference or gardening. Plans to plant additional field scabious to provide pollen sources for *Andrena hattorfiana* have been shelved partly because the factors controlling population size for *Andrena* on site are not fully understood. The use of Shetland sheep on part of the site has created a more open sward amongst marram-dominated fixed dune, reduced the amount of scrub and created additional suitable habitat for a range of thermophilic invertebrates. Control quadrats are being established in grazed areas, so that the long-term effects of different management regimes can be demonstrated. However, one of the most important factors affecting the distribution and abundance of species at Gear is the amount of disturbance. This disturbance is harmful to a range of birds including skylark and to plants such as shore dock (which may be badly trampled), but can be beneficial to thermophilous and psammophilous invertebrates that require open trampled areas for basking, hunting and burrowing (Kirby, 1992). Some plants may also benefit from trampling; for instance, the nationally scarce dune fescue *Vulpia fasciculata* was recorded for the first time here on eroded pathways near the holiday complex.

Conclusion

This complex project attempts to bring together ecological survey and monitoring work with a detailed programme involving the control of recreational access and habitat management by grazing regimes. The emphasis so far has been on setting baseline data, discovering new species and establishing natural trends before management procedures are established. This paper attempts to highlight the complexity of habitat and species management within a dynamic dune system while accommodating management for recreation and access; the central theme has been to find out as much as possible about the ecology of the site combined with the minimum interference necessary to ensure the continued presence of key species and habitats.

References

Anderson, P. and Romeril, M.G. (1992), 'Mowing experiments to restore a species rich sward on sand dunes in Jersey, Channel Islands, GB', in R.W.G. Carter, T.G.F Curtis and M.J. Sheehy-Skeffington (eds.), *Coastal Dunes*, Balkema, Rotterdam, 235–50.

Biodiversity Steering Group (1995), *Biodiversity: The UK Steering Group Report. Volume 2: Action Plans*, HMSO, London.

Boorman, L.A. (1989), 'The influence of grazing on British sand dunes', in F. van der Meulen, P.D. Jungerius and J.H. Visser (eds.), *Perspectives in Coastal Sand Dune Management*, SPB Academic Publishing, The Hague, 121–25.

Cornwall Archaeological Unit (1997), *Gear Sands, Perranzabuloe: An Archaeological Assessment*, report to the St Agnes Newquay Countryside Service, Cornwall County Council.

Daniels, R.E., McDonnell, E.J. and Raybould, A.F (1998), 'The current status of *Rumex rupestris* Le Gall (Polygonaceae) in England and Wales, and threats to its survival and genetic diversity', *Watsonia*, **22**, 33–39.

Delius, J.D. (1965), 'A population study of Skylarks *Alauda arvensis*', *Ibis*, **107**, 466–92.

Dijk, H.W.J. van (1992), 'Grazing domestic livestock in Dutch coastal dunes: experiments, experiences and perspectives', in R.W.G. Carter, T.G.F Curtis and M.J. Sheehy-Skeffington (eds.), *Coastal Dunes*, Balkema, Rotterdam, 235–50.

Eversham, B.C. (1994), 'Using invertebrates to monitor land use change and site management', *The British Journal of Entomology and Natural History*, **7** (Supplement 1), 36–45.

Haes, E.C.M. (1996), 'A Survey of the Solitary Mining Bee *Andrena hattorfiana* at Gwithian Towans and Gear Sands, Cornwall', Spalding Associates (Environmental) Ltd, unpublished report for Cornwall County Council.

Haes, E.C.M., Neil, C.J. and Spalding, A. (1996), 'Survey of the Population of the Silver-Studded Blue Butterfly on Gear Sands', Spalding Associates (Environmental) Ltd, unpublished report for Cornwall County Council.

Holyoak, D.T. (1995), 'Report on a Survey of the Vegetation and Flora at Gear Sands, May 1995', Cornish Biological Records Unit, unpublished report for Cornwall County Council.

Holyoak, D.T. (1998), 'The Rare and Scarce Bryophytes at Gear Sands with Suggestions for their Conservation', Spalding Associates (Environmental) Ltd, unpublished report for Cornwall County Council.

Houston, J. (1997), 'Conservation management practice on British dune systems', *British Wildlife*, **8**, 297–307.

Kirby, P. (1992), *Habitat Management for Invertebrates: A Practical Handbook*, JNCC/RSPB, Sandy.

Liddle, M.J. (1973), 'The effects of trampling and vehicles on natural vegetation', PhD thesis, University College of North Wales, Bangor.

Miles, P.M. (1987), 'Terrestrial sand-dune fauna at Ynyslas, Cardiganshire', *Nature in Wales*, **2**, 75–79.

Moore, N.W. (1953), 'Population density in adult dragonflies (Odonata – Anisoptera)', *Journal of Animal Ecology*, **22**, 344–59.

Neil, C.J. (1997), 'Early Gentian *Gentianella anglica* on Gear Sands. Summary Report', Spalding

Associates (Environmental) Ltd, unpublished report for Cornwall County Council.

Pollard, E., Hall, M.L. and Bibby, T.J. (1986), *Monitoring the Abundance of Butterflies 1976–1985*, NCC Research and Survey in Nature Conservation, No. 2, Nature Conservancy Council, Peterborough.

Radley, G.P., Crawford, I.C. and Waite, A.R. (1987), *No. 13 National Sand Dune Vegetation Survey Site. Report No. 4 Penhale Dunes, Cornwall*, Nature Conservancy Council, Peterborough.

Rich, T.C.G., Holyoak, D.T., Margetts, L.J. and Murphy, R.J. (1997), 'Hybridisation between *Gentianella amarella* (L.) Boerner and *G. anglica* (Pugsley) E. F. Warb. (Gentianaceae)', *Watsonia*, **21**, 313–25.

Sidaway, R. (1988), *Sport, Recreation and Nature Conservation*, Research Study 32, Sports Council, London.

Sidaway, R. (1995), 'Recreation and tourism on the coast: managing impacts and resolving conflicts', in M.G. Healy and J.P. Doody (eds.), *Directions in European Coastal Management*, Samara, Cardigan.

Sothern, E.J., Randerson, P.F., Williams, A.T., and Dixon, J. (1985), 'Ecological effects of recreation at Merthyr Mawr Dunes, South Wales', in P. Doody (ed.), *Focus on Nature Conservation No 13*, Nature Conservancy Council, Peterborough, 217–38.

Spalding, A. (ed.), (1997), *Red Data Book for Cornwall and the Isles of Scilly*, Croceago Press, Praze-an-Beeble.

Spalding, A. (1998), 'A Survey of Skylarks at Upton Towans, 1998', Spalding Associates (Environmental) Ltd, unpublished report for Cornwall County Council.

Stubbs, A. (1997), 'British bee-flies', *British Wildlife*, **8**, 175–79.

Tunmore, M. (1998), 'A Survey of Skylarks at Phillack Towans, 1998', Spalding Associates (Environmental) Ltd, unpublished report for Cornwall County Council.

Williams, A.T. and Randerson P.F. (1989), 'Nexus: ecology, recreation and management of a dune system in South Wales', in F. van der Meulen, P.D. Jungerius and J.H. Visser (eds.), *Perspectives in Coastal Dune Management*, SPB Academic Publishing, The Hague, 217–27.

SPATIAL DATA COLLECTION METHODS IN DYNAMIC ENVIRONMENTS

E. EDWARDS and A. KOH

GeoTechnologies, Bath Spa University College, England

ABSTRACT

The progressive development of digital camera technology is yielding new opportunities for near-real-time mapping when used in conjunction with state-of-the-art soft copy photogrammetry. Aerial photography is a tried and tested means of data acquisition for coastal studies, but airborne digital photography is a new technique which may ultimately supplant the use of traditional aerial photographs and provide the final link in the transition to an all-digital environment. The use of Global Positioning Systems (GPS) is emerging as a promising solution to the problem of reliable spatial data collection, providing latitude, longitude and altitude data for ground control, for aerial image rectification and for aircraft navigation. For many environmental management applications, the 100m positional accuracy obtained from mapping-grade GPS receivers used in autonomous mode is insufficient, but when differentially processed and integrated with high-resolution multispectral fly-on-demand digital imagery, the data provide an extremely valuable and timely source of information. This paper demonstrates the potential of these technologies for producing orthophoto maps of coastal sand dune systems. Digital imagery was acquired using a Kodak DCS 460 CIR (colour infrared) camera and dGPS data were collected using two Magellan PRO Mark X CM hand held receivers. The imagery was processed using VirtuoZo (soft copy photogrammetric software) to produce digital elevation models and orthophotos of the dune system.

Introduction

Coastal sand dunes are increasingly perceived as areas worthy of sustainable development; they are a cost-effective, natural means of coastal defence, provide a range of habitats for a diverse community of highly specialized animal and plant species and in some areas are important in providing communities with a clean supply of fresh water.

In the economic climate of today there is a need for efficient, objective and quantitative approaches to coastal zone management and planning which will generate maximum returns on investment. Geographic Information Systems (GIS) are making significant contributions to everyday life and growth in this area is matched by corresponding advances in Remote Sensing (RS) and differential Global Positioning Systems (dGPS). These rapidly evolving technologies provide coastal managers with state-of-the-art tools for collecting objective data for baseline survey and continuous monitoring purposes.

Aerial photography, a well-established form of remote sensing which has been operational for many years, is still widely used for coastal studies today but it is a relatively expensive data format which does not lend itself to modern computerized approaches to data handling for resource management and planning. In contrast, airborne digital photography is a new remote sensing technique that integrates seamlessly with other digital technologies such as GPS and GIS.

The concept of using digital cameras for aerial photography is in its infancy, and therefore has not been tested extensively, but several workers, including King et al. (1994) and Bobbe and McKean (1995), have demonstrated its potential as an alternative means of imaging for landcover analysis. In comparative assessments of film cameras and digital cameras, Bobbe (1997) has reported good compatibility between the spectral response of the CIR digital camera and CIR film. Bobbe also reported that the CIR digital camera performed better, on several occasions, under adverse lighting conditions when a Hasselblad 70mm camera failed to expose CIR film. This latter point is significant in that the use of a digital system could greatly extend the window of opportunity to fly aerial missions, especially in regions where cloud cover is frequent. Now that much of the technology is in place there is a need for further research to prove the concept of this new data-gathering facility.

The Aerial Digital Photographic System (ADPS)

The Aerial Digital Photographic System used for this research was developed at Bath Spa University College and comprises 3 principal components: a digital camera (Kodak DCS 460 CIR); an aircraft mounting plate; and an intervalometer to manage the framing sequence. The assembly is self-contained, transportable, and can be easily mounted in a light aircraft within minutes (Plate 1). The system is easy to use since it can be operated like a film-based small-format camera. The ADPS is compatible with images acquired from current earth observation satellites, since the spectral sensitivity spans the entire range of visible light and extends well into the near infrared.

The original system, designed around the Kodak Digital Science 420 camera, has been described fully elsewhere by Koh and Edwards (1996a). A major new development is the inclusion of the Kodak Digital Science 460 CIR camera. This differs from the Digital Science 420 camera in the array size of the CCD (charge coupled device). Whereas the 420 has an array size of 1524 (H) by 1012 (V) the 460 has an array size of 3060 (H) by 2036 (V) giving a photocoverage four times larger than that of the 420. This configuration produces a 24-bit colour image of 18.6Mb whereas the 420

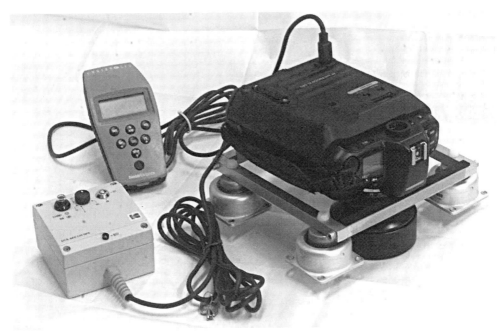

Plate 1. *The ADPS – Kodak Digital Science camera mounted on a shock-resistant chassis, intervalometer and Trimble Ensign GPS receiver.*

produces a 24-bit colour image of 4.4Mb. Graham and Mills (1997) have observed that although the nominal spatial resolution of the DCS 460 is the same as that of the DCS 420, the image quality is far superior due to the improved CFA (colour filter array) interpolation algorithm incorporated within the DCS 460. In addition, the 460 CIR has been designed to integrate information regarding the spatial position of the camera at the point of image capture into the image data file. This information is in the format of a standard NMEA (National Maritime Electronics Association) sentence derived from Global Positioning Systems. Increased accuracies of the camera position can be achieved using real-time differential global positioning (Koh and Edwards, 1996b).

Aerial Photographic Mission

Colour infrared, aerial digital imagery of the dune site at Holywell Bay in Cornwall, southwest England was acquired in May 1996 using the ADPS configured with the Kodak DCS 460 CIR camera, fitted with a 28mm focal length Nikon Nikkor lens and a minus-blue filter. This was mounted vertically on a chassis with integrated shock- and vibration-absorbers in a Partanavia aircraft. The mission was flown at an altitude of 4000 feet above ground level (AGL) at a true ground speed of 80 knots. The flight plan entailed a two line block, each line consisting of 10 images with around 60 per cent forward overlap and 35 per cent side overlap for stereo coverage. The nominal ground resolution was 40cm so that each image covered an area on the ground of approximately 1.2 × 0.8km. Flight navigation was effected using a hand-held GPS receiver connected into an active antenna mounted on the top of the fuselage.

Ground Survey

Differential GPS was used to acquire ground control for registration of the imagery since it was not possible to identify sufficient control points from existing maps owing to the dynamic nature of the subject matter. Two Magellan PRO Mark X CM receivers were used to collect latitude, longitude and altitude data. One receiver was set to log at a known location or 'base station', to provide correction data for differential post processing. Position data for targets including fence lines, paths, path intersections, gates, rocks and any other static features which would be visible on the images were collected using the 'rover' receiver. The data were processed in pseudorange mode using Magellan M Star post processing software.

Plate 2. *ADPS CIR mosaic of Holywell Bay coastal dune system.*

Image Processing

The images were acquired using Twain-compliant software, Adobe Photoshop, where the 6Mb compressed format image (TIF-EP) is resampled to produce a 36-bit (12 bits per channel) file; this is further resampled to 24 bits as the best 8 bits are chosen for each channel. The-24 bit image can then be stored in TIFF format giving an 18.6Mb file containing the 8-bit data for each of the three bands (near infrared, red and green), together with the meta-information in the header file.

An uncontrolled mosaic of the images was constructed in Adobe Photoshop and (a greyscale version) is shown in Plate 2. This type of image can provide a rapid and cost-effective source of data giving an excellent overview of the status of the dune field and providing early warning of dune vulnerability to managers. Visual analysis can supply qualitative data which are adequate for many applications, such as the presence of walkways and car parking facilities and the degree of fragmentation of the dune front, while digital image analysis can yield semi-quantitative data such as the relative areas of bare sand and vegetation. It is also possible to count features such as the number of paths through the dunes and the number of path intersections.

In a colour infrared image such as this it is also possible to differentiate between

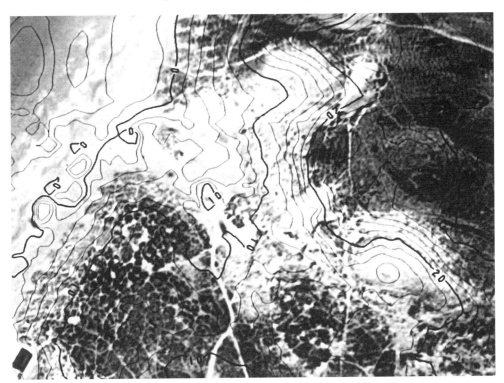

***Plate 3.** Orthophoto with 2m contours overlaid.*

types of vegetation units; for example, marram returns a grey-red signal and can be seen clearly in the foredune ridges with small areas of deep red vegetation where patches of shrubby vegetation occur. The area of permanent pasture at the top of the photograph returns a bright red signal and ploughed fields appear grey. A golf course is clearly visible with the fairways and greens appearing clearly in bright red tones due to the differing mowing regimes and grass species composition. Bunkers appear white, reflecting the sandy substrate. To obtain planimetric information from the imagery, errors inherent in any type of photography were resolved using soft copy (VirtuoZo) photogrammetric techniques. Each photo model (stereo pair) was processed to produce a digital elevation model (DEM). The DEM was then applied to the image base to produce an orthophoto which is rectified for image distortions due to camera lens distortion and aircraft attitude as well as for topographic effects. Plate 3 shows part of the orthophoto of Holywell Bay with 2m interval contours overlaid.

Conclusion

Aerial Digital Photography has great potential for providing rapid, low-cost data which can generate much of the information required by coastal managers. The integration of differentially corrected aircraft position GPS data can provide the spatial control needed to rectify the imagery for mapping purposes and for integration into Geographic Information Systems. With the rapid advances in technology and the explosive growth in the GPS and digital imaging market, the cost of acquiring these new technologies is falling, so that they may now be included in the suite of tools available for management at the coastal zone.

References

Bobbe, T. (1997), 'Applications of a colour infrared digital camera system as a remote sensing tool for natural resource management', *Proceedings of the First North American Symposium on Small Format Aerial Photography*, October 14–17 1997, 71–79.

Bobbe, T. and McKean, L. (1995), 'Evaluation of a digital camera system for natural resource management', *Earth Observation Magazine*, March, 46–48.

Graham, R., and Mills, L. (1997), 'Experiences with airborne digital photography for photogrammetry and GIS', *Proceedings of the First North American Symposium on Small Format Aerial Photography*, October 14–17 1997, 17–36.

King, D., Walsh, P. and Cuiffreda, F. (1994), 'Airborne digital frame camera imaging for elevation determination', *Photogrammetric Engineering and Remote Sensing*, **60(11)**, 1321–26.

Koh, A. and Edwards, E. (1996a), 'Integrating GPS data with fly-on-demand digital imagery for coastal zone management', AGI Marine and Coastal Zone Management Special Interest Group, *Managing the UK Marine and Coastal Zone Environment*, Proceedings of a 1-day meeting, Manchester, 24 July 1996, AGI.

Koh, A. and Edwards, E. (1996b), 'Differential Global Positioning Systems for coastal zone management', in E. Ozhan (ed.), *ICZM in the Mediterranean and Black Sea: Immediate Needs for Research, Education – Training and Implementation*, Proceedings of the International Workshop, 2–5 November 1996, METU, Ankara, 171–85.

MODELLING THE HABITAT OF THE GRAYLING BUTTERFLY (*HIPPARCHIA SEMELE* L.) ON THE SEFTON COAST SAND DUNE SYSTEM, UK, USING A GEOGRAPHICAL INFORMATION SYSTEM (GIS)

JON DELF

Liverpool Hope University College, England

ABSTRACT

Results obtained from observations of the Satyrid butterfly, *Hipparchia semele*, indicate that behaviour is influenced by a small number of key habitat variables. Combining these data with aerial photography and topographical information relating to the Sefton Coast sand dunes, a three-dimensional habitat model was created using the IDRISI Geographical Information System. It is suggested that this methodology could be of value in monitoring *H. semele*, or any other species with similarly specific habitat requirements.

Introduction

'Geographic Information Systems (GIS) are extremely valuable in determining biotopes and landscapes for conservation' (Samways, 1994, 191). To support this statement the example is used of the GIS analysis of red panda habitat in Nepal (Yonzon et al., 1991) where a number of thematic maps were combined to identify that 60 per cent of core habitat was at risk from human pressure. There are indeed a number of other examples of GIS being used at the landscape scale to study habitats and populations of a variety of vertebrate species (e.g. salmon [Guest, 1995]; red squirrels [Gurnell et al., 1996] and short-eared owls [Aspinall and Humble, 1997]) but how applicable could this tool be for invertebrate conservation management? Samways (1994) goes on to suggest that similar approaches using GIS could be useful for the distribution mapping of selected invertebrates but that it may be more applicable to determine biotopes in order to predict where certain species, assemblages or communities may occur. Other guidelines with respect to the conservation of invertebrates also suggest that in many cases their occurrence cannot necessarily be predicted by information concerning plant species or communities alone; invertebrates often spend much of their lives in microhabitats and their needs may be supplied to a fairly major extent by the vegetation structure or physical features of sites (Kirby, 1992).

Although Lepidoptera are one of the most studied of invertebrate groups, they have not received the same level of investigation using GIS as other groups. Where studies have been initiated, focus has tended to include the most destructive species, as in an example involving the spruce budworm *Choristoneura occidentalis* and gypsy moth *Lymantria dispar* from the United States (Williams and Liebhold, 1995). A few studies have used GIS to address the conservation of Lepidoptera, such as that by Weiss and Weiss (1993) to model the habitat of the bay checkerspot butterfly *Euphydryas editha bayensis* based upon its thermal range, and that by Clayson (1996) to identify suitable habitat for the high brown fritillary *Argynnis adippe* in Cumbria based upon altitude and bracken/woodland edge characteristics.

The current study, however, assesses the possibility of using a raster-based GIS, IDRISI (Eastman, 1992), to model the habitat of the Grayling butterfly *Hipparchia semele* using secondary data and, in particular, those spatial data relating to vegetation structure and physical features of the habitat. In the UK, *H. semele* lives and breeds on a wide range of arid, well-drained, unimproved grasslands (Heath et al., 1984). It typically inhabits sparsely vegetated dry heaths, chalk and limestone areas and exposed coastal dunes in particular (Dennis, 1992). The Sefton Coast provides a convenient example of a diverse coastal dune habitat with an established *H. semele* population.

The aims of the research so far have been:

- to explore the use of the IDRISI GIS with respect to describing the patterns of abundance and distribution of this particular animal in relation to its habitat;
- to design a GIS model that would facilitate the visualization, interpretation and analysis of its habitat; and
- to construct an IDRISI GIS model using available secondary data and rules based upon field observations.

Methodology

Primary data concerning the distribution and behaviour of *H. semele* adults were collected using a combination of a standard butterfly monitoring technique (Pollard and Yates, 1993) together with an adaptation of a methodology (Shreeve, 1990) to monitor the behaviour of selected individuals. The observations were confined to a 9km transect of the Sefton Coast, from Formby (SD 2709) at the southern end to Ainsdale (SD 2912) at the northern end, and took place between the end of June and the beginning of September 1994. The full data set for the transect over this period comprises 670 butterfly encounters and observations on the behaviour of 12 individuals. Butterfly positions were digitized in IDRISI and grouped into three datasets representing morning, noon and afternoon individuals.

Secondary data sources consisted of 1km × 1km digitized (24-bit, true colour image) aerial photographs of the dune system and Sefton Coast engineers' topographic sheets (non-standard drawings at 1:5000 scale) showing contours at 1m height intervals.

For an initial attempt at habitat modelling, two 400 × 400m areas (sites 1 and 2) were chosen which demonstrated established *H. semele* populations throughout the period of monitoring. This size of area was dictated primarily by computer memory limitations at the time (6–30 megabytes per km^2) but the choice was also justified by considering such areas as likely to contain entire habitat requirements and which, when viewed from an 'organism-centred' perspective, could be considered a reliable compromise between the home range of an adult butterfly (based upon flight data) and its regional distribution (McGarigal and Marks, 1994).

The digitized aerial photographs were preprocessed to produce 256 colour images where each pixel represented 1m^2 of ground cover. The engineers' drawings of the area were digitized using a flatbed scanner and then cleaned and height coded using image-processing software. Both sets of secondary data were then imported into IDRISI and the contours were 'rubbersheeted' to fit the aerial photographs using features such as slacks and dune ridges to act as control points. All datasets were registered spatially using British Ordnance Survey co-ordinates and, following interpolation of the height data in IDRISI to create digital elevation models (DEMs), the photographic and topographic data were combined to create three-dimensional views which were used to assess visually the degree of success in correlating the spatial co-ordinates of the two types of data.

The DEMs were analytically hillshaded using IDRISI's SURFACE application (using sun azimuths and elevations for 10am and 5pm on Monday 22 August 1994). This process was used to simulate clear sky insolation of the surface at these particular times. The same application was also used to calculate slope angles and aspects of the dune surface.

Supervised classification of the aerial photographs for the two sites was performed using IDRISI's MAXLIKE application (Maximum Likelihood procedure) to create four broad habitat types, viz: 'sand', 'marram grass', 'grass' and 'scrub'. This procedure resulted in less than 2 per cent of pixels in each area being unclassified.

Landscape pattern and structure in each of the sample habitats were examined using the FRAGSTATS Version 2.0 patch analysis program (McGarigal and Marks, 1994) which returns statistics such as number and shape of patches present, edge/area ratios, juxtaposition of patches, nearest neighbour relationships and core area positions, in addition to a number of other metrics. During this analysis it was assumed that any areas greater than 1m^2 were vegetation changes and areas larger than 50m^2 were core areas of vegetation.

Arbitrary 'territories' 20m in diameter were then placed around the positions of individuals on the butterfly distribution data sets by the process of 'buffering' and these data were then used to extract statistics about suitable slope, aspect, height, insolation and vegetation category from the secondary data sources.

Results and Discussion

Actual observations of butterflies along the transect indicated that, in general, individuals were most abundant in areas dominated by marram *Ammophila arenaria* and short grass sward. Distribution tended to be parallel with the coast, particularly between the main mobile dunes/semi-fixed dunes and the intermediate-stage mesotrophic grassland further inland. No sightings occurred in front of the foredunes or in the pine woodland at the back of the dune system. Activity was found to be correlated with weather conditions, presence of sunlight and time of day (butterflies tended to occur further inland later on each day). From the observation of individuals on the Sefton Coast, it was noted that the adult butterflies were relatively sedentary and spent on average 74 per cent of their time basking on a variety of substrates but mainly, particularly during periods of greatest insolation, on bare sand. This compares favourably with the findings of others such as Dennis (1992), Findlay et al. (1993), Shreeve (1990) and Tinbergen (1972). However, flight is also vital for finding a mate, laying eggs and escaping predators, and incoming solar radiation is likely to be a major contributing factor to enable such activity. Size of sand patch did not seem particularly important. Feeding on nectar sources occurred sporadically throughout the day and often occurred 50m or more away from basking sites. Only one pair of individuals was observed copulating and no females were seen ovipositing. Most of the butterfly data therefore related to male and female insects that were primarily concerned with basking, flying or feeding.

Patch analysis showed that there were differences in structural diversity and fragmentation between the two sites, particularly with respect to edge:interior ratio (which was significantly higher for site 1). The correlation between this 'edge-effect' and butterfly distribution was examined using Cramer's *V* correlation coefficient. When compared directly to the distribution of individuals only a mild correlation was found (Cramer's $V = 0.5$ for 196 degrees of freedom) whereas if the aspect of slopes facing the sun for that time of day was also taken into account then the strength of the correlation improved significantly (Cramer's $V = 0.9$ for 196 degrees of freedom).

Combining the above results with information from published research, a number of criteria were established which seemed to be important in governing the behaviour of *H. semele* on the Sefton Coast dune system and which could be used in generating the GIS model (see Table 1).

Figures 1a and 1b give some idea of the model's output and illustrate the topography of the two sites for which models were produced. They show the dune surface in orthographic perspective over which has been draped a coloured map of butterfly probability predicted by the GIS. To assist in the interpretation of the output as monochrome illustrations, some areas predicting a high probability of encountering adult butterflies have been indicated using arrows. Annotated circles have been superimposed above the topographic surface, but from the same orthographic perspective, to

Table 1. Examples of factors affecting butterfly distribution used as criteria in the modelling process

Criterion	Considerations with respect to GIS
1 *H. semele* is most active on sunny dry days and least active in rain	A limitation on the use of the current model as basking sites only have been included
2 Most of the adults' time is spent basking regardless of the time of day but the butterflies are most active in the early afternoon	Used to weight the importance of insolation in predicting butterfly distribution and results in different model outputs with respect to the time of day
3 Sunny/light areas are preferable to shady areas	Used to weight the importance of topography and insolation in creating the model and data relatively easily obtained from the secondary data-sets
4 Settling sites follow the sun throughout the day	As for 3 and also results in the current model being time-dependent
5 Bare sand is the favourite settling surface	Used to weight the importance of this substrate determined by image analysis
6 Areas close to vegetation are preferred	Used to determine the importance of some of the metrics obtained from the use of FRAGSTATS
7 Males are territorial and return to the same hilltop site	Omitted from the original model
8 Females tend to settle near foodplants	Apart from a general weighting towards 'grass' and 'marram grass' features obtained from image classification, detailed preferences omitted from the original model and currently subject to further investigation

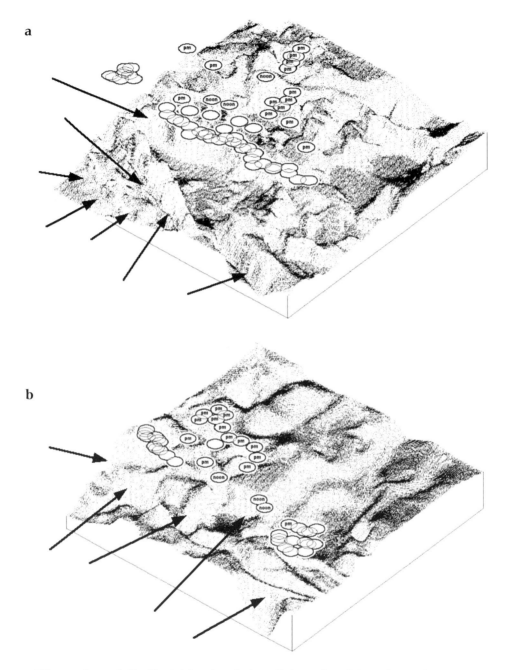

Figures 1a and 1b. *Site 1 (above) and Site 2 (below) viewed from the south west showing probability outputs (➡ = areas of highest probability) for the distribution of* H. semele *compared with distributions observed during fieldwork (**noon** = observations around mid-day; **pm** = afternoon observations; unlabelled circles = morning observations).*

show the relative positions of actual butterfly sightings at various periods throughout the day.

Conclusions and Considerations

Although the results of the behaviour monitoring provided more data on the species' preferences than could be realistically incorporated into the model at this stage, there were also important omissions with respect to certain details of adult behaviour (e.g. oviposition sites), and information on aspects of the larval and pupal stages, which would need to be incorporated to create a holistic model for this butterfly. Resources are currently being sought to enable these omissions to be corrected.

Technical considerations include the fact that IDRISI's analytical hillshading capabilities and image classification characteristics are not considered to be 100 per cent accurate and the fact that it was not possible to obtain completely coincidental sets of aerial photographs, topographic data and butterfly observations. These shortcomings would also need to be addressed in cases where a more detailed model was required.

Despite these limitations the study produced an insight into the possibility of representing *H. semele's* habitat requirements derived from secondary data which it is hoped may prove useful in the conservation of the species at this site and elsewhere.

Acknowledgments

I am indebted to Andrew Goodwin who carried out the 1994 fieldwork surveys and developed the original GIS modelling techniques during the completion of his final-year Environmental Studies dissertation at Liverpool Hope University College. I should also like to acknowledge the assistance given from Sefton Metropolitan Borough Council who provided the aerial photographs and engineers' drawings which were essential for creating the GIS model.

References

Aspinall, R. and Humble, A. (1997), 'Ear today, gone tomorrow? A GIS-based approach to wide-area conservation is pointing the way to habitat-protection schemes', *Mapping Awareness*, **11(4)**, 32–35.

Clayson, J. (1996), 'On a wing and a prayer? GIS helps target fieldwork for a rare butterfly', *Mapping Awareness*, **10(9)**, 24–26.

Dennis, R.L.H. (ed.) (1992), *The Ecology of Butterflies in Britain*, Manchester University Press, Manchester.

Eastman, J.R. (1992), *The IDRISI User's Guide*, IDRISI Project, Clark University, Massachusetts.

Findlay, R., Young, M.R. and Findlay, I.A. (1983), 'Orientation behaviour in the Grayling butterfly: thermoregulation or crypsis?', *Ecological Entomology*, **8(2)**, 145–53.

Guest, M. (1995), 'Salmon stakes', *Mapping Awareness*, **9(7)**, 34–36.

Gurnell, J., Clark, M. and Feaver, J. (1996), 'Habitat forming', *Mapping Awareness*, **10(4)**, 36–40.

Heath, J., Pollard, E. and Thomas, J. (1984), *Atlas of Butterflies in Britain and Ireland*, Viking,

London.

Kirby, P. (1992), *Habitat Management for Invertebrates: A Practical Handbook*, Royal Society for the Protection of Birds/Joint Nature Conservation Committee/National Power, RSPB, Sandy.

McGarigal, K. and Marks, B. (1994), *FRAGSTATS: Spatial Pattern Analysis Program for Quantifying Landscape Structure*, Oregon State University.

Pollard, E. and Yates, T.J. (1993), *Monitoring Butterflies for Ecology and Conservation*, Chapman and Hall, London.

Samways, M.J. (1994), *Insect Conservation Biology*, Chapman and Hall, London.

Shreeve, T. (1990), 'Microhabitat use and hindwing phenotype in *Hipparchia semele* (Lepidoptera, Satyrinae): thermo-regulation and background matching', *Ecological Entomology*, **15(2)**, 201–13.

Tinbergen, N. (1972), *The Animal and its World 1. Field Studies*, Allen and Unwin, London.

Weiss, A.D. and Weiss, S. (1993), *Estimation of Population Size and Distribution of a Threatened Butterfly: GIS Applications to Stratified Sampling*, Environmental Systems Research Institute Inc., New York.

Williams, D.W. and Liebhold, A.M. (1995), 'Potential changes in spatial distribution of outbreaks of forest defoliators under climate change', in R. Harrington and N.E. Stork (eds.), *Insects in a Changing Environment*, Academic Press, London.

Yonzon, R., Jones, R. and Fox, J. (1991), 'Geographic Information Systems for assessing habitat and estimating population of red pandas in Langtang National Park, Nepal', *Ambio*, **20**, 285–88.

APPENDIX

A SAND DUNE BIBLIOGRAPHY FOR THE ATLANTIC BIOGEOGRAPHICAL REGION

General

Bakker, T.W.M., Jungerius, P.D. and Klijn, J.A. (eds.) (1990), 'Dunes of the European coasts: geomorphology, hydrology, soils', *Catena Supplement*, **18,** Catena-Verlag.

Boorman, L.A. (1977), 'Sand-dunes', in R.S.K. Barnes (ed.), *The Coastline*, John Wiley and Sons, Chichester.

Brown, A.C. and McLachlan, A. (1990), *Ecology of Sandy Shores*, Elsevier Science Publishers, Amsterdam, The Netherlands.

Carter, R.W.G., Curtis, T.G.F. and Sheehy-Skeffington, M.J. (eds.) (1992), *Coastal Dunes: Geomorphology, Ecology and Management for Conservation*, A.A. Balkema, Rotterdam.

Doody, J.P. (1991), *Sand Dune Inventory of Europe*, Joint Nature Conservation Committee, Peterborough.

Drees, J.M. (ed.) (1997), *Coastal Dunes, Recreation and Planning: Proceedings of the European Seminar, Castricum, The Netherlands*, European Union for Coastal Conservation, Leiden, The Netherlands.

García Novo, F., Crawford, R.M.M. and Díáz Barradas, M.C. (eds.) (1997), *The Ecology and Conservation of European Dunes*, University of Seville, Seville.

Gehu, J.M. (1985), *European Dune and Shoreline Vegetation*, Council of Europe, Nature and Environment Series No. 32, Strasbourg.

Gimingham, C.H., Ritchie. W., Willetts, B.B. and Willis, A.J. (eds.) (1989), *Coastal Sand Dunes, Proceedings of the Royal Society of Edinburgh. Section B (Biological Sciences)*, **96**.

Grootjans, A.P., Jones, P., van der Meulen, F. and Paskoff, R. (eds.) (1997), *Ecology and Restoration Perspectives of Soft Coastal Ecosystems, Journal of Coastal Conservation* (special feature), **3(1)**.

Healy, M.G. and Doody, J.P. (eds.) (1995), *Directions in European Coastal Management*, Samara Publishing, Cardigan.

Jeffries, R.L. and Davy, A.J. (eds.) (1979), *Ecological Processes in Coastal Environments*, Proceedings of the 19th symposium of the British Ecological Society, Oxford, Blackwell.

Jones, P.S., Healy, M.G. and Williams, A.T. (eds.) (1996), *Studies in European Coastal*

Management, Samara Publishing, Cardigan.

Maarel, E. van der (ed.) (1993), *Dry Coastal Ecosystems*, Ecosystems of the World, 2A, Elsevier, Amsterdam.

Maarel, E. van der (ed.) (1994), *Dry Coastal Ecosystems*, Ecosystems of the World, 2B, Elsevier, Amsterdam.

Maarel, E. van der (ed.) (1997), *Dry Coastal Ecosystems*, Ecosystems of the World, 2C, Elsevier, Amsterdam.

Meulen, F. van der, Jungerius, P.D. and Groot, R.S. (eds.) (1989), *Discussion Report on Coastal Dunes, Landscape Ecological Impact of Climatic Change on Coastal Dunes in Europe*, Dutch Ministry of the Environment.

Meulen, F. van der, Jungerius, P.D. and Visser, J. (eds.) (1989), *Perspectives in Coastal Dune Management: Towards a Dynamic Approach*, SPB Academic Publishing, The Hague.

Meulen, F. van der, Witter, J.V. and Ritchie, W. (eds.) (1991), 'Impact of climatic change on coastal dune landscapes of Europe', *Landscape Ecology*, **6(1/2)**.

Packham, J.R. and Willis, A.J. (1997), *Ecology of Dunes, Salt Marsh and Shingle*, Chapman and Hall, London.

Ranwell, D.S. (1972), *Ecology of Salt Marshes and Sand Dunes*, Chapman and Hall, London.

Great Britain

Atkinson, D. and Houston, J.A. (eds.) (1993), *The Sand Dunes of the Sefton Coast*, National Museums and Galleries on Merseyside, Liverpool.

Beebee, T. and Denton, J. (1996), *The Natterjack Toad Conservation Handbook*, English Nature, Peterborough.

Brooks, A. and Agate, E. (1986), *Sand Dunes: A Practical Conservation Handbook*, British Trust for Conservation Volunteers, Oxfordshire.

Doody, J.P. (ed.) (1985), *Sand Dunes and their Management*, Nature Conservancy Council, Peterborough.

Doody, J.P., Johnston, C. and Smith, B. (eds) (1993), *Directory of the North Sea Coastal Margin*, Joint Nature Conservation Committee, Peterborough.

Moulton, N. and Corbett, K. (1999), *The Sand Lizard Conservation Handbook*, English Nature, Peterborough.

Radley, G.P. (1994), *Sand Dune Vegetation Survey of Great Britain. Part 1: England*, English Nature, Peterborough.

Radley, G.P. and Woolven, S.C. (1990), *Research Report No. 122: A Sand Dune Bibliography*, Nature Conservancy Council, Peterborough.

Ranwell, D.S. (ed.) (1972), *Hippophae Rhamnoides on Selected Sites in Great Britain*, Nature Conservancy Council, Petersborough.

Ranwell, D.S. and Boar, R. (1986), *Coast Dune Management Guide*, Institute of Terrestrial Ecology, Huntingdon.

Rodwell, J.S. (in press), *British Plant Communities, Volume 5. Maritime Vegetation and*

Communities of Open Habitats, Cambridge University Press, Cambridge.

Steers, J.A. (1948), *The Coastline of England and Wales*, Cambridge University Press, Cambridge.

Willis, A.J., Folkes, B., Hope-Simpson, J.F. and Yemm, E.W. (1959), 'Braunton Burrows: the dune system and its vegetation. I and II', *Journal of Ecology*, **47**, 1–24 and 249–88.

Denmark

Aagaard, T., Nielsen, N. and Nielsen, J. (1995), *Skalligen – Origin and Evolution of a Barrier Spit*, Institute of Geography, University of Copenhagen.

Brandt, E. and Christensen, S.N. (1994), *Danske klitter oversigtlig kortsaegning*, Miljøministeriet Skov-og Naturstyrelsen.

Feilberg, A. and Jensen, F. (1992), 'Management and conservation of sand dunes in Denmark', in Carter, Curtis and Sheehy-Skeffington, 1992, 429–38.

Ovesen, C.H. (1998), *Coastal Dunes – Management, Protection and Research,* Report from a European Seminar, Skagen, Denmark, August 1997, National Forest and Nature Agency, Copenhagen.

France

Barrère, F. (1992), 'Dynamics and management of the coastal dunes of the Landes, Gascony, France', in Carter, Curtis and Sheehy-Skeffington, 1992, 25–32.

Favennec, J. and Barrère, P. (eds.) (1997), *Biodiversité et Protection Dunaire*, Lavoisier, Paris.

Germany

Dijkema, K.S. and Wolff, W.J. (1983), *Flora and Vegetation of the Wadden Sea Islands and Coastal Areas,* Report 9 of the Wadden Sea Working Group, The Netherlands.

Ireland

Bassett, A. and Curtis, T.G.F. (1985), 'The nature and occurrence of sand-dune machair in Western Ireland', *Proceedings of the Royal Irish Academy*, **85B**, 1–20.

Jeffrey, D.J. (ed.) (1977), *North Bull Island, Dublin Bay: A Modern Coastal Natural History*, Royal Dublin Society, Dublin.

Nooren, M.J. and Schouten, M.G.C. (1976), *Coastal Vegetation and Soil Features in South-East Ireland*, Doctoraal Verslag, Nijmegen.

Quigley, M.B. (ed.) (1991), *A Guide to the Sand Dunes of Ireland*, European Union for Coastal Conservation, Dublin, Ireland.

Quinn, A.C.M. (1977), *Sand Dunes: Formation, Erosion and Management*, An Foras Forbartha, Dublin.

Young, R. (ed.) (1977), *Planning for the Use of Irish Sand Dune Systems*, An Foras Forbartha, Dublin.

The Netherlands
Bakker, T.W.M., Klijn, J.A. and van Zadelhoff, F.J. (1979), *Duinen en duinvalleien,* Centrum voor Landbouwpublikaties en Landbouwdocumentatie, Wageningen.

Boot, R.G.A. and van Dorp, D. (1986), *De plantengroei van de Duinen van Oostvoorne in 1980 en veranderingen sinds 1934,* Stichting Het Zuidhollands Landschap, Rotterdam.

Dijk, H.W.J. van (1992), 'Grazing domestic livestock in Dutch coastal dunes: experiments, experiences and perspectives', in Carter, Curtis and Sheehy-Skeffington, 1992, 235–50.

Dijkema, K.S. and Wolff, W.J. (1983), *Flora and Vegetation of the Wadden Sea Islands and Coastal Areas,* Report 9 of the Wadden Sea Working Group, The Netherlands.

Doing, H. (1989), *Landscape Ecology of the Dutch Coast,* Stichting Duinbehoud, Leiden.

Dorp, D. van, Boot, R. and van der Maarel, E. (1985), 'Vegetation succession on the dunes near Oostvoorne, The Netherlands, since 1934, interpreted from air photographs and vegetation maps', *Vegetatio,* **58,** 123–36.

Grootjans, A.P., Lammerts, E.J. and van Beusekom, F. (1995), *Kalkrijke Duinvalleien op de Waddeneilanden,* KNNV, Utrecht.

Maarel, E. van der, Boor, R., Dorp, D. and Ryntjes, J. (1984), 'Vegetation succession on the dunes near Oostvoorne, The Netherlands: a comparison of the vegetation in 1959 and 1980', *Vegetatio,* **58,** 137–87.

Portugal
Braun-Blanquet, J., Braun-Blanquet, G., Rozeira, A. and da Silva, P. (1972), 'Résultats de trois excursions géobotaniques a travers le Portugal septentrional et moyen. IV. Esquisse sur la végétation dunale', *Agronomia Lusitana,* **33,** 217–34.

The following references are Symposium Volumes of the Associação Eurocoast-Portugal, each one having several papers or summaries of papers on different aspects of dunes, mainly on their dynamics.

Associação Eurocoast-Portugal (1991), *Seminário sobre a Zona Costeira e os Problemas Ambientais-Aveiro,* Eurocoast-Portugal, Porto.

Associação Eurocoast-Portugal (1994), *Segundo Simpósio Internacional Littoral '94,* Lisboa, Eurocoast-Portugal, Porto.

Associação Eurocoast-Portugal (1995), *2° Encontro de informação sobre dinâmica, Conservação, Protecção e Uso da Zona Costeira,* Figueira da Foz, Eurocoast-Portugal, Porto.

Associação Eurocoast-Portugal (1996), *2° Seminário sobre a Zona Costeira de Portugal (Ordenamento, Gestão e Aproveitamento da Zona Costeira de Portugal) Resumos das Comunicações,* Eurocoast-Portugal, Porto.

Carvalho, G.S., F.V. Gomes. and F.T. Pinto (1997), *Colectânea de Ideais sobre a Zona Costeira de Portugal,* Eurocoast-Portugal, Porto.

Carvalho, G.S., F.V. Gomes and F.T. Pinto (1998), *Dunas da Zona Costeira de Portugal,*

Seminário, Leiria Eurocoast-Portugal, Porto.

Dijk, H.W.J. and Murkmans, W. (1998), *The Portuguese Coast: Values and Threats*, European Union for Coastal Conservation.

Scotland

Countryside Commission for Scotland (1980), *Final Report: Highland Beach Management Project 1977–1979*, Countryside Commission for Scotland, Battleby.

Countryside Commission for Scotland (1981), *Beach Management Project, First Interim Report: 1980*, Countryside Commission for Scotland, Perth.

Countryside Commission for Scotland (1981), *Beach Management Project, Second Interim Report: 1981*, Countryside Commission for Scotland, Perth.

Dargie, T.C.D. (1993), *Sand Dune Vegetation Survey of Great Britain. Part 2: Scotland*, Joint Nature Conservation Committee, Peterborough.

Mather, A.S. and Ritchie, W. (1977), *The Beaches of the Highlands and Islands of Scotland*, Countryside Commission for Scotland, Battleby.

Ranwell, D.S. (ed.) (1986), *Sand Dune Machair 1, 2, 3*, Institute of Terrestrial Ecology, Huntingdon.

Ritchie, W. and Mather, A.S. (1984), *The Beaches of Scotland*, Countryside Commission for Scotland, Battleby (Summary of 18 regional reports 1969–81).

Ritchie, W. and Kingham, L. (eds.) (1997), *The St Fergus Coastal Environment: Monitoring and Assessment of the Coastal Dunes at the North Sea Gas Terminals St Fergus*, Aberdeen University Research and Industrial Services Limited, Aberdeen.

Ross, S. (1992), *The Culbin Sands – Fact and Fiction*, Centre for Scottish Studies, University of Aberdeen.

Whittington, G. (1996), *Fragile Environments: The Use and Management of Tentsmuir NNR, Fife*, Scottish Cultural Press, Edinburgh.

Spain

Aboal, I.L. (1982), *Espacios naturales protegibles. Coloquio Hispano-Francés sobre Espacios Litorales, 1981*, Ministerio de Agricultura, Madrid.

Bonnet Fernandez-Trujillo, J. (1989), 'Aspects of conservation and management of the sand-dune areas in Spain', in van der Meulen, Jungerius and Visser, 1989, 269–75.

Costa, P. and Pacheco, T. (1990), *Guia natural de las costas espaftolas*, ICONA, ser. Materiales, Madrid.

Diaz, T.E. (1974), 'La vegetación del litoral accidental asturiano', *Rev. Fac. Cienc. oviedo*, **15–16,** 369–545.

Flor, G. (1980), *Las dunas costeras de Cantabria: valores singulares geológicos*, Actas de la I Reunin de Geologia Ambiental y ordenación del Territorio, Santander.

Flor, G. (1983), 'Las formaciones dunares eólicas del litoral asturiano', *Astura*, **1,** 9–19.

Guitión, P. (1989), 'Ecosistemas litorales del noroeste de la Peninsula Ibérica: complejos de vegetación psamófila e higrófila', unpublished PhD thesis, Fac.

Bioloxia Universidade de Santiago de Compostela.

Izco, J. (1992), 'Diversidad y originalidad ecológica y floristica del litoral cantabro-atlàntico español', *An. Real Acad. Farm.*, **58,** 483–508.

Izco, J. (1993), 'Dry coastal ecosystems of northern and northwestern Spain', in van der Maarel, 1993, 329–40.

Izco, J. and Sanchez, I.M. (1994), 'Fuentes de impacto ambiental en el medio litoral: aplicación al la Ria de Betanzos (Galicia, España)', *Fitosociologia*, **27,** 97–106, Società Italiana di Fitosociologia.

Joven, M. (1993), *Spanish Coastal Dunes and Wetlands – Student's Report*, European Union for Coastal Conservation, Leiden.

Loriente, E.(1974), *Vegetación y flora de las playas y dunas de la provincia de Santander*, Instituto de la Cultura de Cantabria, Santander.

Ministerio de obras Públicas (1979–1981), *PIDU, Plan Indicativo de Usos del Dorninio Público Litoral (A Corufta, Lugo, Oviedo, Pontevedra, Santander)*, Ministerio de obras Públicas, Dirección General de Puertos y Costas, Madrid.

Obeso, J.R. and Aedo, C. (1992), 'Plant-species richness and extinction on isolated dunes along the rocky coast of Northwestern Spain', *Journal of Vegetation Science*, **3(1)**, 129–32.

Wales

Dargie, T.C.D. (1995), *Sand Dune Vegetation Survey of Great Britain. Part 3: Wales*, Joint Nature Conservation Committee, Peterborough.

Ranwell, D.S. (1958), 'Movement of vegetated sand dunes at Newborough Warren, Anglesey', *J. Ecol.*, **46**, 83–100.

Ranwell, D.S. (1959), 'Newborough Warren, Anglesey. I. The dune system and dune slack habitat', *J. Ecol.*, **47**, 571–602.

Ranwell, D.S. (1960a), 'Newborough Warren, Anglesey. II. Plant associes and succession cycles of the sand dune and dune slack vegetation', *J. Ecol.*, **48**, 117–42.

Ranwell, D.S. (1960b), 'Newborough Warren, Anglesey. III. Changes in the vegetation on parts of the dune system after the loss of rabbits by myxomatosis', *J. Ecol*, **48**, 385–95.